W0018639

# Wireless Transceiver
# Systems Design

Wolfgang Eberle

# Wireless Transceiver
# Systems Design

 Springer

Wolfgang Eberle
Katholieke Universiteit Leuven
Interuniversity Microelectronics Center (IMEC)
Leuven
Belgium

Library of Congress Control Number: 2007939952

ISBN 978-0-387-74515-2        e-ISBN 978-0-387-74516-9

© 2008 Springer Science+Business Media, LLC
All rights reserved. This work may not be translated or copied in whole or in part without the written permission of the publisher (Springer Science+Business Media, LLC, 233 Spring Street, New York, NY 10013, USA), except for brief excerpts in connection with reviews or scholarly analysis. Use in connection with any form of information storage and retrieval, electronic adaptation, computer software, or by similar or dissimilar methodology now known or hereafter developed is forbidden. The use in this publication of trade names, trademarks, service marks and similar terms, even if they are not identified as such, is not to be taken as an expression of opinion as to whether or not they are subject to proprietary rights.

Printed on acid-free paper.

9 8 7 6 5 4 3 2 1

springer.com

# Preface

During the last 30 years, wireless[1] in communications has grown from a niche market to an economically vital consumer mass market. The first wave, with the breakthrough of 2G mobile telephony focused on speech, placed wireless communication in the consumer mass market. In the current second wave, services are extended toward true multimedia, including interactive video, audio, gaming, and broadband Internet.

These high-data rate services, however, led to a separate IP-centric family of wireless personal (WPANs) and local area networks (WLANs) outside the 2G/3G mobile path. Since diversity between data- and voice-centric solutions and the competition between standardized and proprietary approaches is today more blocking than enabling effective development of successful products, a third major wave is unavoidable: a consolidation of both worlds in portable devices with flexible multistandard communication capabilities enabled for quality-of-service-aware multimedia services.[2] At the same time, the dominance of wired desktop personal computers has been undermined by the appearance of numerous portable and *smart* devices: laptops, notebooks, personal digital assistants, and gaming devices. Since these devices target low-cost consumer markets or face wired competition, time to market is crucial, designed-in flexibility is important, low-power operation is a key asset, yet device cost shall be at a minimum.

This book approaches this *design tradeoff* challenge from the perspective of the *system architect*. The system architect is concerned both in an efficient design process and in a competitive design result. Already with the advent of the second-wave high-rate communications, traditional design techniques, relying on early partitioning and substantial engineering margins per design domain, hit the red-brick wall of design efficiency and product cost. We present solutions to the two central design challenges: how to efficiently design a second-wave WLAN system and how can we prepare already for the upcoming challenges of flexible multi-standard terminals? Consequently, we illustrate true crossdisciplinary electronic system-level design with examples in algorithm–architecture codesign, mixed-signal algorithm/architecture codesign, and crosslayer system exploration. Three recurring themes distinguish this work: preference for scalable and hence reusable architectural concepts, proof of concept through actual design and experimental verification, and consequent analysis of design steps and their development into a methodology.

---

[1] Traditional broadcasting-only services such as radio or TV distribution are excluded.
[2] In IMEC, the term M4 for multimode multimedia was coined for this communication and service paradigm. Outside, the term 4G is often used but sometimes also 5G.

The research described here is based on the Ph.D. dissertation of the author which took place mainly between 1998 and 2003 at the Interuniversity Microelectronics Center (IMEC), an independent large-scale research center for micro- and nanoelectronics and its applications in Leuven, Belgium.

Our research resulted in the world's first two low-cost and high-performance OFDM baseband ASICs for wireless LAN, practical solutions for mixed-signal acquisition, digital front-end nonideality compensation, and application-driven transceiver design.

Concerning design methodologies, we developed a practical digital and mixed-signal design flow, contributed to behavioral modeling, cosimulation and design technology, and extended the multiobjective optimization paradigm to a practically applicable design–calibration–run-time approach as well as showed its suitability in the mixed-signal architectural and crosslayer domain.

This book is not focusing on the latest modifications and adaptations in the quickly emerging world of standards around WLANs. Instead, it aims at capturing the design challenges, decision-taking processes, and crossdisciplinary aspects of designing complex transceiver systems in the wireless domain and proposing generic ways of addressing these challenges. This is at the heart of the tasks of a *system architect* which in the era of system on chip (SoC) coincides more and more with the one of a *chip architect*.

# Acknowledgments

This book has emerged from my Ph.D. dissertation at IMEC/KU Leuven. So, most acknowledgments go into that direction. However, the way that Ph.D. project started and the way it moved on is an integral part of the story. To understand the motivation behind the Ph.D. work and ultimately behind this book, it may be necessary to know more than the purely technical aspects of designing wireless LANs. Hence, I want to disclose some of it to the reader of this book.

Going for a Ph.D. is always striving for the creation of something new: groundbreaking results and major contributions to the state of the art. Definitively, but it is not also about crossing traditional research boundaries, challenging discussions, and fruitful work with other people. So it would not be poor if we only looked at the endpoint of the story and not at the travel itself?

To begin with, this thesis was not born as the child of a "normal" Ph.D. student's life, which starts with a supervisor and a nicely defined focused topic, with a scholarship, all in a familiar environment, and a birth after 4 years. Rather, it has been a 6-year journey, starting with a sudden move to Belgium. What followed turned out to be a tricky balancing act between industrial projects and scientific publications, between teaching and project organization, a walk across different groups, and meandering along the chasms of analog and digital, systems and components, implementation and methodology. It has been an adventurous quest, an all-inclusive trip with pain and joy, and the simple conclusion that I would not want to miss anything from it, after all.

Particularly, I will never know how to thank all the people that made this journey fruitful and interesting. Still, I will make a try, so let us dig into the roots of how it all started...

*Back in Germany*

It would have never happened this way, I had not followed the vivid lectures of Prof. Robert Maurer[†] on RF communications and microwave design back in Germany. All of a sudden, my eyes opened up for the wide world of wireless communications, of microwave and RF, of analog and digital. Add to this the overwhelming experience of 3 years part-time work in a great atmosphere at the Institute for Biomedical Engineering, which made engineering put in a perspective to the realms of biology and medicine. Thanks to them the ground was paved for my interest in personal wireless communications systems.

*IMEC comes into the picture*

My diploma in reach (not yet in my pocket), a thesis on wireless LAN for biomedical applications[3] just behind me, and an invitation for an interview based on a brief e-mail sent before Christmas 1996, I arrived in Leuven on a cold day, even with snow (quite exceptional these days) in the corners of the parking. While having no experience in microelectronics and applying for a job as "wireless expert in ASIC design," destiny appeared in the form of Bert Gyselinckx, Marc Engels, and Ivo Bolsens who put an incredible amount of trust into me and offered me to join the WISE group. Fortunately, I soon got company on my journey through the wonders of wireless LANs, when Liesbet Van der Perre joined a few weeks later. I still remember her saying "Proof is the bottom line for everyone" and our first bottles of champagne for the Festival ASIC.

*Well, where is the Ph.D. starting in the end?*

The inspiring talks of Prof. Hugo De Man illustrating the challenges of system-level design and expressing the crucial need for crossdisciplinary research grew the idea for a broad, system-level and both design- and methodology-oriented thesis: mixed analog/digital exploration and design for wireless broadband transceivers. A start had been made already with the digital VLSI designs of Festival and Carnival. As the topic suggested, revolution instead of evolution was the way; hence I moved with my new focus into the mixed-signal and RF applications group to complete this work. I owe invaluable appreciation to my promoters Prof. De Man and Prof. Georges Gielen for their inspiration, scientific criticism, and support during my Ph.D., and to Marc Engels, Stéphane Donnay, Gerd Vandersteen, Piet Wambacq, Liesbet Van der Perre, and Bert Gyselinckx for their patience and help, for suggestions, reviewing, and proofreading.

*You never work alone*

Research is a path that you do not walk alone and it particularly means that you never work alone either. My thanks go to all people I enjoyed working with and learning from: Veerle and Alain for telling me what digital design is about *en voor een warme ontvangst die mijn taalkennis Nederlands zonder twijfel bevorderde*. Thanks also to Geert, Mario, and Mustafa during the busy days of Festival and Carnival tape-outs. For the algorithmic side of life, my thanks go to Patrick, Steven, Luc, Frederik, Huub, and Andrew; for occasional demo highlights and nightmares to Roeland (2×), Tom, Mike, Bhasker, Peter, and Maryse. Moreover, thanks to Patrick, Luc, Radim, and Erik for being incredibly patient with respect to OCAPI bells and whistles. Thanks to all *former* MiRA people for providing me a nice welcome in 2000 and a steady home and especially for the open discussions and their CAD support to Michaël, Petr, and Gerd. Particular thanks to Björn, Joris, Hideki, and Guido. Who did I forget? Indeed, Boris, for having a great time, tough discussions, and for bringing me always down to earth when dreaming in

---

[3] In fact, this work centered around body area networks which, since 2003, have been receiving significant scientific attention. In those days of 1996, it may not have been the right time yet but this is another story.

the Pareto space. In this space, I enjoyed adventures together with Bruno, Sofie, Francky, and Gregory. Many thanks go also to Wendy and Yves at the VUB ELEC lab for real measurements. I also enjoyed a great time as their daily thesis advisor with Mario, Ludwig, and Roeland. I guess I would better stop here and now before I have addressed everyone in DESICS individually. So, last but not least, many thanks to Annemie, Karine, and Myriam for their warm help and support.

Numerous contacts with industry provided me with an amazingly rich amount of feedback. Particular thanks for vivid discussions and permanent challenging, for their technical support, for bottles of champagne and their appreciation of our work goes to National Semiconductors. Thanks also to project partners at Motorola Genève,[4] Infineon Technologies in München and Villach, ST Microelectronics in Pavia, and Sony Japan for their interest in and feedback on my work at IMEC. Finally, I would like to express my gratitude to everyone at Resonext Communications[5] for a great time in the hot phase of a startup endeavor, particularly to Jess, Karl, Farbod, Patrick, and Radim. I am very grateful to Marc, Bert, and Stéphane at IMEC and David and Morteza at Resonext for offering me this unique opportunity.

*Friends at work*

Besides work, there are other good reasons to spend time together with your colleagues. So, floorball comes into my mind. Having started with only a handful of people in the *good old* VSDM days in late 1999 – it must have been one of those rainy autumn sunsets – things have grown and this sport still gathers quite some people every week. This was a welcome activity complementing office life and providing a good balance for the, sometimes, unavoidable deadlines, weekend and evening work. I have made numerous friends with people working throughout whole IMEC and outside of IMEC, a truly unforgettable experience and one of the reason why I simply love this place and its people.

My greatest gratitude goes to my mom and dad, who always trusted in me, helped me on whatever way I had chosen and supported my sudden decision in January 1997 to leave for Belgium: Von Herzen-vielen, vielen Dank für Eure Unterstützung. I would like to devote this book to both of them.

Leuven                                                              Wolfgang Eberle
October 2007

---

[4] Now Freescale Semiconductor.
[5] In the meanwhile, Resonext Communications have been acquired by RF Micro Devices (RFMD).

# Contents

# 1 Introduction

*You see, wire telegraphy is a kind of very, very long cat. You pull his tail in New York and his head is meowing in Los Angeles. Do you understand this? And radio operates exactly the same way: you send signals here; they receive them there. The only difference is that there is no cat.*
Albert Einstein, 1879–1955.[6]

*A problem well stated is a problem half solved.*
Charles F. Kettering, 1876–1958.[6]

This work reflects the research process and its results that the author has obtained concerning functional and architectural system design and design methodology for high-data rate wireless local area networks (WLANs). In this context, functional design aims at achieving a particular transformation of inputs to outputs independent of implementation and hence cost constraints. Architectural design focuses on the efficient realization of input/output transformations constrained by available technology constraints. We approach the entire design process from the perspective of a system architect.

Our main contribution is a structured approach to wireless transceiver design, which aims at enlightening the traditional black box of system design into a structured gray, ultimately white box. The traditional approach of early partitioning of design tasks according to *disciplines*, such as analog or digital, functional or architectural, physical or higher OSI layers, has over the years resulted in a huge design efficiency gap and intolerable accumulating design

---

[6] Clearly showing Einstein's inventiveness and spirit, this explanation raises more questions than it answers. For sure, Kettering would have said that Einstein's view on it was great but unfortunately of no worth to get on to a pragmatic solution. Ever striving to get his new ideas transferred into reality, he pushed development as a cooperative team effort. But, with Kettering's words, both would have agreed that "research is a high hat word that scares a lot of people. It needn't. It is rather simple. Essentially, research is nothing but a state of mind – a friendly, welcoming attitude toward change."

margins. Unlike in *traditional design*, we treat functional design, architectural design, and design methodology in a *joint, crossdisciplinary* manner (**Fig. 1.1**).

Our approach leads, first of all, to a better understanding of the design tradeoffs involved in wireless design; in a second step, this awareness enables us to propose better functional and architectural solutions and a more efficient design flow. We use high-data rate WLANs as a real-world driver application to illustrate our approach; yet, most solutions were conceived to scale such that the underlying principles can be reused in the design of future multistandard multimedia transceivers.[7]

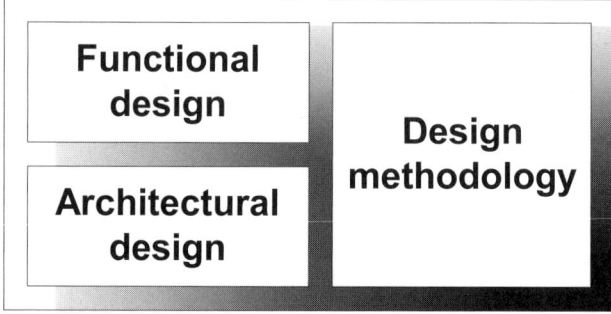

**Fig. 1.1.** Three components of the design of integrated systems are jointly treated

Particular contributions to the state of the art were achieved in each of the three domains. On the functional side, novel algorithms were conceived for the acquisition, mixed-signal front-end compensation, equalization, and tracking process in packet transmission. On the architectural side, a baseband processor architecture for OFDM has been proposed that efficiently accommodates the functionality mentioned before, introduces some energy–performance scalability features, and integrates neatly both with analog front-ends and the MAC/DLC layer. Finally, design technology has been developed to support the efficient exploration and design of functional, architectural, mixed-signal, and crosslayer design.

This introductory chapter is structured as follows. Section 1.1 sets the scene with a description of current wireless broadband communications and its evolution to integrated multimode multimedia devices. Section 1.2 focuses on the design process for such a wireless local area transceiver, which results in the situation and motivation of our work. Section 1.3 sets clear objectives and details the methodology we followed to approach them. Section 1.4 gives an outlook on the contents and structure of this work and puts forward our contributions to the state of the art.

---

[7] In the most general case, we require a multiband, multimode, and multifunction transceiver [Wiesler01]. Since standardization often covers both band and mode require-ments, we refer to the first two criteria as multistandard.

## 1.1 Context

Wireless communications today has become a mature field, yet it still faces tremendous growth rates and continuously enters new markets. A mix of new service ideas together with technology awareness and exploitation of the cost benefits in an economy of scale seems to have created a perpetuum mobile.

Widespread usage appeared when the introduction of 2G mobile telephony pushed mobile phone usage from 10 million subscribers in 1990 to an estimated 700 million in 2001 and an expected 2 billion in 2007 [Rappaport02]. Originally, the breakthrough of 3G deployment was expected for 2004, which, after about 10 years of development, will enable the transition from voice to data traffic and from data rates of 9.6 kbps for 2G to 384 kbps for high mobility and 2 Mbps for reduced mobility [Ojanpera98]. However, a combination of technical challenges and difficulties in clearly positioning this technology in the marketplace has not yet led to widespread deployment by now. These mobile standards establish a cellular layer in the range-data rate plane (**Fig. 1.2**), exhibiting a clear trend toward higher data rates.

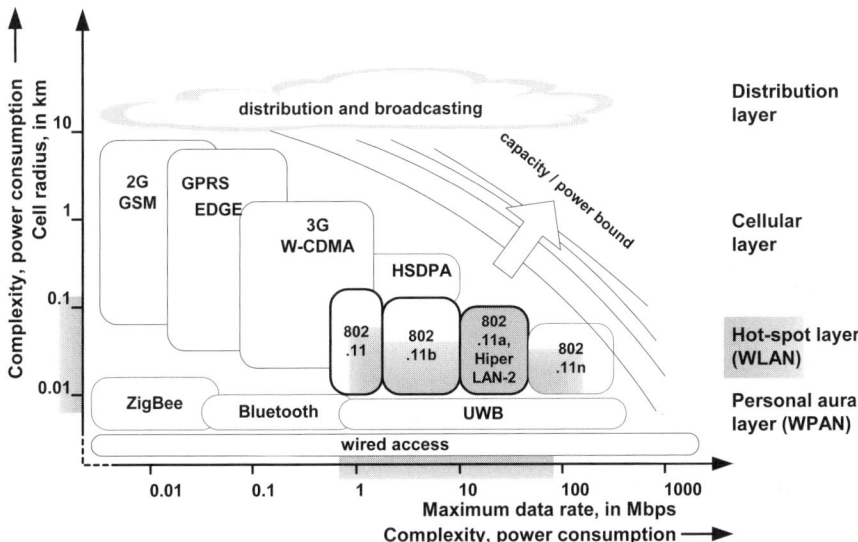

**Fig. 1.2.** With higher data rates and flexibility at moderate power consumption, wireless LAN has been establishing a separate layer that fills the gap between short-range personal area networks and classical cellular solutions. *Shaded areas* denote our involvement around 802.11a/b/n

A similar evolution toward higher data rates has taken place in the field of distribution and broadcasting, e.g., through a family of digital video broadcasting standards [DVB-T, DVB-H] for suburban and rural coverage, and stationary wireless access schemes for urban coverage [Honcharenko97, Eklund02]. For shorter distances up to 100–150 m, WLANs appeared. WLANs emerge as a hot-spot layer to fill the gap for higher data rates and high network density in indoor and campus environments, which could otherwise overload cellular infrastructure. For even shorter distances, notably inside the same room or in the aura of persons, a family of WPANs came up [Gutierrez01] and ultra-wideband (UWB) solutions were proposed. Recently, WiMAX [IEEE802.16] and mobile WiMAX [IEEE802.16e] have been developed to extend area coverage and higher mobility, respectively. Next to the development of solutions for specific applications such as short-range broadband, we notice a general trend – summarized in the terms "beyond 3G" or "4G" – that aims at higher efficiencies, getting closer to the capacity/power bound.

In general, we can observe that both an increase in range/mobility and in data rate lead to higher system complexity and hence power consumption; portable devices avoid the *empty spot* at the upper-right corner of the design space where the gain in system capacity density becomes prohibitively low and/or power consumption exceeds operation specifications of portable devices.

Our research has primarily focused on wireless LANs for indoor and campus operation. They originally emerged from the DECT standard, which from 1992 on was mainly used for cordless telephone applications [DuttaRoy99]. Later on, ETSI and IEEE standardization bodies started separate initiatives for wireless LAN standards: the HiperLAN and IEEE 802.11 families,[8] respectively. Early versions suffered from higher costs and lower data rates compared to their wired Ethernet LAN counterparts and could only move into markets where mobility or convenient ad hoc installation was a key; with the *no new wires* paradigm, the medical care, campus networks, conferences, and warehouse logistics could be entered [Rose01]. A breakthrough was only possible through placement in a consumer market. The idea of *networks for home* was born but has put an even higher pressure on manufacturers to design low-cost and easy-to-use WLAN products within a sufficiently short time to market [DuttaRoy99, Rose01]. Today, everyone can buy WLAN terminal and access points in supermarkets; more and more hot-spot locations appear where people can enjoy wireless LAN access while sitting in a café or walking in a shopping mall. Companies change their infrastructure and WLAN is on the verge to become a standard component of cell phones like Bluetooth became for short-range communication.[9] While sales grow, chipset prices are dropping from $20 in 2002 over $8 in 2003 to about $4 in 2004 and expected $2 in 2006 [Wheeler03, Merritt03].

So, what is next? Parallel evolution paths have established a multilayer structure (**Fig. 1.3**). Next-generation 4G communication systems will benefit from vertical handover between these layers [Mohr00], selecting the most appropriate end-to-end route through the available networks to provide the user with desired quality

---

[8] For a more detailed view, we refer to Chap. 3.

[9] Intel: "a cheap way of connecting," Boingo Wireless: "a standard component of cell phones," and Qualcomm: "we are on standby when requests come from carriers."

of experience (QoE).[10] Due to the enormous difference in requirements across the range-data rate plane, the appearance of a single, flexible standard is very unlikely. Instead, future terminals will have to flexibly cope with multiple transmission techniques as diverse as single carrier and multiple carrier, single antenna and multiple antenna, CDMA or TDMA.

This integration of multiple modes and standards into a portable device represents a major increase in design complexity. For 2G and 3G systems, a 10-year cycle between research and successful market appearance seems appropriate [Raivio01]. Still, the introduction of 3G faced such a large amount of, *also technical*, problems that the original term "beyond 3G" [Steele00] has been deliberately replaced by the less biased 4G or even 5G [Raivio01]. It appears beneficial to learn from the difficulties that the introduction of 3G mobile faced and compare them to the important points for the success of wireless LAN:

- Achieving substantially higher rates than the previous generation and sufficiently high for relevant applications and services is crucial.
- Providing a sufficiently open communication system relying on a layered approach such as IP is beneficial to meet time-to-market constraints and foster development of services.
- Delivering low-cost devices on-time, preferably from multiple manufacturers, is a key concern.
- Exploiting shared unlicensed frequency spectrum which avoided the high financial cost and risk that have been introduced by governments when auctioning spectrum for 3G operation.

We can conclude that the current wireless communications scene offers a tremendous potential for growth in consumer markets; the required increase in data rates and the move toward more flexible multistandard devices, however, results in a severe performance/cost dilemma. Moreover, the current explosion in standardization activity and the needs for multistandard devices push for a fundamental increase in exploration and design efficiency to meet time-to-market constraints. Essentially, the gap between *abstract* service and communication concept definitions and their *concrete* implementation in products has never been as wide as today.

## 1.2   Motivation and Objectives

In fact, Sect. 1.1 sketched the wireless evolution over the last 15 years until 2005. Obviously, the *initial* motivation for this research work must be found at its beginning. Hence, *back* to the concrete situation in 1997.

---

[10] The term *QoE* is defined in Chap. 3; it could be *roughly* described as quality of service (QoS) perceived by the user.

As early as 1996, Harris Semiconductor[11] had introduced its first PRISM chipset for spread-spectrum-based wireless LAN at 1–2 Mbps which turned the company later into the leading provider of 802.11b chipsets [Harris96]. The InfoPad multimedia terminal at UC Berkeley had a similar focus with a 1–2 Mbps peak data rate for the downlink but a low 64–128 kbps asymmetric uplink only [Truman98]. However, 1–2 Mbps neither could compete with the wired 100-Mbps Ethernet solution nor could these data rates support true multimedia or compete in the business and office market. There was only very little preliminary work existing on the combination of OFDM in a WLAN context [McDermott97]. Their focus on discrete board-level design, initially 40-GHz carrier frequencies and picocells, using only 16 subcarriers and a spectrally inefficient DQPSK modulation left considerable room for our research.

This situation represented a common starting point for two study paths at IMEC: one was to explore theoretically algorithms and techniques for a further increase in data rate [Vandenameele00, Thoen02a] and the other was to look for solutions to close the design gap between theory and practice. We adhere to the second approach. We jointly assumed, for overall infrastructure cost reduction, a microcellular approach for indoor and campus environments relying on moderate RF frequencies in the 2–6 GHz range, and the need for a significant increase in data rates into the 100-Mbps range.

Not uncommon to new trends and markets, skepticism was present in 1997 on whether wireless LANs with high-data rates could be designed for sufficiently low cost and power consumption. The recent spread-spectrum-based 1–2 Mbps products and the 11-Mbps 802.11b proposal being in an early standardization phase in these days, turned down chances for a high-rate proposal based on a fairly unknown technology such as OFDM. Over time, with prior advancement in the design, three of these fundamental roadblocks appeared at the application side:

1.  The wireless world in 1997 was – to a large part – focused on single-carrier transmission schemes. Industry was yet to be fully convinced that digital solutions could really enhance and replace significant parts of their analog transceiver designs. Despite promising forecasts for OFDM [Bingham90], many feared the complexity of large-size FFTs such as in the upcoming DAB or DVB-T work [Bidet95]; another challenge to overcome was the need for novel solutions for practical problems such as channel estimation, signal acquisition, and synchronization [Meyr98]. Proof was needed that an efficient digital solution would be feasible in the wireless LAN context.
2.  Once digital complexity was mastered, skepticism focused on the analog/RF front-end part of the wireless LAN. The need for high-performance front-ends to cope with OFDM would unavoidably lead to

---

[11] Harris Semiconductor was spun off as Intersil Corp. in August 1999 and divested its WLAN business to GlobespanVirata in August 2003.

high-cost and high-power consumption [Martone00]. At that time, evidence was yet to bring that digital compensation techniques had the capability of relaxing front-end specifications and increasing the performance at negligible additional cost.

3. While these first two steps may have improved performance, a link analysis revealed that the transmitter, more notably the power amplifier, represented a major power consumption bottleneck [Raab02]. This is partly due to the fact that transmitters have been *traditionally* designed for worst-case conditions and basic class-A power amplifier topologies are the rule. We strive for solutions to reduce the average power consumption following three approaches: scenario analysis, power control [Ebert99], and DSP-controlled power amplifier usage [Asbeck01].

Inherent in all three suggestions, we find the need for crossdisciplinary design: at the algorithm–architecture, analog–digital, and crosslayer system–component level. Unfortunately, joint treatment of multiple domains imposes a significant increase of design complexity to the designer. This can only be resolved in a structured way through methodology and, eventually, design technology and computer-aided design (CAD) tool support. In particular, two challenges can be identified that, together with the previous three, establish a total of five challenges that we will address:

4. Design technology for architectural design and design refinement at the physical layer,[12] be it algorithm–architecture or analog–digital, is required for the DSP core and the suggested digital compensation techniques are needed both at transmit and receive side. While appropriate solutions have become available, e.g., for HW/SW codesign [Bolsens97], the mixed-signal threshold has not been sufficiently reduced despite some positive signs as early as 1994 [Halim94]: both the SRC and MEDEA consortia keep mixed-signal design listed as a significant research gap [SRC00, MEDEA02, MEDEA05].

5. Multimode capability and the move toward multistandard solutions increase the need for application and design space exploration. Already for WLAN, it could be shown that some modes in the standard are redundant [Doufexi02], while inflexibility at other places diminishes performance [Asbeck01]. Again, methodologies were lacking that enable a system-wide, mixed-signal exploration to establish a balanced, performing and cost-efficient, flexible design [SRC00, MEDEA02, MEDEA05].

It becomes obvious why, given this amount of unanswered questions already for plain OFDM, we decided to center our work at applying OFDM for WLAN. Exten-

---

[12] The physical layer of the OSI networking layer scheme is meant here, *not* physical design.

sions such as multiple-antenna techniques, adaptive loading, or advanced coding techniques have not been treated in our work, but are currently studied by others.

Importantly, most questions arose around the *joint* design of the analog or digital signal processing functionality. A look at the OSI networking layering scheme for wireless LAN explains this fact (**Fig. 1.3**). Established in the telecommunications field with seven layers ranging from the application down to the physical layer, the IP-centric data communications world separated the stack into an application-oriented top and a transport-oriented bottom part. Contrary to the telecom world, both parts are treated in a largely independent way, such that wireless LAN itself only introduces new layers at the L1, L2a, and L2b layers. For the rest, it relies on existing IP and other standardized application protocols.

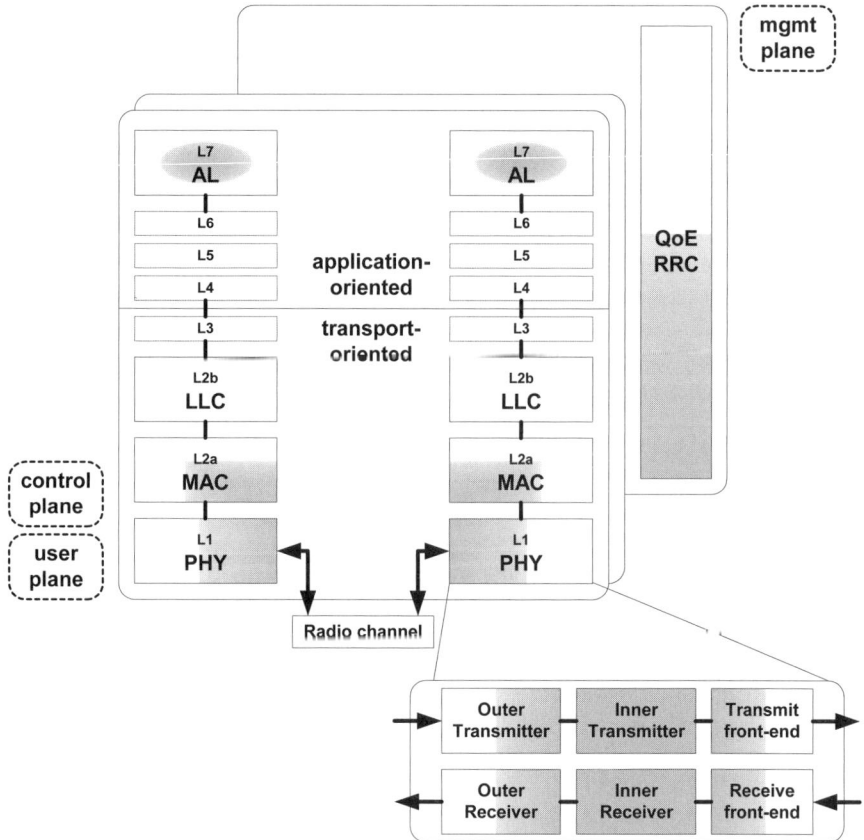

**Fig. 1.3.** In the OSI layering scheme, we treat both aspects in the user/control plane as well as in the management plane. Our main focus is on the lower layers (L1–L2a) with the inner transceiver as a starting point. In the system exploration, we also take into account different service and application requirements (L7). The *shaded areas* denote our involvement on QoE RRC, AL, MAC, and PHY

The OSI scheme distinguishes between the user and the control plane, which are strongly layered, and a nonlayered management plane. In packet-oriented wireless LAN, user and control plane share equal or similar semantics and are hence treated together; they take care of user information transfer and the setup and maintenance of connectivity. The management plane includes radio resource control (RRC) and exchange of resource information between layers and network resources; it is there where total user QoE is enabled.

When we map the questions to the appropriate layers and planes, three foci are recognized: the L1 and L2a layer user/control plane for digital and analog signal processing, the management plane for QoE-aware crosslayer RRC, and the application layer which provides QoE constraints based on application and user requests.

Finally, we can summarize the objectives of our work: we will propose adequate solutions to the five previously mentioned challenges. In the context of OFDM-based 2G wireless LAN and similar future flexible broadband transmission schemes, this means:

- Developing functionality and scalable architectures enabling a low-cost low-power implementation including the mitigation of all relevant practical nonidealities
- Establishing a mixed-signal exploration and design flow that enables an efficient process for this design task, ultimately speeding up time to market

## 1.3   Approach

Having stated the objectives of this work is the necessary entry point for the discussion of how we are approaching[13] a solution.

A key aspect for an efficient development of new architectures and methodologies is a concrete functional application driver with accompanying product constraints. In our case, the driver is wireless LAN and the constraint is the implementation in low-cost low-power terminals. All results will be judged regarding their usefulness for this driver. Moreover, aware of the flexibility requirements for future systems, we also assess the scalability of our solutions; in general, we will favorite *scalable and modular* solutions.

---

[13] This section is sometimes titled "methodology." We prefer *approach* over methodology since *approach* actually refers to a particular mindset rather than the term *methodology* that, originally, relates to a body of practices, procedures, and rules. This distinction is taken up again in Chap. 2.

As a system designer, addressing this goal with a pure top-down approach seems intriguing. However, limited knowledge on the application and constraints prevents him or her from doing so. Motivated by the classical divide-and-conquer approach, we propose here to start from a core problem and, incrementally, add more and more context to it. Which context is added depends on a selection process that determines its criticality to the final system. This results in a mixture of a top-down selection, bottom-up modeling, and top-down decision process, on which we will elaborate more in Chap. 2.

A fundamental difference between engineering and *art* in general is that engineering should develop and follow a methodology through which an increase in design efficiency is obtained in repeated or similar derived designs. Hence, we carefully analyze each design step and place it in a design flow. Design steps which appeared critical in design time or error-prone in the hands of a human designer were taken up for the development of CAD technology.

Clearly, from these assumptions, we expect different results than from the approach in *canonical* dissertations[14] on algorithms, architectures, *or* design technology. When comparing our approach to those commonly applied in dissertations, we recognize three major differences.

First, the system scope will not lead to optimal component or algorithm design *in the classical* sense under narrow conditions. Instead, an operational system will be the result and favoring scalable and modular solutions will guarantee its reuse with extended requirements. Consequently, noncritical components will not be optimized further than necessary.

Second, the focus on the numerous system-level challenges prohibits spending significant effort on, e.g., circuit-level techniques. Instead, scalable and modular architectural solutions will be proposed with clear interfaces and boundary conditions such that, e.g., circuit-level techniques can still be applied independently.

Finally, we try to establish a mixed-signal design flow and prove its benefits in a world still deeply divided into analog *or* digital *and* algorithms *or* architectures. Intentionally, our effort was concentrated on a *crossdisciplinary* design methodology only; in all other cases, we suggested interfacing with experimental or commercially available tools. Our goal is *not* to provide design technology within one domain.

---

[14] Remember that a Ph.D. dissertation lies at the base of this work. The majority of Ph.D. theses address localized problems where deep exploration of a particular problem or class of problems is desired.

## 1.4    Preview of Contents and Contributions

This section introduces the structure and conceptual flow of this book. This introductory chapter represents the first of, in total, eight chapters. The discussion of a variety of design and design methodology aspects at different applications and abstraction levels calls for a particular structure for this text to enable quick access both to general ideas and detailed information (**Fig. 1.4**). We aimed at a clear, formal separation of functional/architectural design on the one hand and design methodology/technology on the other hand.

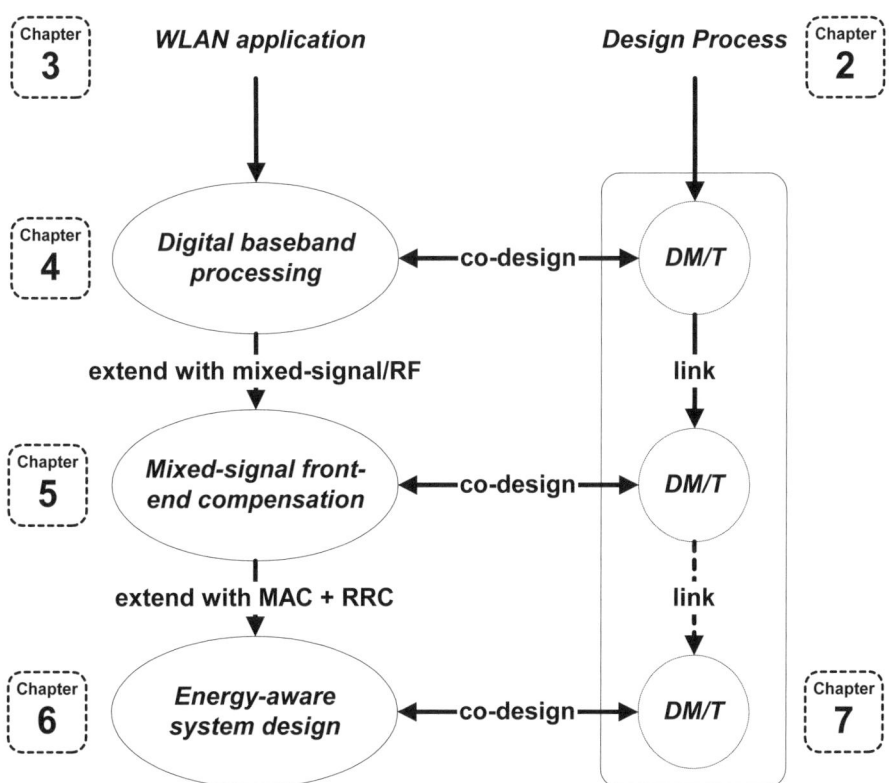

**Fig. 1.4.** Our research has been subdivided into two parallel tracks: an application-oriented design track covered in Chaps. 3–6, which is embedded in a design methodology/technology (DM/T) context covered in Chaps. 2 and 7

Chapters 3–6 focus on the design aspects for the WLAN driver case, extending its functional and architectural scope from a general specification over a single design aspect toward the system level. These four chapters are embedded in a design process by Chaps. 2 and 7; the first introduces its requirements and philosophy, the latter details the developed design technology (DT) and our experiences during its application to the WLAN design driver. As illustrated, design and design methodology (DM) were, in practice, codeveloped.

Chapter 2 focuses on the design process and methodology challenges apparent to the design of integrated wireless communications systems. It motivates and situates the functional and architectural research work in Chaps. 3–6 in a design flow. We stress that only a structured, crossdisciplinary approach in design can reduce the design efficiency gap and pave the path for wireless terminals with appealing performance–cost properties. Two techniques have been addressed in particular: first, the consequent codevelopment of design and design technology to increase design efficiency; second, the systematic multiobjective exploration and partitioning of a design into design-time and run-time aspects to reduce product cost for the same performance:

- The importance of a crossdisciplinary approach to wireless system design has been motivated and illustrated in [Eberle02c, Eberle02e, Verkest01a, Bougard03a].

Chapter 3 introduces WLANs as particular application driver, which is used throughout the entire book to analyze actual design issues and prove both our proposed design and design technology solutions against a practical case. Since system design, as a *true* engineering discipline, is mainly concerned with living up to its numerous constraints, we will briefly review constraints coming from standardization, business, application, technological, and physical constraints. We motivate three system-level assumptions: a microcellular approach to wireless LAN, the need for a significant increase of peak data rates from existing 1 Mbit s$^{-1}$ to about 100 Mbit s$^{-1}$, and the importance to include flexibility and quality-of-service (QoS) awareness for multimedia applications:

- The microcellular networking scheme appears in [Eberle97a, Eberle97b]. A substantial increase in data rate by two orders of magnitude is recommended in [Engels98, Gyselinckx98]. Multimedia applications as a main driver are proposed in [Deneire00a, Deneire00b] and importance and implications of power and QoS awareness are addressed in [Eberle02e].

Chapter 4 presents the complete functional and architectural concept for the digital signal processing core of the OFDM wireless LAN modem. The data rate extension to about 100 Mbit s$^{-1}$ required innovations in algorithmic and architectural design. Moreover, an innovative C++-based design flow was applied. Each design step is motivated and placed in the context of this design flow. Finally, observations from the two ASIC designs and an FPGA implementation are analyzed:

- We proved that a cost-efficient solution for the core-FFT functionality is viable and that the FFT is *not* the design bottleneck for OFDM wireless LAN [Eberle97b, Eberle99a, Vergara98a, Gyselinckx99].
- A functional low-cost transceiver concept for OFDM was introduced in [Eberle99a, Eberle01b]; a high-performance transceiver concept was added in [Eberle01a]. Importantly, both concepts distinguished themselves from the state of the art by a significant increase in data rate and the integration of the complete synchronization *and* equalization functionality of the receiver. Particular aspects on acquisition and synchronization were further published in [Eberle00c, Fort03a].
- Both low-cost and high-performance transceivers were implemented and experimentally verified as ASICs in digital 0.35- and 0.18-μm CMOS technology using a novel C++ design flow [Eberle00a, Eberle01a, Eberle01b]. The chosen distributed multiprocessor architecture has an excellent scalability. Details and an extension to a digital-IF front-end were published in [Eberle98b, Eberle00b].
- A patent on the transceiver architecture was filed in 1999 and has been granted in 2004. It covers in particular the partial reuse, data format conversion, and acquisition architecture [Eberle99b].

Chapter 5 extends the work in the previous chapter toward the analog receiver front-end. This widening in focus is crucial since analog nonidealities have proven to have a more detrimental effect than the already harsh indoor radio channel. In particular, we address predominantly digital and hence scalable techniques to mitigate gain, offset, mismatch, and phase noise problems. These improvements can be used either to relax front-end specifications, and hence cost, *or* to increase performance at equal cost. Specific attention in algorithmic and architectural development was required since the wireless LAN packet transmission scheme requires estimation and compensation of these nonidealities being accomplished at low protocol overhead. We will show that this combination of requirements renders most existing solutions unusable:

- Front-end compensation techniques have been published in [Eberle00c, Eberle02a, Eberle02b, Eberle02c, Eberle03] for automatic gain control and DC offset compensation; in [Fort03b], an approach for efficient interaction between automatic gain control and synchronization is described. Our AGC approach has also been described as a concrete application case for design-time/run-time multiobjective optimization in a patent filing [Eberle02d].
- The impact of the transmit/receive transfer function, including radio channel, analog, and digital filtering, on timing synchronization has been studied in [Debaillie01a, Debaillie01b, Debaillie02], leading to practical design criteria for filters and timing synchronization.

- In [Eberle02a, Côme04], we proposed an elegant architectural extension to the OFDM core that embeds all compensation techniques; the result is a practically relevant and consistent acquisition and tracking architecture.

Chapter 6 raises the abstraction level from subsystem to system design, particularly to the design of the radio link. Application scenarios for wireless LANs are very variable with changing environments, variable payload traffic, and QoS requirements. Traditionally, systems were statically designed to cope with the worst-case link and traffic conditions, with unrealistically high-power consumption as a result. Consequently, the power amplifier was identified as the main power bottleneck. We show how design-time scenario exploration can lead to better component choices and how run-time QoS-driven component usage can significantly reduce the average power consumption:

- QoS-driven multiobjective system exploration and scenario-/application-aware design are described in [Eberle02e, VanDriessche03] with main focus on the transmitter; an extension including receiver scalability appears in [Bougard03a].
- Practicality of transmit chain power reduction has been studied for the power amplifier in [Eberle02c, VanDriessche03, Bougard03a]; compatible digital signal adaptation methods in the transmitter have also been studied.
- The design-time/run-time approach for a QoS link between analog component steering knobs and higher-layer QoS decisions is described for the power amplifier case in a patent filing [Eberle02d].

Chapter 7 presents the design flow and design methodology developed and applied in Chaps. 4–6. While particular methodology aspects have been introduced already there, this chapter falls in two parts: first, we describe design technology developed to efficiently implement methodology; second, we situate these techniques in an integrated flow for mixed-signal transceiver design. This flow extends the traditional focus of *detailed design refinement* to two threatening methodology gaps in *efficient design space exploration* and *crosslayer design*.

- The application and extension of C++-based design technology for digital ASIC design from the architectural to the synthesis-entry level are described in [Eberle98a, Eberle01b, Verkest01a].
- A digital reuse case involving two ASICs and one FPGA designs is analyzed; some observations have been published in [Verkest01a, Eberle01b].
- The development of MATLAB/C++-based simulation technology for front-end and mixed-signal design and its integration with the digital design flow is detailed in [Vandersteen01, Wambacq01, Wambacq02a, Wambacq02b]; particular behavioral modeling techniques for strong

      mixed-signal interaction are addressed in [Eberle03].
- Our design-time/run-time approach for mixed-signal systems is described in [Eberle02d] and illustrated in [VanDriessche03, Bougard03a].

Chapter 8 concludes with a review of the major achievements and a comparison of the actual design results to the objectives. Forthcoming from our research, a selection of promising topics for future research is presented.

# 2 The System Design Process

*Everyone takes the limits of his own vision for the limits of the world.*
Arthur Schopenhauer, 1788–1860.[15]

Research in microelectronic integration for wireless LANs is obviously the right way to go to obtain low-cost low-power solutions that meet business expectations for a consumer market. So, why not start designing them ad hoc, right now?

Maybe because a few questions remain, such as do we have a clear specification of *what* we want to obtain as a result? Do we know *how* we get from our expectations toward a prototype implementation or a final product? Can we repeat this process?[16] And finally, what does *design* actually mean?

*Designing*[17] as a human activity is not easily captured in a single definition, but a combination of *creating or executing in a highly skilled manner, formulating a plan or devise something*, and *having something as goal or purpose* [AHD00] seems to bring together all important aspects: each design needs a clear goal to achieve; it requires the concept of a method, path, or process that leads to this goal; and it relies on particular skills to reach the goal while staying on the intended path.

---

[15] Schopenhauer's words – "Jeder Mensch hält die Grenzen der Wahrnehmung für die Grenzen der Welt." – obviously go beyond simple perceptions of working style; but has not there always been a fight between old and new methods of accomplishing things? Schopenhauer mentions also that this often includes that new thoughts are first ridiculed, then attacked, and finally become common thinking – "Ein neuer Gedanke wird zuerst verlacht, dann bekämpft, bis er nach längerer Zeit als selbstverständlich gilt."

[16] Obviously, products are designed everyday. A fundamental question is how much experience from previous designs plays a role to reproduce a design success and whether and how some of this experience can be transferred from an experienced human designer to other designers, into a methodology, or into tools.

[17] The word *design* has its origin in the Latin word *designare* which means *to devise (to invent), to mark out for something*.

A particular design activity relies on a requirements specification as input on *what* to achieve and is basically given freedom in *how* to achieve it. Obviously, the most efficient process of achieving the specification is desired. Traditionally, requirements for an electronic system could be rather easily and early partitioned into a few discrete disciplines such as analog and digital, mechanical and electronic, low-power and high-power parts, etc. Little margins on cost, power consumption, and performance for portable, consumer-oriented devices such as WLAN do not allow such an upfront partitioning; instead, a multiobjective system-level optimization is preferred. The definition of a system[18] directly explains that codesign and optimization at a larger scope, taking into account more components together, offer a potential benefit in better functionality or lower cost. However, it also complicates the design process since the design space has been enlarged. **Fig. 2.1** illustrates the difference between vertical intratechnology and horizontal intertechnology exploration. The figure is an extension of Kienhuis' abstraction pyramid [Kienhuis99].

If we assume that the same design cost can be spent, then we might reduce design cost along the vertical design space exploration axis and shift it to the horizontal exploration axis. First, this requires increased efficiency in design refinement and synthesis within a particular technology or discipline. During the last years, progress has been made on this aspect. Second, horizontal exploration requires understanding and interaction between *traditionally* different skills, hence interdisciplinarity. Kienhuis [Kienhuis99] stated that an increase in abstraction level leads to lower exploration cost. However, he assumed the *availability* of efficient and sufficiently accurate behavioral models. This assumption neither does hold for analog and microwave/RF nor will be applicable for digital designs processed in nanotechnology. Also, intertechnology exploration leads to heterogeneous modeling approaches, which largely increase the complexity of horizontal design space exploration. Colwell [Colwell04] stated this as "there's always cost in tying conceptually separate functions together." But, is there an alternative and are these functions actually *conceptually separate*?

This chapter is organized as follows. Section 2.1 describes design as a process. We introduce terminology and the scope of design methodologies and design technology. Section 2.2 focuses on microelectronic system design. Requirements and common approaches are described that allow the identification of areas of mature design technology and methodology gaps. Finally, we synthesize our particular design approach for WLAN. Section 2.3 analyzes the consequences of such a design flow based on different levels of crossdisciplinarity. Codesign of design technology and application is addressed in particular. Section 2.4

---

[18] A system can be defined as a set of interrelated elements perceived as a whole, performing a useful function by interaction with human or other environment. A system is composed of components, typically in a hierarchical way. The interrelation makes the system more useful than the sum of its isolated components. The system reacts to or interacts with its environment.

summarizes our findings and directs the reader to the implications of our design rationale in Chaps. 4–7.

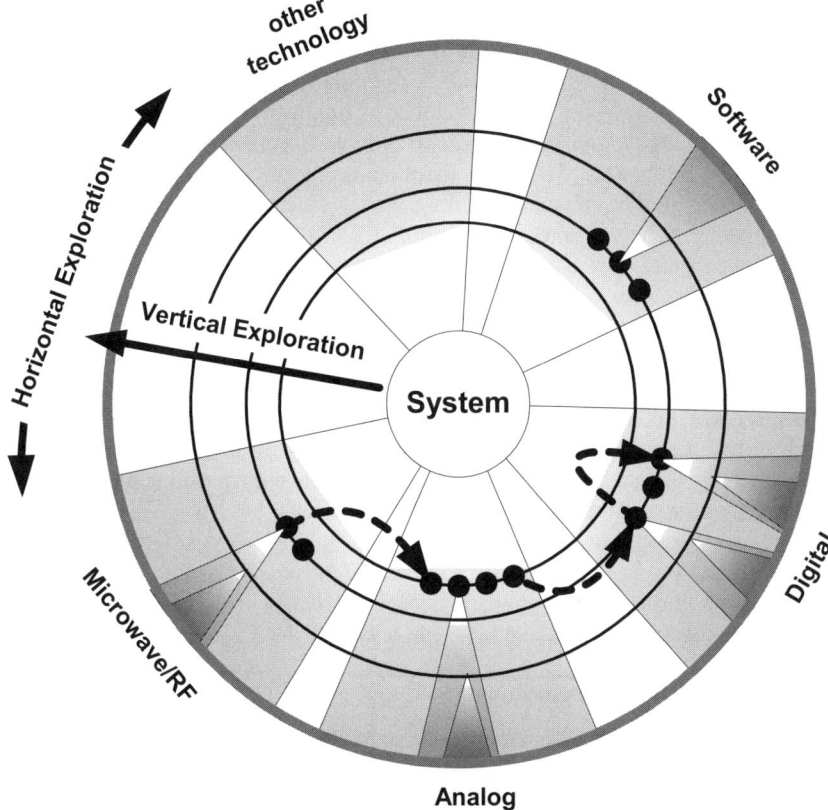

**Fig. 2.1.** Extension of the abstraction pyramid to a multisegmented abstraction circle which captures both intratechnology (*vertical*) and intertechnology (*horizontal*) exploration. Horizontal exploration within one technology may start a trajectory of changes across multiple technologies to meet overall specifications

## 2.1 Design

Both artists and engineers rely on creativity, rooted in knowledge, and craftsmanship, as means for innovation. In principal, they both start with a blank page and a few requirements. However, in contrast to the fine arts, design in the engineering context has become far more constrained a priori by *nontechnical* requirements and cost aspects [Rissone02]. The amount of constraints allows a distinction between two different design goals: the design of a new product and the design of a derivative of an existing product. The first refers to the *blank page* situation in which creativity and exploration is central, the latter expects a quick transition from a changed specification to an upgraded product.

Hence, the purpose of engineering design is the translation of initial requirements into a product or object with these desired properties in a predictable, efficient manner. This identifies the need to describe design as a process [Jacome96]. In contrast to the era of the *renaissance engineer*, where the limited amount of knowledge still allowed a single engineer to be aware of all engineering aspects, design requirements and design processes have become so complex and diverse that engineering is only viable in a team context today. A division of engineering into a large number of specializations and subdisciplines has been the result. Consequently, it becomes more and more difficult for each engineer to identify his or her role in the design process. Managing and enabling smooth transitions of design information between engineers have become *the* major challenge.[19] Lately, the immersion of application and service aspects into the design process, for example the interaction between service cost and service quality, has forced a shift from a product- and hence object-centric approach to a goal-centric approach [vanLamsweerde03]. An example is the embedding of sophisticated control into devices to meet user-specified conflicting goals such as long operation time *or* high-quality content in a portable. This adds environment and content awareness to the design process, making it even more complex.

We will now try to understand the design process in more detail. First, we introduce a clear terminology and structure of the design process. Next, we investigate its development, deployment, and use.

## 2.1.1   Design as a Process

An indisputable fact, design represents a process and shall lead to a particular goal. It is a philosophical question whether we embed the desired characteristics of the process, such as design time or design cost, in the goal or whether we establish them separately as properties of the process. *Design philosophy* may include abstract and hardly quantifiable requirements for the design process such as[20]:

- Maximize the probability of success
- Develop all critical design support in-house
- Use only advanced, stable, and proven CAD tools
- Identify major risk factors first

---

[19] "This made me think that there is no bottom, no top, and every year more steps are added as our field gets chopped into more manageable segments. I look down the stairs [of the design process] and see them disappearing into blackness. Above me the steps ascend into fog." [Lucky03a]

[20] These are quotes originating from various microelectronics company websites.

Clearly, this is input to but can hardly *be* the subject of an engineering thesis. Instead, we focus at a design process that leads us from an initial specification $S[0]$ to a final specification $S[m]$ through a number $m$ of discrete design review steps. We expect from this process that it enables a design team to maximally reach the design goal, i.e., it optimizes the use of capabilities given the initial requirements and constraints.

**Fig. 2.2** outlines the design process in a generic way. Starting from general design principles, we first have to come to a clear initial requirements specification. This specification includes desired and undesired properties as well as constraints. The result is a definition of the *maximum* design space in which a solution for the particular design has to be found. The properties of the requirements and the design space are input to the choice of a particular design methodology. The combination of methodology and design technology, i.e., computer-aided design tools that support the designer, allows establishing a design flow. This flow describes how methodology and tools are applied in the *actual* design, which is the last step in the design process. Note that we will first analyze this ideal *top-down* process. Practical issues are considered in Sect. 2.1.2.

### Requirements Specification

The capturing of initial requirements in a requirements specification is a crucial but often underestimated point in practice. A significant amount of project or product failures can be traced back to an *insufficient* specification [Bell76]. The most important reasons for this are:

- Positive requirements are more difficult to estimate than constraints [PageJ88]; this is obvious since constraints assume a preselection of technology from which then concrete data are available. Definition of requirements requires much more creativity.
- Elicitation of primary objectives is difficult given the complexity and heterogeneity of current products [vanLamsweerde03]; conflicting objectives have become the rule, not an exception.
- A requirements specification shall establish precisely what a product must do without describing how to do it. Unfortunately, distinguishing *what* from *how* is a dilemma in itself [Davis88]. The distinction is necessarily a function of perspective. Hence, requirements differ with the degree of abstraction and as the point of view changes, for example, from the designer to the user of a product.

Several approaches have emerged to deal with this problem, mainly instantiated in the field of software engineering, but largely applicable to system design in general. This ranges from general guidelines such as *structured design* [PageJ88], guidelines for requirements specification documents [IEEE830, IEEE1233], the unified modeling language to capture multiparadigm specifications [UML,

Rumbaugh98], or the KAOS[21] approach in software engineering and support for automated reasoning [vanLamsweerde03].

Essentially, all these approaches treat design as an *object-oriented* process with a particular object instantiation as the endpoint of design. Hence, the design process is focused on managing the *what*-aspect. Recently, a *goal-oriented* view has been introduced [vanLamsweerde03]. This approach tries to embed the *why*-decisions in a design process.[22] This allows tracing back decisions in a design process.

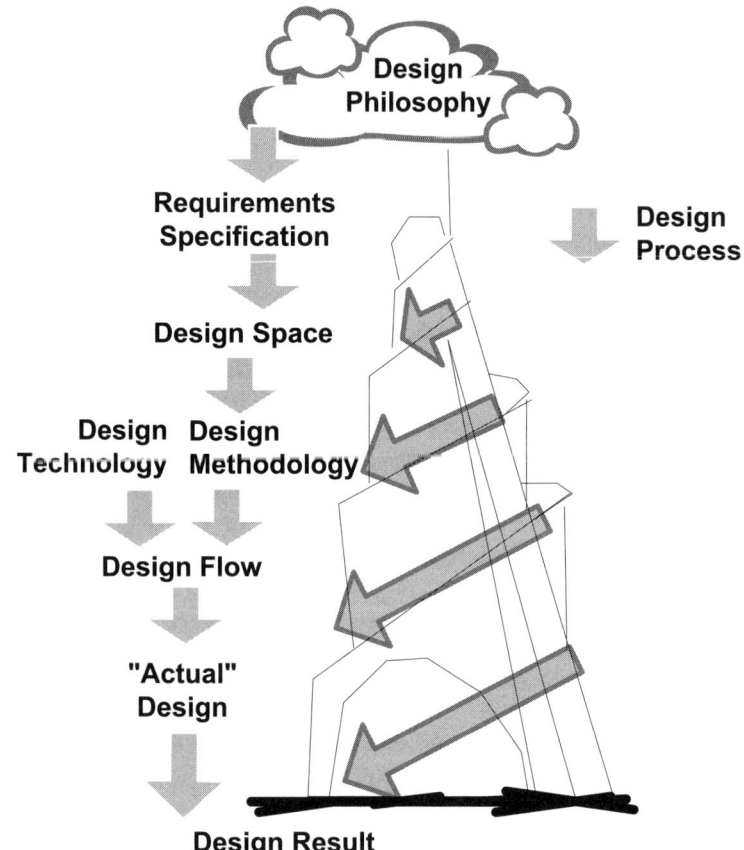

**Fig. 2.2.** The design process structures a design from the hardly quantifiable design philosophy that specifies the goals down to the actual design result

---

[21] KAOS, keep all objectives satisfied.
[22] We can ask the question whether we need a fully formalized specification process; regarding the *what*-aspect, this appears useful. However, most misunderstandings are correlated with mistakes on the *why*-aspects, i.e., with the derivation of wrong specifications. Formalizing the *why*-process appears as an even more challenging task.

With this initial notion of requirements specification, we can reformulate our definition of a design process as a process of deriving a *complete* specification, i.e., a set of properties that uniquely characterize a particular design object, starting from an *initial* requirements specification [Jacome96].

## Design Space

The initial requirements specification implicitly defines the initial design space. Depending on the viewpoint and abstraction level at which the requirements are captured, the application design space is divided into a *problem design space* and a *solution design space*. The problem design space starts from the problem (*goal*) and encapsulates all possible solutions to the problem without constraints. The solution design space is a subspace of the problem design space with particular solution constraints being imposed (e.g., in terms of bounds on cost, latency, etc.). We do not focus on the problem design space in detail. Hence, from now on, the short-term design space denotes the solution design space.

Essentially, the design process aims at identifying the optimal configuration of the solution design space that meets all specifications. The optimal configuration represents a collection of entities that, based on their configuration, perform a specific set of tasks. This equals the definition of a system [Papalambros00]. Hence, the derivation of the complete specification or configuration in the design process completely defines the system.

Essentially, this represents a decision-oriented process that falls into an exploration/ optimization and a mapping/refinement step. The first allows traversal and comparison of different configurations in the design space based on requirement metrics. The second narrows down the design space based on a design decision. A design process that includes an evaluation criterion is a decision-making process. The result is an arbitrarily complex, partially ordered, sequence of generation design steps (mapping and refinement) interleaved with test and validation steps (exploration and optimization).

## Design Methodology, Design Technology, and Computer-Aided Design

This definition of the design process is neither constructive nor predictive in terms of cost. Hence, we need a methodology[23] that establishes a quantitative and qualitative capturing of the progress in the design process, a clear sequence of preferably finite design steps of predictable complexity. Moreover, the effort in implementing and learning the methodology should result in, e.g., better, faster, or more consistent design results compared to an ad hoc approach. Essentially, a methodology must go beyond the description of a single method: it has to be

---

[23] A method is a means or manner or procedure to accomplish something in a regular and systematic way [AHD00]. A methodology establishes a body of methods and principles particular to a particular domain. Interestingly, the authors also state the misuse of the term methodology as a pretentious substitute for method.

general enough to be applicable in a *significantly large* domain, i.e., its deployment and applicability must have been evaluated.[24]

Efficiency and reproducibility advocate the use of computerized automation techniques for those design steps that have been mathematically formalized and do not require human reasoning. This results in design technology[25] or tools. The degree of automation allows a differentiation between electronic design automation (EDA) and computer-aided design (CAD) tools. In practice, no formal distinction is applied but CAD emphasizes that technology only supports the designer while EDA aims at full automation of design steps. A design methodology embraces typically both techniques. The application of design tools to accommodate a particular sequencing of design steps is called a *design flow* or *tool flow*.

### 2.1.2 Application and Rationale of the Design Process

**Fig. 2.2** illustrates that the *actual design process* has a much wider scope than the *actual application design*, which is perceived by the designer as *the* design task. The design process includes both selection and/or development of design methodology and technology and its deployment toward the application designer.

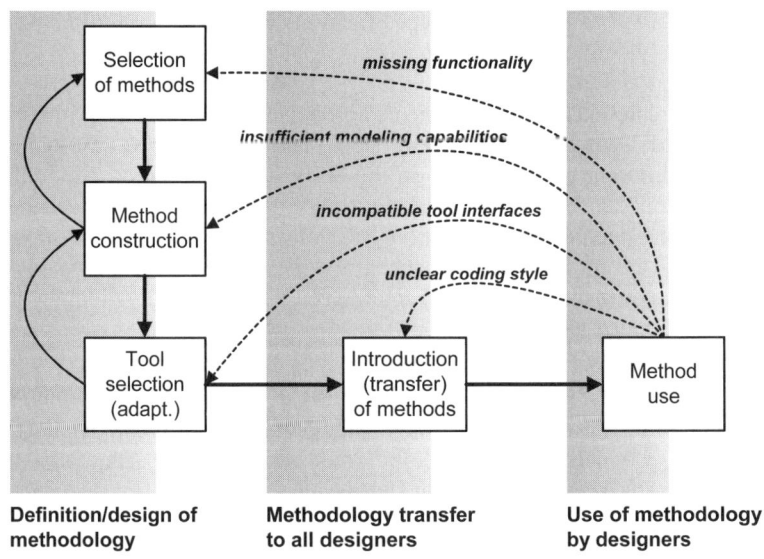

| Definition/design of methodology | Methodology transfer to all designers | Use of methodology by designers |

**Fig. 2.3.** Design methodology is developed in a metaprocess called *method engineering* and essentially consists of the three phases of development, deployment, and use

---

[24] Reference [AHD00] states that misuse of the term methodology as a pretentious substitute in case of nongeneralizable methods has unfortunately become a widespread phenomenon.

[25] Technology is the systematic treatment of an art, craft, or technique [AHD00].

For a successful deployment and use of design methodology, it is important to place this theoretical idea of a design process in a real designer's environment. Applicability *and* acceptance by the designer are the essential proof for the rationale and success of a methodology. **Fig. 2.3** illustrates this transfer of methodology through the three phases of development/definition, deployment, and use. These three aspects are addressed individually.

*Methodology Development*

As we have illustrated, the development of a methodology is itself part of the design process. This partial development, usually called *method engineering*, is itself driven by another process. Kumar and Welke [Kumar92] defines method engineering as a method for designing and implementing domain-specific methods. Hence, method engineering represents a so-called *metamethod*[26] with the goal of instantiating a general metaconcept into an explicit and concrete domain-specific methodology. The derivation from the generic to a particular instantiated concept is achieved through the application of additional constraints and requirements. Practical examples are the limitation to tools from a particular CAD vendor, the usage of a particular software coding style, or the experience of the design teams.

These examples show that methodology development does not necessarily result in writing your own tools. Instead, it consists of methodology selection, optional method construction, and tool selection (**Fig. 2.3**). Depending on requirements, an entire design methodology can be acquired from a CAD vendor along with the corresponding tools. An example is a register transfer level (RTL)-based design flow for synchronous digital designs starting from VHDL and resulting in gate-level netlists.

In general, however, design methodologies require adaptation to *local* requirements since design teams can rarely find a perfect match based on existing methods and tools [Tolvanen98]. This process is called *local method development*. We can distinguish between the development of new methods and the adoption of existing methods to differing constraints or assumptions. Tolvanen [Tolvanen98] classifies the method selection into three principles:

1.  The textbook approach is driven by technical rationality [Schön83], i.e., the idea of reuse. This assumes that similar problems can be approached with a proven solution. However, this often excludes a formal assessment of similarity. Quite often, a number of *potential* techniques are selected and tried out in an ad hoc manner: this is essentially a trial-and-error method.

---

[26] The Greek *meta* means "beyond, after." Interesting from a linguistic point of view, method stems from the Greek *methodos*, which is already an agglomeration of *meta* and *hodos* (the latter meaning way, journey). Thus, the word *method* alone addresses already the metacharacter.

2.  The contingency approach tries to classify problems into domains and reuse again domain-specific solutions [Kumar92]. The approach may result in suboptimal results due to the low granularity in classification but delivers fast results due to maximum reuse.
3.  The method engineering approach requires an exploration and optimization strategy and hence a domain-specific development of techniques. Although a costly approach due to the method development cost, it may be the only technique to find the optimal solution in a complex constrained design space.

Obviously, these approaches can be combined at different levels of decision.

## Deployment of a Design Process

Design methodology is supposed to enable and support the application designer, but is rarely developed by the application designer. Current approaches for method selection and development actually do not provide adequate support for learning and creation of methodical knowledge. Although local (in-house) methods develop over time and methods must be seen as one part of the organizational knowledge, this knowledge is rarely captured in an adequate way [Tolvanen98]. This gap reveals:

- An unawareness of what the in-house design methodology embraces (due to a lack of global view on the design process)
- A hesitation to adopt new, disruptive design methodologies and tools
- A resistance in formalizing the in-house methodology due to cost or effort

This prohibits a fast transfer or modification of the design methodology, e.g., in case of the integration of new team members or the partial change of the design flow. It also prevents the export of design methodology to third parties.

[Ciborra94] and Stolterman [Stoltermann92] have described that the case of transferring skills from a skilled designer to a less proficient one fails often due to a lack in formalization and capability of transferring know-how into methodology.

Stolterman [Stoltermann92] linked the capability of adopting disruptive new techniques to the capability of an organization for radical learning. Typically, individual designers prefer an incremental way of learning. Hence, adoption of new, disruptive methodologies cannot be delegated to individuals but requires the support of the organization to bridge development, transfer, and use phase.

In practice, new methodologies are often introduced through pilot projects in which a number of designers are trained and apply the methodology to a *lightweight* design project. This early feedback often allows a priori corrections and adaptations of the methodology before actual use in large projects. Note, however, that this assumes that the methodology *scales* adequately with the design size. This is not evident for complex system designs.

### The Design Process in Use

Once the design process is used by designers, the practical application gives rise to adaptations. We can distinguish four levels of feedback (**Fig. 2.3**). Ranking in severity from low to high, e.g., missing coding style can be locally supported through additional clarification or training. On the contrary, missing functionality may require the overall adaptation of the design flow and involve the acquisition of different or additional tools from external sources. Only local, domain-specific know-how will allow the appropriate selection of the feedback mechanism and hence minimize the cost of adaptation of the design process.

Ideally, the initial specification should never be modified. However, limitations in the accuracy of technological estimates (e.g., updating performance/cost models when more accurate information becomes available) or nontechnical constraints (e.g., fulfilling the time-to-market constraint that may vary due to changing market interests or competition) frequently require iterative adaptations of the initial specification [Jacome96]. It is essential though to avoid changes that require fundamental changes in the design process and the underlying technology. Consequently, for the technological side, it is important to identify less trustworthy estimates as early as possible in a *risk assessment* step and investigate them prior to other parts.

### Rationale and Conclusions

Referring to the need of a design process, we cannot end this analysis without addressing a measure for its success. Although need and success seem to be cornerstones of a design rationale, they only form part of it, mainly the *reasoning* component. A rationale must be seen as the sum of at least three different forms of knowledge: reason, aesthetics, and ethics.[27]

Aesthetics denotes the ability to judge whether the process as a whole is considered appropriate, i.e., *a good design*. Stolterman [Stoltermann92] proposes the confrontation of the process with the designer in practice to answer this question. Interestingly, his studies revealed a so-called *hidden rationality of practice*, which appears familiar from a phenomenological point of view but deserves a deeper analysis to allow conclusions. This hidden rationality of practice is constituted by two apparent contradictions:

- *A fundamental mistrust in new methodologies*
  Methods that show a simple and consistent picture of the design process appear *irrational* to designers. Designers evaluate their intrinsic structure

---

[27] I. Kant used the slightly different terms of cognitive powers, feeling of pleasure and displeasure, and power of desire in 1790. See also [Stoltermann92].

and come very quickly to the conclusion that a method is not realistic enough. The designers' own complex and disparate view of the process appear to them a more attractive and pragmatic approach. Again, the fact that they cannot describe their own approach in engineering terms makes it acceptable; basically, the missing structure prevents a deep analysis and hence prevents early rejection based on facts. The result is often a fundamental, initial mistrust with new, unfamiliar methodologies.

- *A dilemma between art and engineering*
  In fact, designers have difficulties in identifying the crucial skills that characterize good design; often the only answer is through experience [Stoltermann92]. Thus, system design appears sometimes as an art. System designers themselves and even more those who develop methods would, however, *if they could choose*, like to change the process to be like an engineering process. This either results in uncompleted approaches to formalize design steps or the early denial of any substantial generalization and documentation of *how* a design process looked like.

- *Design granularity: Creativity vs. problem solving*
  Commonly, design is seen as a sequence of problem-solving steps that fix a malfunctioning reality. This impression often results from a too narrow insight of the designer toward the design process. Indeed, when assigned a small task with strict requirements without knowing *why* and *how* these requirements were derived, the designer's reaction becomes understandable. The small granularity prevents the exploitation of the designer's creativity. On the contrary, some designers may want to remain focused in their particular field of knowledge and feel overloaded by too much context. These opposing cases prevent an efficient information exchange within a design team when not addressed particularly.

Finally, we want to summarize a number of critical points on which a candidate design process needs to be evaluated. They form cornerstones of our design rationale:

- An encapsulation of the entire design from requirements specification over methodology development to the actual design is obtained.
- Effort invested in the requirements specification pays off in reduced iterations over the specifications and leads to a more predictable process.
- Methodology is conceived as a whole from development over transfer to the actual use. The importance of local method development is crucial to the successful use of the methodology by designers.
- Developing design methodology is a metaprocess that requires the instantiation of a generic idea into a specific flow on the one hand and on the other hand the generalization of a particular design solution into a reusable method.

- Design has traditionally an object-oriented focus with the product as center. Newer concepts favor a goal-oriented approach that enables a traceback of design decisions.

We will come back to the rationale behind these points frequently in the following chapters, in particular in Chap. 7.

## 2.2 Microelectronic System Design

The goal of electronic system design is to minimize production cost, development time, and cost subject to constraints on performance and functionality of the system [Ferrari99].

Progress in IC fabrication technologies has enabled integration of complete electronic systems on a chip (SoC) or in a package (SiP) [Shen02]. This evolution toward embedded solutions is driven by the size and production cost advantage of integrated solutions and has surpassed the previous *PC era* as an industrial driver [Schaumont01a]. For example, in our particular case, we consider the complete integration of a wireless LAN transceiver – analog, RF, and digital, hardware and software, discrete and integrated components – using an advanced packaging technology with integrated multichip modules (MCMs) [Donnay00]. This move from general-purpose components to heterogeneous application- or domain-specific systems has been recognized by MEDEA [MEDEA02] and ITRS [Edenfeld04], which both introduced system-specific drivers, for example for wireless communications.

Hence, electronic system design moved from a board-level to a chip- or package-level approach. This also means that the task of designing and partitioning the system has moved from selecting discrete components and connecting them on a board into the microelectronics design domain. We refer to this approach as *microelectronic system design*. Characteristics of microelectronic system design are heterogeneous integration and performance/cost awareness.

This section is structured as follows. First, we state the challenges in microelectronic system design that come along with complexity and heterogeneity. Next, we analyze the key methodologies that have been proposed for this design domain. Finally, we synthesize a specific design methodology for our wireless LAN design case.

### 2.2.1   The Challenge of Complexity and Heterogeneity

Difficulties in merging microelectronics with system design have first been revealed in digital design. Digital design so far could benefit from technology scaling according to Moore's law [Moore65]. The result was higher performance at reduced chip size and reduced cost per function. This paved the way for the integration of more functionality onto a single chip. However, it is today clearly recognized (and addressed in detail in Chap. 6) that performance comes at a power cost, maximum performance is not always required, and hence performance can be traded against energy or peak power. Besides this performance/cost tradeoff aspect, we see that this evolution reveals the four major problems that even gain in importance when moving to a truly heterogeneous microelectronic system.

### Technology vs. Design Productivity Gap

The MEDEA roadmaps consistently referred in 1999 [Borel99], in 2002 [MEDEA02], and again in 2005 [MEDEA05] to the increasing gap between the available logic transistors on a chip and the capability of designers to make use of them in application design (**Fig. 2.4**). Due to this gap, the importance of design chains has become as important, if not more, than technology innovation since a lack in design efficiency prevents the full exploitation of a technology advantage and leads to a delay in time to market. Obviously, current EDA support is not sufficient. We can think of two reasons for this fact: either the capabilities of design automation tools were too limited or they were not adequately embedded in the actual design process.

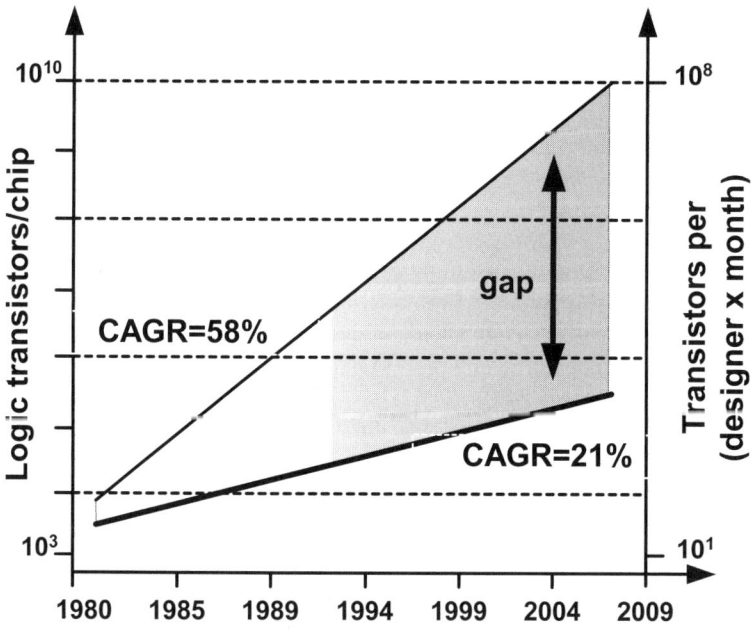

**Fig. 2.4.** The technology vs. design productivity gap has been increasing for a long time. Disruptive design methodologies are needed to close the gap

*Delays and Costs Due to Complexity-Induced Functional Errors*

Advancing technology allows the integration of increasing functionality onto a single chip. However, failures in SoC designs are causing a significant number of redesigns: 48% of the designs fail on first silicon, 20% still on second spin, and 5% require even more than three spins.[28] Failing on first-time-right silicon may lead to missing the market window and hence to a significant loss of profit. Importantly, a large number of these designs actually failed due to functional or specification errors. This reveals a test and verification gap. The yield reduces with increasing heterogeneity in the same design, such as analog and digital on a SoC or discrete and integrated RF components in a SiP. With 20% of mixed-signal SoCs today and a projection of 70% for 2006,[29] the severity of this problem rises dramatically. The urgency for design support in the integration of heterogeneous components is also expressed in the ITRS 2003 roadmap [Edenfeld04].

*Lacking Support for Power- and Energy-Aware Design*

Unfortunately, performance and power consumption in Moore's law scale at the same rate. As a consequence, thermal dissipation problems add tremendous design complexity and slow down the usage of the available chip area. Both ITRS and MEDEA identify low-power design and system power management as key challenges.

*An Incoherent Marketplace for System-Level Design EDA*

In 1993, the EDA industry introduced electronic system-level (ESL) design automation solutions for the first time at a large scale [Goering03]. The very diverse approach with many niche solutions such as graphical code generation, HW/SW codesign, architectural modeling, behavioral synthesis, etc., resulted in fading interest soon. ESL design obtained a bad reputation from that moment on. Around 2000, with the advent of SystemC, a similar movement started with significant EDA industry support to address high-level system modeling in C-based language styles. At the beginning of 2004, a diversification over several approaches and even a withdrawal of major EDA players from the ESL market can be observed.[30] Since then, we have mainly seen a HW/SW digital-driven initiative. ESL design tools for truly heterogeneous systems including analog/RF, power management, sensor/actuator support, etc., appear in 2006 to be far away from the designer's toolbox.

### 2.2.2  State of the Art in Electronic System-Level Design

Techniques for ESL design have to overcome the four problems aforementioned. For an understanding of the state of the art, we first have to introduce the concepts

---

[28] Several quotes from Cadence Design Systems, Simplex, and at the International Symposium on Quality Electronic Design, March 27, 2001; based on a study of Collet International, 2000.

[29] Estimate as of Cadence Design Systems, 2003.

[30] M. Santarini, "Panel debates viability of ESL tools market," EEdesign, March 5, 2004.

of *function*, *architecture*, and *interface* which allow the principles of hierarchy and abstraction.

A system implements a set of functions. A function is an abstract view of the behavior of the system. It is the input/output characterization of the system with respect to its environment. It has no notion of implementation associated to it. Contrary to this, we can describe an architecture that by itself has only a notion of implementation but no particular functionality. The notion of function and architecture depend on the level of abstraction. Higher abstraction levels require a partitioning of the design into visible and hidden or, equivalently, detailed and nondetailed aspects. This introduces general design principles such as hierarchical composition and the concept of interfacing to describe the interaction with a component whose content is not detailed, a so-called *black-box* component model.

## A Chain of Dependencies

Besides the automation of particular design refinement tasks through, for example, automatic synthesis techniques, we can order existing methodology trends into a sequence of depending methodologies with decreasing complexity:

1. multiobjective optimization to take into account concurrent constraints on performance, power, cost, etc., which requires
2. horizontal design space exploration and codesign techniques to compare and validate heterogeneous design solutions; they rely on
3. behavioral and gray-box modeling for an efficient comparison at a manageable level of detail; this calls for
4. design capture, verification, and reuse at higher abstraction levels to speedup the design and improve design robustness

Obviously, the least complex tasks have been addressed first in time and have, so far, been most successful. An example is methodologies for the move of digital design to higher abstraction levels. For verification purposes at design time, a so-called silicon virtual prototype (SVP) is built. Over time, the abstraction level of this prototype has moved from the classical layout, transistor, gate level to the register transfer level [Mehrotra03, Dai03] and finally to the transaction level[31] [Gordon02, Schlebusch03]. Preferably, a SVP is prepared at all levels but verification of system-level features is achieved most quickly at the highest abstraction level. The design cost reduction due to reusing existing components, the so-called IP,[32] increases also with the abstraction level.

---

[31] The concept of transactions actually covers a large range of abstractions and is hence not very well described. An example is the modeling of an actually time-distributed physical bus access as a single transaction with a lumped timing and energy consumption. However, we could also go further and represent several related bus accesses (a burst access) again as a single transaction. The goal of transaction level modeling is finally representing the detailed interaction by its relevant properties.
[32] Intellectual property.

Further abstraction is possible through *platform-* and *interface-based* design techniques. The *Y-chart* approach combines separate functional and architectural specifications through a formal mapping step into an implementation [Gajski83, Ferrari99, Kienhuis99]. The separate architecture specification allows the mapping of different functionality to the same architecture, which is denoted as *platform-based design* [Keutzer00]. A formal separation of data manipulation into communication/transfer and computation/behavior constitutes *interface-based design* [VanRompaey96, Rowson97, Lennard00].

### Unequal Progress Across Different Technologies

Importantly, major advances have been limited so far to the digital world. While hardware/software solutions have been elevated to the codesign level [Bolsens97], analog/RF design, despite slow but steady progress, struggles at the synthesis, reuse, and modeling steps [Chang96b, Vassiliou99, Miliozzi00, Gielen00]. As a consequence, the differences in modeling effort, model complexity, and model quality create significant problems at the codesign level. Similarly, considerable state of the art has been established for power-aware digital design techniques [Singh95, Benini99]. At the analog side, low-power design happens too but at a much slower pace and in a much less structured and automated way [Abidi00].

### Lack of Techniques to Handle Multiobjective Optimization Complexity

The addition of power consumption as a major additional constraint is particularly serious for the design of wireless, portable consumer devices. The performance-driven state-of-the-art built up in the PC era cannot be reused in the domain of portable wireless devices. General-purpose off-the-shelf microprocessors (0.1 GOPS $W^{-1}$) have to compete with custom-designed, far more efficient ASICs (100 GOPS $W^{-1}$) spanning up a design space gap of three orders of magnitude [Claasen99]. Domain-specific architectures with the optimum tradeoff between performance/power and design cost are needed [Schaumont01a].

At the same time, well-known single-objective optimization strategies cannot handle anymore the additional concurrent constraints. Multiobjective optimization strategies have to be introduced and tailored. Moreover, adequate figures-of-merit need to be defined [Papalambros00, Allais43, Pareto06, vanLamsweerde03]. Exploration and optimization require parameterized but not entirely open white-box models. Preferably, gray-box models[33] are used that exhibit the relevant parameters to be optimized to the user while hiding irrelevant details [Büchi97, Melgaard94].

### Lack of an Executable Specification as Starting Point

The majority of formal design approaches relies on the availability of an executable specification to apply transformation or refinement methods [Lieverse99, Kienhuis99, Catthoor98a, Catthoor98b]. Specifically in the

---

[33] Gray-box modeling originates from system identification, e.g., Bohlin, 1984.

telecommunications domain, this assumption is not met. First, standardization does not provide a complete specification, which requires an exploration of the function design space. Second, multiple concurrent constraints often force a redefinition of the application requirements to obtain an overall feasible design. Both cases require modifications of the executable specification. The problem of functional iterations has been termed *algorithm selection* [Potkonjak99]. It is essentially the complementary problem to the implementation platform selection. However, it has rarely been addressed in a methodic way.[34]

### 2.2.3    Synthesis of a Future-Proof Design Methodology

The previous section indicated a significant gap between the capabilities of the current state of the art in microelectronic system design and the expectations and requirements from both a product design and process technology point of view. In fact, this situation will aggravate with the advent of next-generation multistandard multimedia communication devices. Hence, for our design flow, we neither can rely on selecting existing design methods nor should develop new methods that only fit the design of WLAN transceivers. Instead, we aim at a future-proof design methodology.

First, we stress an integrated view on design space exploration and optimization. Second, we recommend transition-aware modeling and codesign. Finally, we emphasize the importance of a *comprehensive* design process rationale including nontechnical factors as a measure of success.

#### An Integrated View on Design Space Exploration and Optimization

Design methodology for next-generation communication devices will need to address diverse flexibility requirements [Rappaport02] and diverging interests of multiple players ranging from operators over device manufacturers to users [Pereira01]. Devices need to meet a multitude of discrete standards. In fact, we cannot expect a convergence to one *flexible* standard,[35] but we will need cooperative interaction, compatible services, and standard-compliant devices. This adds the application design space as a third dimension to the existing functional and architectural exploration design space (**Fig. 2.5**). Efficient exploration techniques are needed for all three subspaces together with a common change management based on multiobjective optimization principles.[36]

---

[34] In fact, Potkonjak and Rabaey [Potkonjak99] state: "this area is currently more art than science or engineering and designers almost exclusively rely on intuition instead of accurate quantitative procedures. Very likely that this goal will remain elusive for years to come."

[35] Citing Goodman in [Wickelgren96]: "Flexible standards is an oxymoron. If it's too flexible, it's not a standard."

[36] Note that our view goes beyond the definition of system-level design as a third discipline between algorithmic exploration and implementation [Meyr01].

Efficient design solutions will employ interface- and platform-based architectural concepts together with generic, scalable functional concepts to allow adaptation to changing application scenarios. One part of this design flow will be executed by the designer at design time, while the other part, notably a subset of instance selection, change management, and optimization procedure, will be refined to an implementation which resides in the actual device for intelligent self-adaptation at run time.

The principles of design space exploration, multiobjective optimization, and design-time/run-time tradeoff will be employed throughout Chaps. 4–7.

### Transition-Aware Modeling and Codesign

We have already stressed earlier that design space exploration and optimization can impossibly be performed at low abstraction levels of design. Hence, we require an adequate vertical design flow, which allows the extraction and embedding of behavioral models, and a fast mapping process to analyze results of a design configuration. Particular attention is given to reduce transition overhead between abstraction levels during specification and simulation as well as the lowering of the entry threshold for new designers. An important role plays the adequate construction and usage of gray-box models (**Fig. 2.6**).

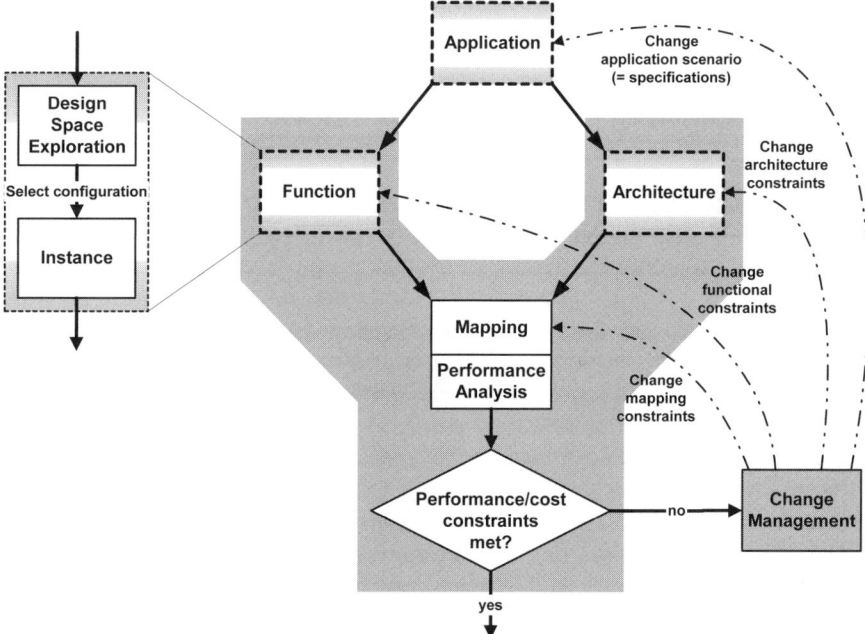

**Fig. 2.5.** The extended Y-chart integrates the exploration of the application, function, and architecture design space with a change management component

An efficient vertical top-down flow for digital design, its extension to mixed-signal system-level design, and behavioral modeling and gray-box techniques for scalable analog components will be presented in Chap. 7.

## A Comprehensive Design Process Rationale

In 1993, a number of CAD solutions were introduced as *the* solution to the ESL design problem. At that time, the approach failed: ESL design became the bad notion of a *buzzword* and it took years more until recognizable success could be achieved [Goering03]. Our analysis of the design process reveals that the complex interplay between many contravening influences – to a large part nontechnical ones – prohibits a single design methodology or tool from solving the problem. We emphasize the difficulty of managing the design process *as a whole* and a systematic, horizontal design space exploration as the key *enabling* techniques of the future.[37]

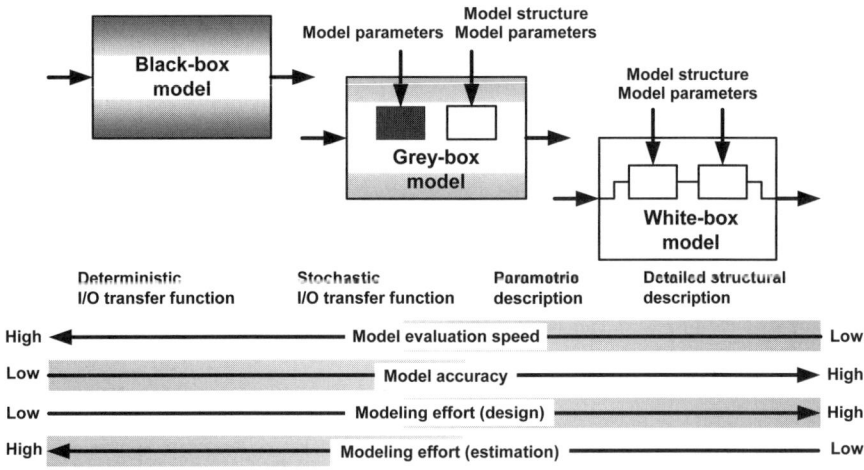

**Fig. 2.6.** Black- and white-box modeling approaches have contradictory properties. Efficient use may require a gray-box model or automated support for model-level transitions

Since the early days, methodology and tool support for ESL design have been steadily but slowly improving. Still, most ESL tool solutions of today, e.g., the CoWare tool suite, address only a part of the embedded system design process. The main focus has been on hardware/software codesign at the digital platform level. Improvements are still required, for example, on the integration with analog/RF and on the mapping of functional code toward the platform.

---

[37] Traditionally, design space exploration and design process have rather been treated as a cost factor, although both *enable* a larger design space and hence are key to a better, if *not to the optimal* solution.

Hence, we live up to the rationale defined in Sect. 2.1.2 and evaluate the developed methodology in a broader, also nontechnical context. More details can be found in the conclusions of Chap. 2 and in Chap. 7.

## 2.3 Crossdisciplinarity

Design has been described as a structured process, yet tying together a multitude of diverse design and technology disciplines. Obviously, a unidisciplinary approach cannot provide a satisfying solution to the intrinsic diversity. We advocate a *crossdisciplinary* approach to design, which embeds all layers of disciplinarity. Importantly, we try to bring together the classically divided processes of *designing the application* and *designing the design methodology* and advocate for a codesign of application and methodology design. The consequences of this approach for the design process and the designers in a design team are analyzed. We stress that the codesign itself requires a *transdisciplinary* approach.

### 2.3.1 Disciplines

The term *disciplinarity* encapsulates the experience and the final state-of-the-art approaches for the design process and its individual techniques within a particular discipline. In classical terms, microwave/RF, analog, and digital design are considered as three different disciplines. The degree of interaction in a project between team members originating from different disciplines leads to the following classification [Rosenfield92, Bannon94, d'Hainaut86]:

- *Unidisciplinarity* means that representatives of individual disciplines meet, but the meeting does not affect their disciplinary identity at all; this often leads to difficulties in understanding since each discipline has its own terminology and standards for information exchange.
- *Multidisciplinarity* states that researchers work in parallel or sequentially from a disciplinary-specific base to address a common problem. A basic requirement is that a minimum level of common terminology is established early to allow a partitioning of the common problem into subproblems with their own specifications. The limited terminology forces the use of a top-down-only approach for partitioning, which leads to a suboptimal partitioning with a high probability. In practice, collaboration is often depending on a system architect linking together team members.[38] An effective *horizontal* design space exploration is not possible.
- *Interdisciplinarity* requires researchers to work jointly but still from a disciplinary-specific basis to address a common problem. Joint work goes

---

[38] Unfortunately, system architects with a bird's eye view on a system-wide scale are scarce.

further than an initial partitioning and involves a combination of top-down and bottom-up process steps. This enables design space exploration.

- *Transdisciplinarity* assumes a shared conceptual framework, which brings together disciplinary-specific ideas and approaches to address a common problem. Consequently, an early partitioning of the problem into multiple disciplines is not required; instead, all divide and conquer strategies are available. Transdisciplinarity requires a significant multidisciplinarity per team member to distinguish itself from an inter-disciplinary approach.

*Crossdisciplinarity* is a term that covers multi-, inter-, and transdisciplinarity. We will not use this term since we want to stress their subtle, but important differences between them at different layers in the design process.

### 2.3.2   Consequences for the Design Process

We can now analyze a design process in terms of its disciplinarity needs (**Fig. 2.7**). Four levels of design can be distinguished from highest to lowest abstraction level: design space definition, horizontal design space exploration, horizontal partitioning, and vertical design space exploration. Vertical design space exploration is the domain of the classical, unidisciplinary designer. Horizontal partitioning consists of taking quantitative design partitioning decisions and derivation of independent specifications for multiple vertical design space exploration trajectories. The required common terminology and quantitative information exchange ask for a multidisciplinary approach. Horizontal design space exploration goes beyond the partitioning decision because it requires an in-depth understanding of tradeoffs between different disciplines. An example is a clear understanding of the model accuracy vs. computational complexity tradeoff across different disciplines and the capability to specify model requirements. This interactivity, which closes the DSE loop, requires interdisciplinary capabilities. Finally, the definition of the design space requires a system-wide reasoning capability,[39] be it qualitative or quantitative. The large complexity at this level requires significant abstraction capabilities and experience for *relevance*. The commonalities of heterogeneous subsystems must be extracted to make them comparable. This is also the level where ad hoc decisions have to be taken to extend the design space with particular technologies for further exploration.

In practice, we can often observe a *clash* between system and circuit designers when it comes to exchanging information about a design because of the different abstraction levels handled by default by both classes of designers. Fairly often, misunderstandings result in either wrong decisions being taken or problems not being recognized early enough. Hence, translation between abstraction levels and reasoning at various abstraction levels is an important skill for both system and

---

[39] System-wide reasoning capability is hard to find since it requires a combination of broad experience and fast mental abstraction skills. The experience part may require some time in practice. Mental abstraction skills, however, can be taught and trained to some extent.

circuit designers. Note that, for the circuit designer, this does not require an extension of his disciplinary scope to other disciplines.

**Fig. 2.7** clearly indicates that different levels of disciplinarity are required. It also shows that systems with limited heterogeneity may not exploit capabilities at all levels.

### 2.3.3 Consequences for the Designers and Design Methodologies

Changes in the design process are reflected both in requirements for the application designers and the design methodologies applied by them.

*Increased Need for Interdisciplinarity in Application Design*

When we consider the particular case of wireless system design, the designers face both a complex, heterogeneous design case and a large number of conflicting design objectives. This requires a detailed horizontal design space exploration,

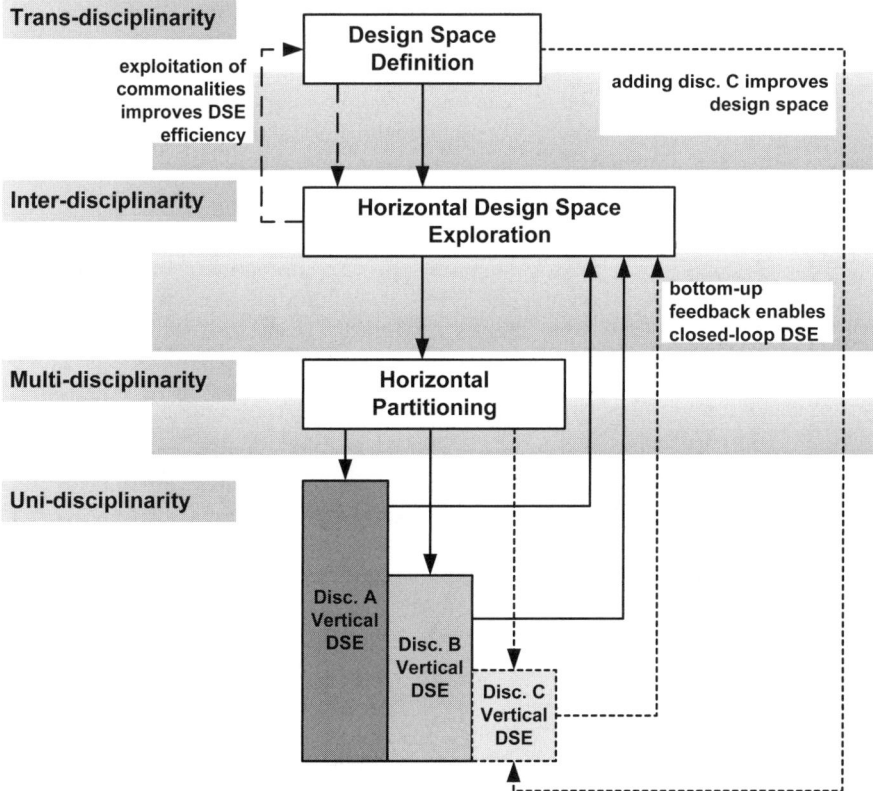

**Fig. 2.7.** Different levels of disciplinarity are required during the design process. Required crossdisciplinarity increases with heterogeneity and abstraction level

such that a close-to-optimal solution is found. Hence, we can state an *increased need for interdisciplinary methodologies and designers*. This conclusion seems to be in contrast with several statements in [Bannon94] where a multidisciplinary approach for the discipline of system development is thought to be sufficient. In fact, there is no contradiction but this evolution toward interdisciplinarity reflects the steadily increasing performance/cost pressure in the wireless and consumer market [Wheeler03].

### Link Between Role Models and Disciplinarity for Application Designers

Muller [Muller03] describes the stepwise development from a unidisciplinary trained designer toward a system architect. This reflects the transition from a specialist in a narrow discipline toward a generalist with a broad experience. Importantly, each generalist should have developed a feeling for the design complexity depth, i.e., for the vertical DSE process (unidisciplinary root know-how). **Fig. 2.8** applies this principle to integrated system design. Designing a system requires – in one way or another – filling the box of competences which is span across several disciplines and across several levels of abstraction. Lacking competences in the picture lead either to local, suboptimal contributions or even to risks since implications of particular technologies remain unknown. Disciplinarity increases with the breadth of know-how. Obviously, human capabilities in general do not allow the combination of deep know-how in a large number of disciplines. Hence, a specialization in vertical or horizontal know-how results. This reflects very well the classification of disciplinarity mentioned above.

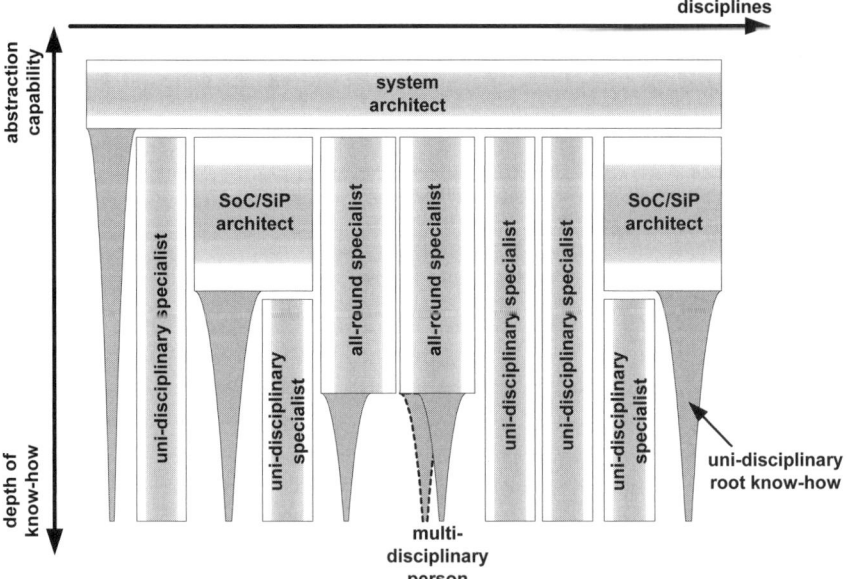

**Fig. 2.8.** Since depth and breadth of know-how are difficult to combine, system design requires a balance of all disciplinary capabilities supporting each other

Opportunities to replace designers' experience by external IP appear mainly at the unidisciplinary level. In SoC or SiP design, this was first limited to hard or soft macros or modules. The extension toward programmable platforms, e.g., embedded processors and bus systems, FPGAs, etc., reflects already a reduction of design complexity at the multidisciplinary level since these components hide partitioning decisions [Chang99].

*Increased Need for Transdisciplinarity in Design Technology Development*

The development of design technology has always been an interdisciplinary task. It requires knowledge in a particular application domain and, for the implementation of the design technology, in software engineering and algorithms, and digital signal processing. The move to system-level design brings multiple domains together along with their individual tool implementation knowledge. Unless embedded in a single conceptual framework,[40] application of methodologies and tools at the system level will become inefficient and unattractive to the designer. A transdisciplinary approach can prevent the creation of a pile of incompatible tools and, instead, provide suggestions for domain-specific design flows and hence guide the development of the essential tools and support for the relevant design step transitions. Importantly, only the combination of application- or domain-specific know-how with insight into the capabilities of design technology can produce a design process as proposed in Sect. 2.1, which truly *enables* a design team.

### 2.3.4   Codesign of Design Technology and Application

The heterogeneity of integrated electronic systems does not allow finding a perfect match of design tools when the design starts without limiting design creativity too much and too early in a deliberate way. While a basic set of tools, mainly at the unidisciplinary level, can be reused due to its generality, higher abstraction layers will require more dedicated approaches to efficiently define, explore, and optimize the system. Also, transitions between selected tools may not fit perfectly the desired design flow.

Traditionally, this resulted in a black-or-white decision [Schaumont01a]: The "pure design technologist" would wait for the design of the optimum tools before starting a design while the "designer-in-the-trenches" would abandon any new tool development and go for the working chip first. Both approaches are undesirable since they either result in tools that come too late for use in practical designs, or vice versa, no progress in tools ever and a nonscalable and hence unpredictable way of designing. Actually, the strong connection between application domain and suited design methodologies allows industry to turnaround this vicious circle. In-house codevelopment of application know-how and domain-specific design methodologies can lead to a competitive advantage over competitors in terms of

---

[40] Note that we do not mean here what CAD vendors call an "integrated framework," i.e., a tool suite with a common GUI, etc.

design efficiency and hence design cost and quality. Similar to the customer-owned tooling (COT) approach at the physical design level, we may call this concurrent process a *customer-owned system design process*.

Consequently, we motivate a joint path for application and design technology in an integrated codesign process. The idea for the codesign of application design and design methodology is not new and has been stressed, for example, in "you have to design the [design] environment plus the chips together" [DeMan00] or "design the design system for your system" [Schaumont01a]. This work illustrates this codesign idea in an explicit transdisciplinary way, which is based on three principles. First, we start with an application goal and *not* with a particular design problem. Second, we develop design methodology only when it proves to be necessary to overcome a specific application design problem in the foreseen path. Third, each developed design methodology needs to be evaluated with the problem that triggered its development and at least one other example that allows conclusions about the generalization of the method.

We illustrate the codesign of design technology along with the application design for three design challenges (Chap. 7):

- A smooth, efficient design flow for digital VLSI design from the system to the gate level; an analysis of reuse cases demonstrates that an initial investment in design methodology pays off
- An extension of this digital flow with an analog/RF component that allows mixed-signal system simulation and hence horizontal design space exploration
- A methodology for design space exploration at the communications link level, at the mixed-signal system level, and at the component level, which allows a partitioning of the system complexity into a design-time and a run-time aspect

## 2.4  Conclusions

This chapter puts the focus on the design task itself, its implications for the designer and the design team but also for the outcome and output of the design project. Today's communications systems have become very complex and depend on close collaboration of many disciplines. The resulting design process is full of interactions and requires careful planning, observation, and steering. An important evolution complicating this process is that, beyond the continuous pressure for short design time and low design cost, performance of wireless devices must increase while energy consumption must decrease. The latter dilemma requires optimization across disciplinary boundaries to obtain the aggressive design targets. A consequence of this evolution is the move toward crossdisciplinary exploration and optimization which call for novel methodologies and tool support to ease the designers' work.

The concrete design examples in the following chapters will illustrate how we translated this methodology into practice. Chapters 4–6 are not object oriented but goal oriented and the accompanying process is driven by elimination of the major design obstacles in terms of performance and/or power. In Chap. 6, this culminates in a methodology that systematically trades off performance and energy cost in a multiobjective optimization approach. Importantly, this approach can be used early in the design process and updated later, leading to an encapsulation of requirements specification, methodology, and actual design into a single process.

In particular, Chap. 7 addresses how particular design challenges were translated into methodologies and supported by tools. There, we show how a concurrent development of design methods for digital design, mixed analog/digital codesign and modeling, and system-wide optimization results not only in an ad hoc design result but also in a generic reusable methodology that is applicable to a wider range of problems.

# 3 Specification for a Wireless LAN Terminal

*Gedanken ohne Inhalt sind leer, Anschauungen ohne Begriffe sind blind.*
*Thoughts without content are empty, intuitions without concepts are blind.*
Immanuel Kant, 1724–1804.[41]

Chapter 2 postulated the need for a fundamental rethinking of design processes in the context of portable communication devices. Demonstrating relevance and impact in a particular application context is the missing component, the *key experience* that would convince the designer. With a wireless LAN OFDM transceiver, we have chosen for an application driver that should help us define clear and relevant goals and a case that illustrates applicability and benefits of the proposed design methodologies.

This approach must be judged from two sides. On the one hand, specifically in the context of a Ph.D. dissertation, some may consider this as limiting research already in a too early phase. Indeed, it may prevent us from a deep exploration in a certain topic. On the other hand, this makes the results more applicable to industry practice and real product design, it introduces relevant boundary conditions as they appear in practice and, instead of exploring one narrow topic, it may force us to look for transdisciplinary solutions.

Hence, this chapter positions wireless LAN in a wireless world that saw its birth in 1894 with Marconi's long-distance radio wave experiments as a first *application with potential* and that has evolved into a heterogeneous mix of communications networks spanning the whole planet Earth and beyond. Moreover, the need for a careful analysis of the service context and the operating environment is stressed.

---

[41] Originally in *Kritik der reinen Vernunft* and often translated as "Experience without theory is blind, but theory without experience is mere intellectual play," this translation does not capture the whole original citation. Kant wrote his first works in Latin but moved later to German, contributing significantly to the recognition of German as a language of science. Basically, he links practice to theory but also the importance of terminology and the relevance of practical context to methodology and thinking.

We will also illustrate the impact of standardization processes on research, since they have become a major instrument, not to say *battlefield* for industry.

Service and operating requirements clearly illustrate the need for higher data rates, for a packet-based system, for low-power portable operation, and for more flexibility, but also place wireless LAN into the indoor and campus environment which suffers from severe multipath radio reflection. These *requirements* motivate the use of a transmission technique called orthogonal frequency division multiplexing (OFDM). While, at first sight, theoretically elegant and simple, some practical weaknesses of OFDM have to be carefully circumvented when it is applied in a packet-based wireless context.

The context analysis of wireless local area network (WLAN) together with the assumption of using OFDM contributes to the requirements specification of the application driver. Our focus on the design and implementation of the portable terminal side is added as an important additional constraint. The requirements specification identifies and motivates the goals of the application. Hence, it constitutes the common starting point for the application- and methodology-related Chaps. 4–7. Since Bell and Thayer [Bell76] have already stated that the most severe problems in project failure occur during the requirements specification phase, we took our inspiration for structuring this process from guidelines [IEEE830, IEEE1233, vanLamsweerde00].

This chapter is structured as follows. Section 3.1 situates wireless LANs in the wireless communications world. Based on a brief history of WLAN technology evolution, we analyze the service context and the operating environment. Finally, we determine the impact of standardization and business concepts on research in the WLAN field. Section 3.2 introduces the OFDM transmission technique for wireless packet-based transmission. Section 3.3 integrates our findings from the two previous sections into a requirements specification for our application driver: a high-rate OFDM-based wireless LAN terminal. Section 3.4 concludes the chapter.

## 3.1   Wireless Local Area Networks

WLANs may still *appear* as the little sister or brother next to the wired local area networking world which has gained an important role in the enterprise, business, and administrative world, driven by the exponential evolution of the Internet traffic. Indeed, the question "Why the sudden interest in radio" is legitime [Riezenman01]. Wireless cannot compete with wired gigabit Ethernet in speed; it is easier to intercept and more difficult to secure; finally, next to the dimension of time and frequency, space is added. Yet, many applications require mobility. Wireless comes without wires to install and maintain. Wireless functionality can be integrated in portable devices that can be used virtually *anywhere* and *anytime*:

Connectivity moves with the user. And not every application needs immense transmission speed in the Gbps range.

As a result, wireless communications have grown from a niche player to a major industry sector [Rappaport02]. Between 1990 and 2004, the number of cell phone subscribers has increased by a factor of 100 to about 1.4 billion.[42] This first revolution has been driven by voice and messaging services and only slowly moving toward higher transmission speeds in the context of 3G [Ojanpera98, Honkasalo02]. In 2000, a second revolution began, driven by Internet-oriented data communications: Wireless LAN is finding its way into the enterprise, public, and consumer markets.

So, could wireless LANs, in particular end-customer terminals, benefit from two, now even three, generations of experience in the mobile communications world? The answer is *no*, if we leave out common sharing of the improvements in microelectronics, battery, display, etc., technology. WLAN and mobile terminals show very little overlap in the requirements plane (**Fig. 3.1**). For the mobile terminal, mobility, range, and operation in environments with many users are crucial, WLAN-equipped laptops aim basically at high-data rates only. However, WLAN is on the way to become an add-on capability in handheld devices such as personal digital assistant (PDA), which shifts the main focus from data rate to low-power operation. Still, little overlap is visible; this little reuse motivates a *fundamental* exploration of the application and design space of wireless LANs.

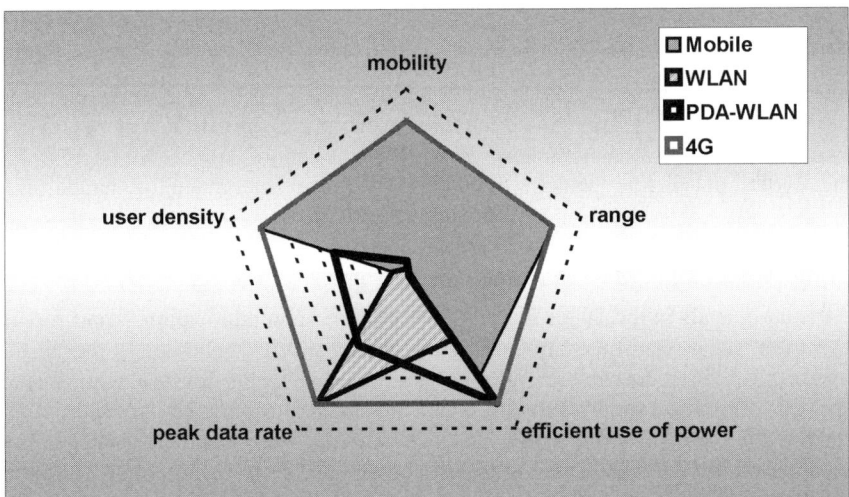

**Fig. 3.1.** Requirements between mobile (2G, 2.5G, 3G) and WLAN terminals differ largely. The move toward PDAs shifts the focus from data rate to low-power operation, yet the overlap remains little. 4G embraces all requirements. *Requirements increase with distance from the center*

---

[42] According to IDC, September 5, 2003.

**Fig. 3.1** also indicates that future 4G systems embrace both the goals of mobile and WLAN terminals. It is important to note that this does not necessarily mean that all goals must be simultaneously met. For example, an occasional medium-power operation for large-range communications is very well possible. An in-depth understanding of the service mix and the probabilities of scenarios becomes *crucial*.

Hence, the reader is invited to a short journey through the early days of wireless data communication, the evolution toward wireless LANs, the *turbulent* situation today with many emerging techniques and standards, and projections toward a unified next-generation wireless world. Our observations will not only result in a set of interesting trends but will also directly lead to service, environment, and terminal requirements for our wireless LAN terminal.

### 3.1.1   Wireless LAN Between Early Radio and 4G

The history of wireless networking stretches far back in time. Yet, about 50 years later than Marconi, the military investigated and extensively used encrypted wireless data transmission first during World War II. In 1971, the first packet-based radio communications network was conceived and installed between seven computer locations on the islands of Hawaii (ALOHAnet). However, with island distances between 50 and 200 km, we would call this today a wireless wide area network (W-WAN). Yet, it already illustrated the concept of a central basestation, the star topology, and the packet-based data communication between computer terminals, which are still predominant in today's WLANs.

The evolution of wireless LANs took a far less straight path than the development of ALOHAnet. Research and technology evolution, governmental regulation, standardization, and business ideas entangled much more. Hence, we *try* to divide the evolution in five phases: early WLANs, the emergence of standardized WLAN, WLAN today, the diversification, and the integration trend toward 4G.

*Early Wireless Local Area Networks*

After early trials in the 1980s with diffused infrared communications[43] and spread-spectrum radio transmission employing analog SAW technology, the FCC's opening of the industrial–scientific–medical (ISM) bands for use with spread-spectrum radio techniques drove research toward RF techniques. In 1990, NCR[44] introduced their first WaveLAN products for the 900-MHz and 2.4-GHz ISM band based on direct sequence spread-spectrum (DSSS) transmission at 1–2 Mbps [Claessen94]; and Symbol Technologies brought the similar SpectrumOne product on the market. The DSSS approach was the starting point for the IEEE 802.11 standards family. Still, nonharmonized regulation in the USA, Europe, and Japan kept wireless LAN research distributed over several frequency bands ranging from

---

[43] IBM Zürich Research Labs, 1979.
[44] Acquired in 1991 by AT&T.

900 MHz to 17–18 GHz [Hollemans94]. In 1991, ETSI standardizes DECT which became a success in cordless voice applications but did not succeed, despite its potential, in the wireless LAN field [DuttaRoy99]. With HiperLAN/1, ETSI started standardization of a 23.5-Mbps GMSK-based true wireless LAN [H1].

### The Emergence of Commercially Viable Standardized WLAN Products

Despite significant effort, in 1995, WLAN was still a niche player. At that time, a 10-Mbps wired Ethernet card was selling for $100 while a 2-Mbps WLAN card cost $500, which revealed a performance/price penalty of 25× for WLANs [Wickelgren96]. Missing were an unambiguous standard to increase interoperability and higher integration to reduce product cost.

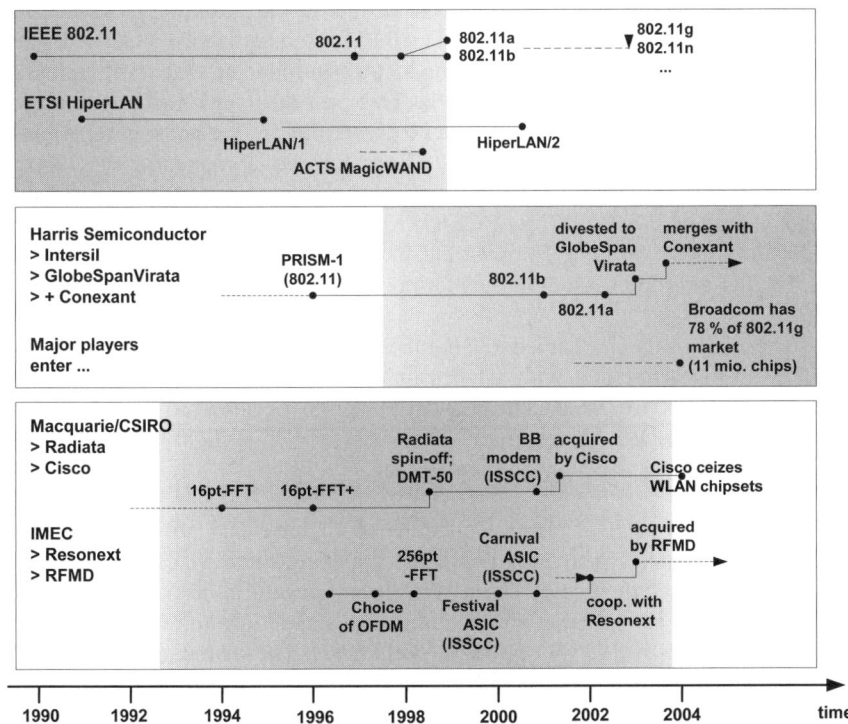

**Fig. 3.2.** The success of WLAN emerged from a fruitful combination of standardization activities, early industry involvement, and a transfer of knowledge from research to industry. Since 2004, WLAN is treated as a commodity item and major manufacturers swept the field. *From top to bottom*: standardization, industry, and interaction research with industry

The IEEE addressed the standardization problem successfully through the 802.11 standard, which was finalized in 1997, with compliant products appearing even earlier [Harris96, Crow97]. Extensions to 11 Mbps [IEEE802.11b], to 54 Mbps in the 5-GHz band [IEEE802.11a], to 54 Mbps in the 2.4-GHz band [IEEE802.11g] followed. In 2006, we are close to the standardization of multiple-antenna extensions for higher throughput [IEEE802.11n]. While IEEE standards were pushed by early industrial product designs, the European standards family ETSI HiperLAN/1 and HiperLAN/2 did not have the same success. HiperLAN/1 was maybe too early in 1995. Research initiatives such as ACTS Magic WAND in 1998 failed to create market success by relying on the newer and more expensive ATM instead of IP. HiperLAN/2 came finally too late in 2001 and proposed a technically better but also much more complex multiple-access control (MAC) [H2-PHY, H2-MAC]. Today, the WLAN world is largely determined through the IEEE standardization process with ETSI and ARIB harmonization as follow-up [vanNee99, Henry02].

This evolution of standards is compared in time to the evolution in research and product design (**Fig. 3.2**). The IEEE 802.11 standardization really triggered product design with Harris Semiconductor, introducing the first integrated chipset[45] for 802.11 in 1996 [Harris96]. This cost advantage made the company the global leader until the market became diversified through new players and the rising business perspectives in 2002. The extension of standards was used by several players to get into the market: Companies such as Atheros or Resonext took the lead in 5-GHz 802.11a CMOS design; Broadcom has become market leader through its 802.11g chipsets and Intel will benefit from the integration of WLAN into microprocessor chipsets for portable computers.

This shake-up of the market led to a number of mergers and acquisitions. Intersil's WLAN business merged with GlobespanVirata and later Conexant, forming a valid counterpart for Broadcom [Souza03]. Cisco decided to leave the chipset business, because WLAN became "a market where chipsets have matured to the point where they have become commoditized."[46]

So far, it seems as if academic research would have had no direct impact on WLAN. Fortunately, two cases prove the opposite. First, Macquarie University and CSIRO started their research on WLAN in picocells at 60 GHz for 100-Mbps data rates in the 1990s. Economic viability and standardization did move them to lower frequencies in 1996 [Osgood97, Skellern97]. They came up with a series of ICs with increasing functional integration until they spun off as an independent company, Radiata Communications. This startup was in 2001 acquired by Cisco in the first phase of acquisitions. The second case involves our own research at IMEC. This research started in 1996 with a spread-spectrum background but, in early 1997, was quickly retargeted to the more promising but so far not considered OFDM technique (Sect. 3.2). During ISSCC 2000, we presented the world's first fully integrated transceiver for 80-Mbps WLAN, the Festival ASIC [Eberle00]. Again, the IEEE 802.11a standardization effort also drove us to investigate this

---

[45] With an incredible amount of luck, I succeeded in getting early samples of this chipset already in 1996 before the market introduction in Europe [Eberle96].

[46] CNET News.com, "Cisco 'winds down' wireless chipsets," February 2, 2004.

spectrally more efficient solution which resulted in the ISSCC 2001 paper on the Carnival ASIC [Eberle01]. End of 2001, we started collaboration with an American startup, Resonext Communications, to bring together excellent baseband and RF knowledge. Little later, Resonext introduced a complete CMOS-based IEEE 802.11a solution and was finally acquired by RFMD which has now become a major supplier of 802.11 chipsets. In the meanwhile, we extended our research to mixed-signal and analog design for WLAN, toward crosslayer design, and toward 4G.

### WLAN Today

During the past 2 years, IEEE 802.11 has become the dominant standard for wireless LAN connectivity. The standardization focus shifted to extensions for service enhancements, increased interoperability and security,[47] which shows that the standard has become matured through practice.

WLAN has entered the enterprise market with industrial solutions.[48] Successful field trials lead to the company introduction of RFID–WLAN solutions in warehouses,[49] and office space becomes equipped with WLAN at a large scale.[50] Microsoft Corp. expects an 18-month return-on-investment period from an installation of 4,000 access points at multiple locations and reports that more than 90% of the users moved over to a far more flexible use of their laptops leading to productivity gain.

Moreover, it has spread from the enterprise world to public and home environments. In 2002, public WLAN hotspots in retail outlets, coffee bars, and restaurants mushroomed.[51] T-Mobile installed 2,600 hotspots in coffee bars in the USA alone. About 350 T-Mobile hotspots in Austria will appear in places where people spend their flexible time; attention is given to the nomadic worker.[52] In the home, wireless has pushed aside phone- and power-line networking [Wheeler03]. The Wi-Fi label for 802.11b devices introduced recognizability for the end user [Henry02].

These evolutions led to sales of about 20 million 802.11 chipsets in 2002 [Wheeler03]. More than 50 companies were developing 802.11 chipsets. In the meanwhile, rising sales have led to a price decay from $20 in 2002 to $4 in 2004 per chipset, putting enormous cost pressure on chipmakers. The trend in 2005 is a

---

[47] For example, 802.11e for QoS, 802.11h for power control, 802.11i for security, and 802.11j for interoperability.

[48] Second-generation IWLAN, SIEMENS Automation and Drives, press release, November 2003.

[49] METRO Group, press release, January 2004.

[50] Intel Corp., "Microsoft Corporation: 35,000 employees go wireless with WLAN," January 2003, http://www.intel.com/business/casestudies/microsoft.pdf.

[51] According to Dataquest, 2003.

[52] T-Mobile, "T-Mobile constructs Austria's biggest WLAN network," press release, February 18, 2004.

further increase in the number of chip sales due to increasing integration of WLAN into, e.g., mobile phones, camera phones, PDAs, etc., while profit margins per sold device are further reduced.

## Diversification

The driving forces behind wireless data communication have not stopped at wireless LAN. A diversification in standards and solutions tailored to particular applications has begun. Besides capacity extensions for wireless LAN such as multiple-antenna techniques [Sampath02], solutions in the wireless personal and metropolitan area network range were conceived to embed wireless LANs. In 1998, work on Bluetooth (IEEE 802.15.1[53]) started as a 2.4-GHz wireless "cable replacement" solution with 1–2 Mbps rates over short distances up to 10 m [Bisdikian01]. Later, the IEEE 802.15.3 committee concentrated on a short-range high-rate wireless personal area network at 2.4 GHz, this time targeting particularly multimedia streaming and rates compatible with WLAN [Karaoguz01]. Both concepts stress ad hoc networking and at least a factor 4 lower power consumption than WLANs as advantages. Proposals for ultra-wideband such as 802.15.3a aim at even shorter distances (4–10 m) yet rates in the order of 100–200 Mbps [Aiello03]. Interestingly, standardization also focused on low-power operation as primary goal with ZigBee and IEEE 802.15.4 for short-range operation with data rates far below 1 Mbps but extremely low-cost and low-power operation for integration in nearly every device at home [Callaway02]. A similar evolution is seen toward wireless metropolitan networks (WMANs) with the standardization of WiMAX [IEEE802.16] and mobile WiMAX [IEEE802.16e]. These standards aim at wider area coverage and higher mobility extensions compared to WLAN. In fact, they partially overlap with mobile services (2G, 3G) when it comes to area coverage and medium mobility and compete with WLANs when it comes to bandwidth and shorter distances.

## Integration

While resulting in very optimized devices for particular application, the emergence of tens of standards is a nightmare for manufacturers, service providers, and customers. Home and office may become crowded with numerous devices of poor interoperability and hence violate the major success criterion for ease of use, besides the question of cost. Service providers have difficulties in scaling, extending, and maintaining their systems and provide support for numerous solutions. Manufacturers will be unable to provide and support products for all relevant standards in time. It is crucial to find a way out of the standards war and the confusion that may be created at the customer through announcements of many different and incompatible technology solutions. This situation existed already twice, during the early years of 1–2 Mbps WLAN with proprietary solutions and later with the first 802.11b products [Parekh01]. At that time, the focus on 802.11b and the introduction of a recognizable label such as Wi-Fi was a

---

[53] Lower layers are identical.

major success factor. With similar goals in mind, the next generation (4G) is seen as an integration phase [Mohr00, Honkasalo02] toward multimode multistandard terminals interacting with multiple layers of networks (Sect. 1.1).

## 3.1.2  Requirements Analysis

Section 3.1.1 has shown that WLANs are used in a multitude of environments today. We concentrate here on the public, office, and home networks which represent the vast majority of WLAN deployment. What makes them suitable to WLAN is the fact that they focus on people in a sitting, working, or studying situation, where they are able to process large amounts of information. Hence, the demand for higher bandwidth services in these situations.

So, how can we get more information on the *right* WLAN device? We have to approach this situation from two sides: a service-oriented and a design-oriented one. While the user expects transparent, easy-to-use, and qualitative services, the WLAN designer looks for architectural and quantitative requirements.[54] A small example of a home networking scenario will illustrate the analysis. Next, we address service, networking, and terminal aspects. Finally, we summarize our findings.

*A Wireless Home Networking Case Study*

Wireless home networking is a particularly demanding case since it covers a large variety of use cases, ranging from watching TV, reading email, downloading files from the Internet, to printing (**Fig. 3.3**). Activities can be indoor or outdoor around the home. Stationary and mobile devices can be used.

Typically, a single WLAN access point would provide wireless connectivity to a wired access network such as ADSL or cable that provides high-speed access to the Internet. Optionally, multimedia data could be received through a terrestrial or satellite DVB receiver. Interoperability should be possible between low-power mobile devices such as PDAs or laptops and stationary desktop devices. Service availability and quality would mainly depend on the capabilities of the device used. Access to particular devices can be shared such as in the simultaneous downloading and email reading example or in the simultaneous printing and downloading case. Connections could be set up with the central access point or in a peer-to-peer manner directly between devices. But what if the user would have bought an access point dimensioned for picocells? Coverage for one room would be fine, but he or she would have to install multiple of them and maintain them. Hence, an analysis of requirements is crucial.

---

[54] Quoting Rose: "Sell the application, not the network." [Rose01]

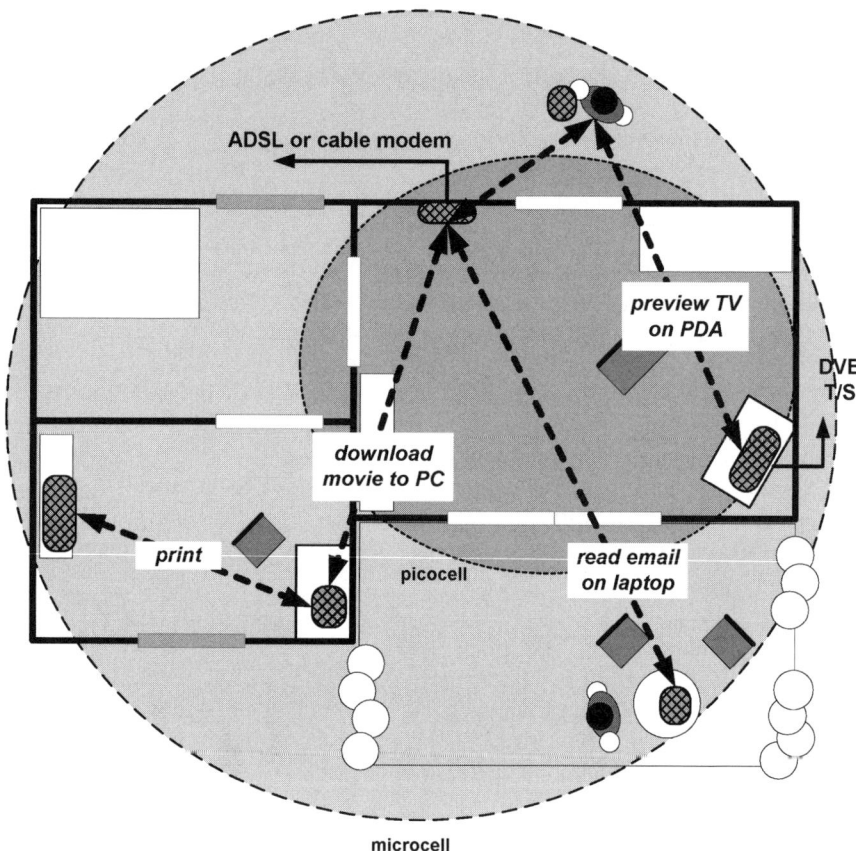

**Fig. 3.3.** Home networking is a demanding case with a variety of applications, devices, and environments

## Application and Service Aspects

Applications and services for WLAN cover such diverse use cases as a few minutes of low-rate email reading, hours of medium-rate streaming video, or high-rate downloading of documents from the Internet. It is important to note that, in a network with multiple users and services, we cannot treat services individually; instead, we have to consider an aggregation of services and its probability of occurrence (Chap. 6). While we can enumerate services in a nearly endless list, we also need more quantifiable metrics to classify them and treat similar services in a similar way. Peterson and Davie [Peterson96] proposed a taxonomy of applications based on time dependency and rate variability, which is very well applicable for the heterogeneous mix of services we encounter in practice [Englund97].

The concept of quality of service (QoS) embraces these service metrics. It is, however, not unambiguous [Fluckiger95]. QoS can be defined at every layer of the OSI protocol stack, but metrics will differ and translation is not evident. We will use the term quality of experience (QoE) for the quality that the user perceives with respect to a particular application and refer to the term QoS in its narrower meaning as link-level QoS between layer L2 and L3.[55] Velez and Correia [Velez02] provide a very general overview of QoS requirements for a multitude of services and Teger and Waks [Teger02] expresse the view on service quality from a user perspective. Englund [Englund97] describes services based on the concept of *use cases* and *scenarios*.

### Networking Aspects

Performance and properties of a network as a distributed system depend strongly on the network nodes (terminals and access points) and the properties of the network links between them. Like our scenario (**Fig. 3.3**), a set of use cases in a particular environment determines the design of the wireless connectivity.

Range is the utmost concern since the radio signal strength decays rapidly with distance in air and even more in the presence of obstacles (Sect. 3.2.1). In the mobile world, a multilayer cellular approach was chosen [Chia91, Coombs99]. The starting point was macrocells with a diameter of a few km, to which a single-access point provides connectivity. In case aggregated service rates exceed the capacity of a single-access point, micro- or picocells are used as an overlay network within a macrocell.[56]

The need for mandatory access points in a mobile phone network does not always hold in WLAN. It is also desirable to set up ad hoc connections between terminals without the presence of WLAN infrastructure [Chen94]. The WLAN network and its nodes should hence be prepared for both modes of operation. Also, WLAN links are expected to cover 100–150 m distance at maximum with a much lower average distance in the 5–20 m range. Since access points[57] form costly infrastructure due to installation cost, their spatial density shall be minimized. Particularly for small office or home usage, a single-access point is advisable. Hence, the overlay principle is not applicable for consumer-oriented WLAN due to cost restrictions.

---

[55] Our terms QoE and QoS match, respectively, the terms *perceived QoS* and *intrinsic QoS* as defined by ITU/ETSI [Gozdecki03]. Note, however, that perceived QoS in ITU/ETSI encompasses required and perceived QoS by the user as well as offered and achieved QoS by the provider. We always use the user perspective.

[56] Note that, in the context of WLAN, the cell-size taxonomy is shifted toward smaller areas. While picocells in cellular systems may cover an entire building, picocells in WLAN target a single room [Prasad99, Skellern97].

[57] Also called *basestations*.

Selection of the appropriate carrier frequency is a tradeoff between available bands with sufficient bandwidth for the desired data rates, the frequency-dependent radio propagation properties, and implementation complexity of the front-end. Bandwidth availability stimulated research in the 40–60 GHz range such as in ACTS MEDIAN [Flament02, Smulders02], but the high cost associated with a picocellular infrastructure and front-ends have shifted interest to lower frequencies. Similarly, Japanese systems at 17 GHz did not succeed and Motorola's Altair was stopped [Hollemans94]. The picocellular approach was also proposed for WLANs at 5 GHz [Skellern97]. Shibutani et al. [Shibutani91] suggested optical backbones and Haroun and Gouin [Haroun03] modulated an 802.11b RF signal on an optical carrier to reduce access point cost. Yet, connectivity remains *wired* leading to an increased installation cost. Systems became practical when the 2.4-GHz band was envisaged. The later move to 5 GHz, despite higher path losses and front-end cost, was motivated by the overallocation of services in the 2.4-GHz band [Lansford01].

With the microcell approach as a primary assumption, we also have to consider MAC schemes that can handle multiple users and aggregated services. A MAC protocol includes authentication, association and reassociation services, and power management to grant different nodes access to the shared radio medium. The origin of WLAN in data networks and the trend toward an all-IP future network [Robles01] have resulted in Ethernet-inspired multiple-access protocols such as CSMA/CA in IEEE 802.11. The collision avoidance (CA) technique is required since the traditional collision detection (CD) approach in wired networks is not reliable in a wireless context. Although easy to implement, CSMA techniques cannot guarantee QoS under high load. Extensions toward scheduled time-division schemes similar to the HiperLAN/2 MAC [H2-MAC] have been introduced or are under consideration. For a detailed comparison between IEEE 802.11a and ETSI HiperLAN/2, we refer to [Doufexi02].

Despite the use of CA techniques, the *unbounded* nature of the radio channel constitutes the *hidden* and *exposed* terminal effects and requires time multiplexing or duplex (frequency-multiplexing) techniques to separate transmission and reception. Yet, the fact that the radio channel in totality appears nonreciprocal to each node, downlink and uplink can be separately optimized [Eberle97b]. Since most of these aspects require a joint physical–MAC layer consideration, we will comment on them in Sect. 3.2.4 and point the reader to particular tradeoffs in Chaps. 4–6.

As an example, Table 3.1 illustrates the variability between practical WLAN scenarios. Parameters such as range and aggregated goodput[58] vary with the scenario. Individual low-rate services are not considered. Access quality and path loss scenarios are classified in best (B), typical (T), and worst (W) case scenarios. Access quality describes the optimality of access point placement; at home, suboptimal placement is likely while cell planning is mandatory for office deployment. The propagation scenario depends on geometry and obstacles in the environment. An interesting early result was that, despite the short range, the data rates for the dense-office scenario could not be provided by a 54-Mbps IEEE 802.11a system [Doufexi02, VanDriessche03], because protocol overhead limited actual goodput under best conditions to about 27 Mbps. This example is part of an in-depth study on service and application aspects specifically prepared for WLAN, since existing published information was very scattered.

**Table 3.1.** Scenarios are a good way to capture requirements variability, from which both the worst-case requirements and the best and typical use cases can be derived

| Scenario | Aggregated goodput[58] (Mbps) | Maximum distance (m) | Access quality | | | Propagation scenario | | |
|---|---|---|---|---|---|---|---|---|
| | | | B | T | W | B | T | W |
| Multimedia home | 18 | 15 | | x | x | | x | x |
| SoHo[59] | 19 | 15 | x | x | | | x | x |
| Ad hoc conference | 19 | 20 | x | | | | x | |
| Office | 23 | 20 | x | x | | | x | |
| Lounge | 27 | 30 | x | | | | x | |
| Short distance, peer-to-peer[60] | 27 | 2 | x | | | x | | |
| Office (dense) | 46 | 10 | | x | | x | x | |

## Terminal Aspects

Services and networking determine ultimately the type of device or terminal that the customer uses to communicate with the network. The PC era has passed its zenith of dominance and faces strong competition by *wireless digital networked devices* [Zimmermann99]. While in the finishing PC era optimization for performance was of utmost importance, size or power consumption did not really matter. The evolution to laptops, cell phones, and PDAs drastically changed the design space for terminals. On the one hand, display and battery size largely determine weight and form factor of portable devices; on the other hand, they influence service types and quality as well as lifetime. Performance optimization is replaced by a tradeoff between service quality and power (*and* energy[61]) consumption. In this context, the research effort at UC Berkeley on a portable multimedia terminal ("InfoPad") in the early 1990s is remarkable [Sheng92, Truman98]. However, InfoPad's assumptions differed from today's WLAN: the wireless link for the InfoPad relied on error-tolerant multimedia data as payload such as speech or video only and was not designed for variable services and QoS requirements.

---

[58] Goodput is the error-free throughput at the interface to the application layer (L7).

[59] The small-office home-office case aggregates office tasks with multimedia consumer services.

[60] This short-distance high-speed link could be taken up by devices according to IEEE 802.15.3a.

[61] The importance of *both* peak power and average energy consumption is detailed in Chap. 6.

WLAN has entered the consumer electronics market where products are primarily differentiated by price and reliability [Sherif02]. Cost falls apart into design or nonrecurrent engineering (NRE) cost and manufacturing cost per device. The mass consumer market aims at a reduction of manufacturing cost through higher integration of functionality and lower component count: WLAN evolved from a discrete board solution with seven ICs in 1996 to a three to five IC solution in 2003 [Harris96], two-chip solutions on PCB have become available in 2004, and two-chip system-in-a-package solutions are upcoming at a cost of \$4 [Struhsaker03, Wheeler03]. WLAN becomes commodity IP such as GPS or digital camera. Contrary to the manufacturing cost, the design cost and design risk for such integrated solution have risen exponentially. Being aware of design cost and risk, our WLAN solutions in Chaps. 4–6 were designed with low cost *and* scalability in mind. The push of next-generation systems for integration of multiple transmission standards and the risk of a design failure will demand software-radio techniques [Mitola95] or, more likely, *software-reconfigurable* transceivers [Brakensiek02].

### 3.1.3  Conclusions

Wireless LAN has gone through a tremendous evolution during the last decade. In fact, this evolution shaped and influenced this work largely. Our early work on digital design for 80-Mbps WLAN anticipated standardization, whereas some of the later work was inspired by or adapted to standardization or feedback from industrial partners. An analysis of this evolution helped us to state requirements for the design of the WLAN terminal. Intentionally, we sketched the complexity of the application and QoS context in which WLANs are already today, since providing adequate QoS under very variable conditions will be *the* challenge in next-generation wireless systems.

So, let us conclude with a few findings on technology, standardization and business, as well as research.

*Technology*

- *A microcell approach is mandatory for the home environment; ad hoc connectivity will become more important with the evolution toward PDAs*. Infrastructure and installation cost for picocells are prohibitive. With more diverse digital devices becoming available, the importance of ad hoc connections will increase. A co-optimized MAC–PHY is required to handle this flexibility at reasonable protocol overhead.
- *60 GHz prohibited WLAN while 2.4 GHz enabled it*. If sufficient bandwidth is available through regulation and the spectrum is not overused, lower carrier frequencies are preferred since they lead to significantly lower implementation costs, particularly in the analog domain.
- *ATM could not compete against IP*. Adoption of IP-compliant protocols is strongly favored compared to non-IP solutions such as ATM for their

protocol modularity, interoperability, and legacy. This complies with a general evolution toward an all-IP network [Robles01] that guarantees the transport for both data, voice, and multimedia services (triple play).

- *Low cost through high integration was a key to the success of WLAN.* Intersil's 1996 chipset resulted in acceptable cost and hence increased its application; the higher visibility was crucial to trigger new applications, more industrial players, and through this also more research.

### Standardization and Business

- *It took about 13 years for WLAN to establish as a significant market and about 18 years if early research is added.* Early products came out in 1990, but a breakthrough in enterprise *and* consumer markets happened in 2003.
- *Standardization needs sufficient product push to show early feasibility and become successful.* Too early, too late, too complex or not fast enough kills standardization as the IEEE 802.11 vs. ETSI HiperLAN case has shown. The even better ETSI HiperLAN MAC could not weigh in against being late and the fear of implementation complexity.
- *Markets and roles are changing faster than ever.* The fast evolution in business makes cooperation with industry difficult for research. It becomes unclear who does what. WLAN evolved from a wholesale system-in-a-box solution in 1995 to a $4 chipset integrated by OEMs in 2004.

### Research

- *A long-term view turns taking risks into taking opportunities.* When we advocated OFDM-based WLAN in 1997, WLAN was essentially synonymous to spread spectrum. Yet, our early studies proved that scalability of spread-spectrum techniques would soon come to an end.[62] Moreover, we decided for increased flexibility, for generic techniques and methodologies that would be reusable in the 4G-integration phase. We also looked from the start at power–performance scalability which is less important for WLAN in laptops, but crucial for WLAN-equipped PDAs.

Finally, the challenge between user and design perspective on WLAN has been addressed. Clearly, the user is at an advantage but designers should learn what it means to "sell the application, not the network" [Rose01]. It is not a threat, but an opportunity to design more intelligent, robust, and self-adapting wireless systems;

---

[62] For classical direct spread-spectrum techniques, higher throughput results in higher signal bandwidth. It was clear that available spectrum is limited (either because it is licensed for fairly high costs such as in the GSM or UMTS cases, or it can become crowded in the case of unlicensed spectrum such as the 2.4-GHz bands). Hence, a solution was required that both offers a better spectral efficiency and allows an intrinsic scalability from low to high bit rates and to varying channel conditions. OFDM meets these requirements.

systems that hide their complexity inside and instead expose natural interfaces to the user. Or, vice versa, when we look at it from a *designer's perspective*.

## 3.2 Orthogonal Frequency Division Multiplexing

In Sect. 3.1.2, OFDM was suggested as a good candidate transmission scheme for wireless LANs at high-data rates.[63] This section addresses the multipath radio propagation problem a bit deeper and explains concepts and challenges of the OFDM technique.

### 3.2.1 Indoor Propagation Characteristics

Practically all scenarios in Sect. 3.1.2 describe communication in fully or largely enclosed environments. This indoor radio channel is characterized by severe multipath propagation. The antenna radiation pattern of the transmit antenna determines in which directions the electromagnetic wave is sent out. The receive antenna faces a superposition of direct, reflected, and diffracted waves depending on the room topology. Attenuation, delay, and phase of each individual wave are functions of the electrical path length traveled and the number and sort of encountered obstacles. The time-domain impulse response of this channel at time $t$ can be modeled by a time-dependent linear finite impulse response $h(\tau,\phi,t)$ in equivalent baseband notation. $h(\tau,\phi,t)$ is a function of the excess delay[64] $\tau$ and the angle of arrival $\phi$ of the propagation paths at the receive antenna [Hashemi93, Janssen96]. Channel properties can be interpreted in the delay, frequency, time, and spatial domain. Since we restrict ourselves here to single-antenna operation and, as we will show later, quasistationary or even stationary temporal behavior can be assumed, we may neglect here time and spatial domain. In-depth treatments for spatial and temporal properties can be found in [Vandenameele00] and [Thoen02a], respectively.

*Power Delay Profile*

For the characterization of the multipath channel, the power delay profile is preferred over the impulse response. The power delay profile is defined as the squared absolute value of the impulse response, given the individual tap weights $\alpha_k$ for a number $N_{path}$ of taps, and describes the time distribution of the received signal power originating from a transmitted Dirac impulse:

---

[63] OFDM is typically seen as a candidate for medium to high bit rates for two reasons. First, the number of subcarriers increases with the frequency selectivity of the channel which is typically not a problem for low-data rates in the range of kbit s$^{-1}$. Second, OFDM offers advanced bandwidth allocation schemes, e.g., for multiuser applications which are often not needed for low bit-rate transmission.

[64] The excess delay is defined relative to the time of arrival of the first propagation path.

$$H(t) = |h(t)|^2 = \sum_{k=0}^{N_{path}-1} \alpha_k^2 \delta(t - \tau_k). \tag{3.1}$$

The first and second moments of the power delay profile, *mean excess delay* $\overline{\tau}$ and *rms delay spread* $\sigma_\tau$, respectively, describe the amount of time dispersion in the channel. They are given by

$$\overline{\tau} = \frac{\int_{-\infty}^{+\infty} H(t)t\,dt}{\int_{-\infty}^{+\infty} H(t)\,dt} = \frac{\sum_{k=0}^{N_{path}-1} \alpha_k^2 \tau_k}{\sum_{k=0}^{N_{path}-1} \alpha_k^2} \tag{3.2}$$

and

$$\sigma_\tau = \sqrt{\frac{\int_{-\infty}^{+\infty} H(t)(t-\overline{\tau})^2\,dt}{\int_{-\infty}^{+\infty} H(t)\,dt}} = \sqrt{\frac{\sum_{k=0}^{N_{path}-1} \alpha_k^2 (\tau_k - \overline{\tau})^2}{\sum_{k=0}^{N_{path}-1} \alpha_k^2}} \tag{3.3}$$

in the nondiscrete and discretized case with $N_{path}$ components. The limited traveled distance in an indoor environment leads to mean excess delays of 10–100 ns and an rms delay spread in the same range [Hashemi93, Janssen96]. An exemplary channel power delay profile with $\overline{\tau} = 87.3$ ns and $\sigma_\tau = 79.6$ ns is given in **Fig. 3.4**. Note that we show there already the discretized version with taps spaced according to the receiver's sampling period (here 50 ns), which is used in end-to-end link system simulation. For simulation purposes, a parameterized stochastic channel model is preferred over the propagation-inspired $h(t)$ impulse response. Practical models work with a finite-length tapped delay line and use an rms delay spread estimated from measurements [Saleh87, Medbo98, Medbo99]. We refer to [Vandenameele00] for a detailed description of the stochastic channel model applied here and for its validation through 2D ray-tracing simulation.

Typically, the impulse response is truncated in time from the point where the individual power of all later paths falls below an *importance threshold*, for example below the required minimum signal-to-noise ratio (SNR). If the rms delay spread is sufficiently small compared to the signal period of a transmission scheme, all paths arrive within one signaling interval. In this case, no interference between adjacent signals appears and single-tap time-domain equalization[65] can be used. As a rule of thumb, the minimum signal period is about one-tenth of the rms

---

[65] Equalization is here reduced to a single-tap gain and phase adjustment.

delay spread; in our example, this would limit signal transmission to about 1.26 Mbaud.[66] We also assumed an importance threshold of 20 dB below the maximum component. Assuming quadrature amplitude modulation (QAM) with $M$ bits per symbol and a maximum probability of bit errors of $10^{-5}$, this would limit us to $M = 4$ or 16-QAM [Chung01]. This would limit the data rate to about 1.26 Mbaud × 4 bits per baud ~ 5 Mbps. Clearly, this makes the power delay profile a key instrument for design of the transmission system.

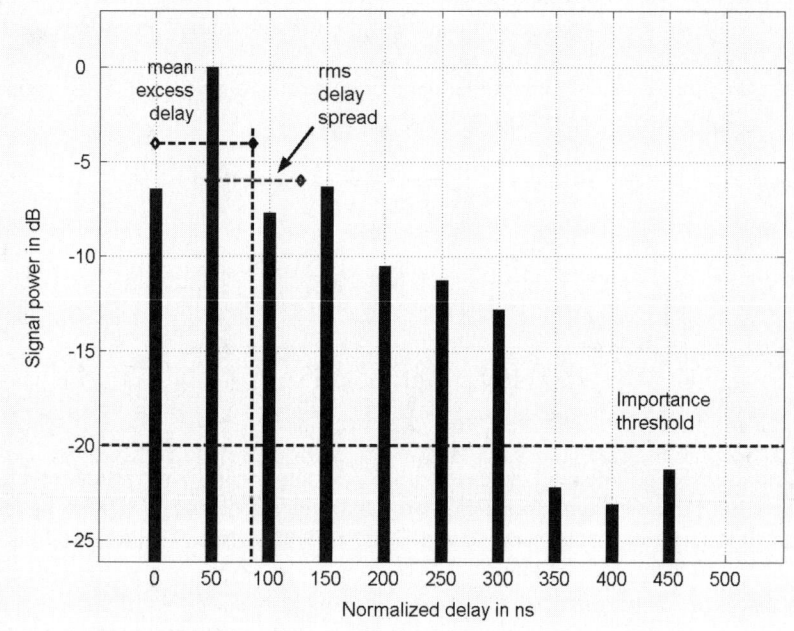

**Fig. 3.4.** Discretized channel power delay profile

## Frequency Response

The frequency response $h(f)$ of the channel is obtained from the Fourier transform of the complex baseband impulse response:

$$h(f) = \int_{-\infty}^{+\infty} h(t) \exp(-j2\pi ft) \mathrm{d}t. \tag{3.4}$$

From **Fig. 3.5**, we can see that the attenuation over a bandwidth of 20 MHz[67] shows a strong frequency dependence. This behavior is called *frequency-selective fading*. For a communication system with this bandwidth, equalization is required

---

[66] The baud rate is defined as the number of signal transitions per second. The actual bit rate can be derived from the baud rate when the amount of signal constellations is known, i.e., the modulation scheme has been defined.

[67] 20 MHz is the bandwidth of IEEE 802.11a/g and HiperLAN/2 WLAN.

to compensate the frequency-selective behavior and restore a flat channel response.

Note that very deep dips, the so-called *spectral nulls*, can occur such as around the 8-MHz bin in the example. The high, local attenuation around a spectral null introduces a low local SNR that may prohibit any successful transmission for a narrowband system.

The coherence bandwidth $B_{coh}$ is a metric that allows the comparison of the frequency selectivity of the channel and the system bandwidth. It is defined as the bandwidth separation over which the autocorrelation of the channel frequency response $h(f)$ decreases by 3 dB. The coherence bandwidth and the delay spread are approximately inversely proportional:

$$B_{coh} = \kappa \frac{1}{\sigma_\tau},\tag{3.5}$$

$\kappa$ depends on the distribution of the paths and it varies between 0.156 and 0.166 for indoor scenarios [Vandenameele00]. For our example, an rms delay spread

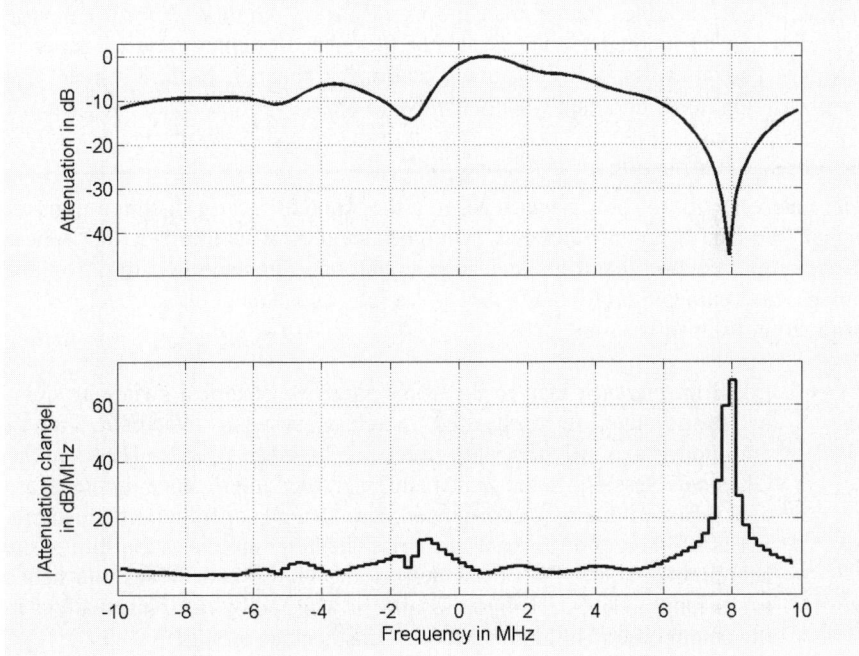

**Fig. 3.5.** Channel attenuation and corresponding attenuation change in dB MHz$^{-1}$ as a function of frequency

$\sigma_\tau = 79.6\,\mathrm{ns}$ results in a coherence bandwidth of about 2 MHz. The coherence bandwidth plays an important role in the initial specification of an OFDM transmission scheme; the subdivision of the total bandwidth over a number of subcarriers can be seen as a sampling process in the frequency domain; for an accurate capturing of the channel, it is required that the subcarrier spacing is smaller than the coherence bandwidth. For 20-MHz bandwidth and 64 subcarriers as used in [IEEE802.11a] and the coherence bandwidth of 2 MHz, we see an oversampling ratio of 2 MHz vs. (20 MHz/64) or 6 to 7.

### Comment on Temporal Variation

Movement of objects introduces a temporal variation of the channel since it changes the overall topology. Higher velocities increase the variability. In an indoor environment, velocities are basically limited to about 5 km h$^{-1}$ or 1.4 m s$^{-1}$, which is the walking speed of moving people [Thoen02a]. Analogue to the concept of coherence bandwidth $B_{\mathrm{coh}}$, the time coherence $T_{\mathrm{coh}}$ can be defined as the duration over which the channel characteristics do not change significantly [Jakes93]. As Thoen et al. [Thoen02b] show, the classical Jakes model derived for moving terminals overestimates the variation in the case of fixed terminals in a changing environment. Yet, in both cases, the coherence time remains bounded to about 20 ms in indoor environments. In Sect. 3.3, we will see that the coherence time is about ten times longer than the usual packet duration. Hence, temporal variation can be ignored for the design of the packet-oriented physical layer in Chaps. 4 and 5: we basically assume a stationary channel. Note, however, that time variation needs attention when we move to crosslayer optimization (Chap. 6).

### Sources of Interference

The receiver will not only have to resolve the wanted linearly distorted transmit signal. We can distinguish between two other sources of interference that appear in the same frequency band as the desired signal: interference originating from the direct environment (cell) and the so-called cochannel interference (CCI) originating from surrounding cells.

- Local interference may result from impulsive electrical switching noise or from other uncoordinated wireless systems. Microwave oven emissions have a detrimental effect on 2.4-GHz WLANs [Blackard93, Ghosh96, Ness99]. In the same band, increased interference problems are also expected due to uncoordinated simultaneous operation of Bluetooth, IEEE 802.11b, and HomeRF systems. The latter motivates the shift to the less problematic 5-GHz band despite the higher path losses and hence lower range. The 5-GHz systems may be affected by radar installations in the neighborhood of airports; though this is much less likely.
- CCI can still be present, although frequency planning avoids adjacent cells with the same frequencies, when insufficiently attenuated signals

from nonadjacent cells can enter [Chen01]. The problem can be relaxed with more available channels.[68]

We focus on the 5-GHz band where we can neglect these interference problems, as there are no prominent interference sources and enough channels are available. As a consequence, we assume that our WLAN operates in a *noise-limited* context instead of an *interference-limited* one as in classical cellular systems [Kalliokulju01].

## 3.2.2   History and Principle of OFDM

Already in 1962, a predecessor of OFDM was considered as an elegant method to treat severe frequency-selective channels [Goldberg62]. Thus, the principle of OFDM as a transmission technique can best be explained through an analysis of its origins. OFDM combines four ideas: multiplexing of data signals in the frequency domain, the use of modulated carriers – hence using multiple carriers instead of a single carrier, exploiting frequency diversity, and introducing orthogonality between the signals.

### Carrier Modulation and Frequency Division Multiplexing

In 1841, C. Wheatstone proposed a time-division multiplexing (TDM) system for telegraphic application; the main driver was to enable several users to share a common *wired* medium. In 1874, E. Gray transmitted a number of tones – first 4, later 8 – simultaneously over the same wire and analyzed them at the receiving end, the equivalent to frequency division multiplexing (FDM). In 1890, M. Pupin experimented with the modulation of signals onto a *carrier frequency*. The advantage of FDM over TDM was that synchronization could be avoided. Note, however, that each tone required an individually tuned radio transceiver; tones were separated by guard bands to prevent unrealistically sharp filters (**Fig. 3.6**a).

### Exploiting Frequency Diversity

It took until the 1950s when the frequency-selective nature of long-range terrestrial radio made conventional single-carrier radio unreliable, similar to what we described in Sect. 3.2.1. There was a particular interest in highly reliable terrestrial HF systems from the military, which led to developments such as KINEPLEX, ANDEFT, and KATHRYN [Cimini85, Bahai99]. These systems were in fact based on FDM techniques. The Collins Radio KINEPLEX TE-206[69] system with four subcarriers[70] that carried each 300 bps may be termed the first

---

[68] Note that, assuming the classical hexagonal cellular layout, no solution exists for the three channels in the 2.4-GHz band, while the eight (12) channels in the 5-GHz band can be effectively used to implement a proper cellular system.

[69] The Collins TE-206 KINEPLEX system from 1959 used four equally spaced tones between 935 and 2,255 Hz. This spectrum was upconverted to a 21-kHz carrier. Two channels were multiplexed on a single tone. The equipment was already transistorized but still it had cabinet size and required a 100-W power supply.

[70] The classical papers mostly used the term *tone*. In this dissertation, we will use the modern term *subcarrier* instead of tone.

FDM system designed for mitigation of frequency-selective channels [Doelz57, CRC59]. "The philosophy of system operation is that when circuit conditions are good, i.e., no multipath or jamming, all channels carry their own individual traffic at capacity. If they deteriorate, redundancy is added [Goldberg62]." This means that already a simple form of coding was applied across the subcarriers in the frequency domain. Note also the difference to the M-FSK transceiver which exploits only a single subcarrier at a time [Ferguson68].

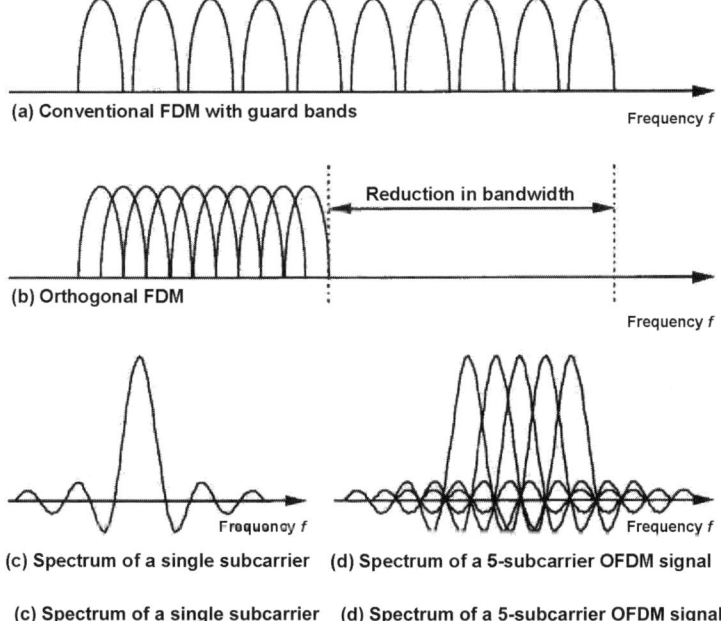

(a) Conventional FDM with guard bands                    Frequency $f$

Reduction in bandwidth

(b) Orthogonal FDM                                        Frequency $f$

(c) Spectrum of a single subcarrier    (d) Spectrum of a 5-subcarrier OFDM signal

(c) Spectrum of a single subcarrier    (d) Spectrum of a 5-subcarrier OFDM signal

**Fig. 3.6.** Orthogonality reduces system bandwidth of a multicarrier FDM system considerably. The *sinc*-shaped spectra of the subcarriers overlap and with increasing number of subcarriers $N_{sc}$, the total OFDM signal approaches asymptotically a rectangular frequency shape, which is the most bandwidth-efficient solution

## The Orthogonality Criterion

In an initially classified[71] paper, Goldberg [Goldberg62] presented the use of a *Fourier transformer* at both transmit and receive side as *the* novel design aspect in a multicarrier system. The system described was the 34-subcarrier AN/GSC-10 KATHRYN digital data terminal [Zimmerman67, Kirsch69, Bello65]. The use of the Fourier transformer had two important consequences:

---

[71] The paper was declassified later and reprinted in 1981.

1.  The *sinc*-shaped spectra coming from each subcarrier on an equally
    spaced frequency grid overlapped largely and reduced the total system
    bandwidth considerably (**Fig. 3.6b**).
2.  Individual filtering and carrier modulation per tone became obsolete; in
    the transmitter, a Fourier transformer summed up signals at baseband and
    a single RF upconversion stage could be used reducing hardware cost.

Chang's paper and patent [Chang69, Chang70] filing in 1966[72] were a major
cornerstone since they derived a general method for synthesizing classes of band-
limited orthogonal time functions in a limited frequency band. The method
showed that the basis function of the Fourier transform,

$$\psi_k(t) = \exp(-j2\pi k \Delta f_{sc} t), \tag{3.6}$$

could be elegantly used to transmit multiple spectrally overlapping tones at a
spacing of the tone sampling frequency $\Delta f_{sc}$. The lowest nonzero subcarrier at $\Delta f_{sc}$
determines the symbol length[73]:

$$T_s = \frac{1}{\Delta f_{sc}}. \tag{3.7}$$

The limiting or *gating* to the symbol length $T_s$ is equivalent to a convolution with a
rectangular pulse of the same duration and introduces a $\sin(x)/x$-shaping of each
individual pulse in the frequency domain (**Fig. 3.6c**). The waveforms of all
subcarriers $k$, $k \in [0, N_c - 1]$, are coherently summed up which creates a spectral
shape that approaches the minimum-bandwidth rectangularly shaped spectrum
asymptotically in the number of subcarriers $N_{sc}$ (**Fig. 3.6d**).

In practice, the *inverse* Fourier transform is used in the transmitter to form the
OFDM time-domain signal, while the receiver employs the Fourier transform to
recover the transmitted signal. The block diagram of the KATHRYN system does
not mention this difference explicitly (**Fig. 3.7**).

For a description of its end-to-end operation, we fall back to the original, clear
words of Goldberg:

> At the transmitting end, the input from various channels is electronically
> commutated to form a time division digital multiplex.[74] The digital stream is then

---

[72] Submitted 1966, published 1969. The corresponding patent was granted in 1970. Note
that it also included the optimum design of the transmit filters for a given amplitude transfer
function $|h(f)|$ of the channel.

[73] In literature, also block length or frame length is used.

[74] Note this particular view of the data input to the Fourier transform as a time multiplex
of *different* sources. In OFDM today, the data stream of a single source is *mapped* to
subcarriers in a particular way.

passed into the Fourier transformer, where it is converted into a frequency division multiplex. [...] At the receiving end, the signal passes through another Fourier transformer where the frequency division spectrum is converted into a time division digital stream which is subsequently commutated out to the proper individual channels. [Goldberg62, © IEEE 1962]

**Fig. 6. "Kathryn" block diagram.**

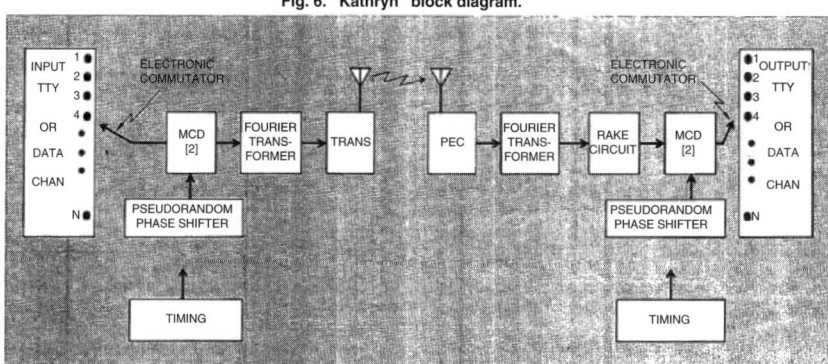

**Fig. 3.7.** End-to-end block diagram of the KATHRYN system from 1959 (© IEEE 1962)

### Digitization and the Fast Fourier Transform

Note that the first systems realized the Fourier transform with an analog implementation, for example with delay lines [Goldberg62]. Also, in the early days of digital design, the computational complexity of the DFT, which scales according to $O(N_{sc}^2)$ with the number of subcarriers $N_{sc}$, was prohibitive. However, digital design allowed to tradeoff datapath against control complexity, which allowed the use of the fast Fourier transform (FFT) instead of the DFT [Weinstein71]. Since then, OFDM techniques could exploit the benefits of digital scaling and mathematical algorithms for their Fourier transform core. The combination of the Winograd Fourier transform and canonical-signed-digit (CSD) is an interesting example of algorithm–architecture codesign [Peled80].

### The Cyclic Prefix

The detrimental *intersymbol interference* that the multipath channel introduced at the transition between adjacent OFDM symbols was already perceived and solved in an analog way in [Kirsch69].

However, digital signal processing allowed Peled et al. to digitize their concept, which added a cyclic prefix[75] to each OFDM symbol (**Fig. 3.8**). This prefix is inserted at the transmitter and removed at the receiver. If its length $N_{cp}$ is chosen sufficiently long compared to the length of the impulse response (Sect. 3.2.1), the cyclic prefix serves as a guard time against intersymbol interference between two

---

[75] Other common names are cyclic extension, guard interval, or guard time.

adjacent OFDM symbols $k$ and $k + 1$. The *duplication* translates a linear convolution into a cyclic convolution.[76] This prevents that the precursor part of the impulse response falls into symbol $k + 1$ and its information is lost for symbol $k$ *and* that the postcursor part of symbol $k - 1$ falls into symbol $k$. Obviously, the length of the cyclic prefix shall be minimized since channel capacity is lost. The related optimization problem linking impulse response, cyclic prefix length, and timing synchronization is treated in detail in Sect. 5.3.

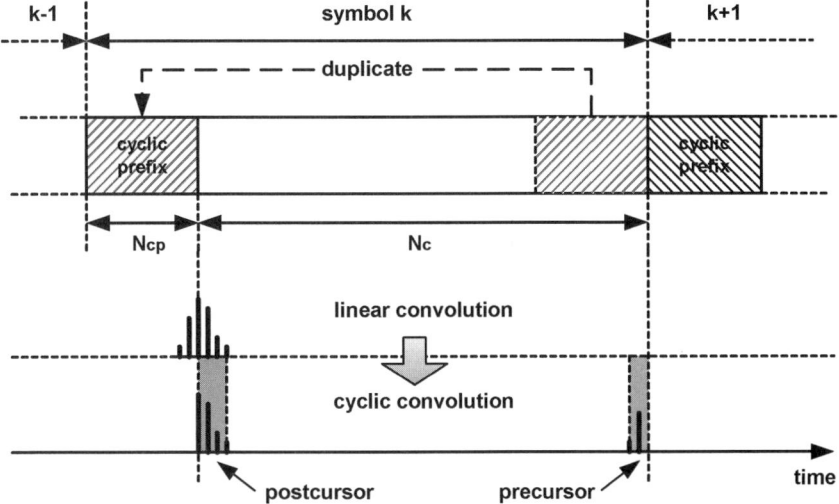

**Fig. 3.8.** The cyclic prefix turns a linear convolution into a cyclic convolution and protects a single OFDM symbol from multipath-induced intersymbol interference

## *Multicarrier Modulation: An Idea Whose Time Has Come*[77]

In 1981, NEC investigated QAM-based OFDM techniques for *fast* data modems but the field did not develop [Hirosaki81]. The breakthrough of OFDM usage may be dated back to the early 1990s with the development of the wire-based digital subscriber line (xDSL) technology[78] in several flavors such as HDSL [Chow91b] or ADSL [ADSL, Chow91a] and V(H)DSL [Chow91a]. At about the same time, OFDM was used to combat severe multipath effects in terrestrial digital audio broadcasting [DAB]. For the DAB case, systematic design of coded OFDM (COFDM) solved the problem that OFDM performance depends on the worst subcarrier [Alard87, LeFloch89]. A little later, development started for an OFDM-based terrestrial digital video broadcasting standard [DVB-T, Sari95]. Data rates evolved from 1.6 Mbps for HDSL to 34 Mbps in DVB-T.

---

[76] Also: circular convolution.

[77] This is the title of a frequently cited article by [Bingham90].

[78] In the xDSL community, the term *discrete multitone* (DMT) is preferred over OFDM.

In 1999, OFDM was applied in IEEE 802.11a-based WLAN, followed by ETSI HiperLAN/2 and the IEEE 802.11g version of WLAN. Still, new frequency-selective environments are discovered for which multicarrier techniques are proposed: OFDM appears in IEEE 802.16 wireless access [Eklund02] and for ultra-wideband short-range communications[79] in IEEE 802.15.3a in the form of *multiband*-OFDM [Aiello03].

### 3.2.3  Mathematical Model

The discrete-time baseband-equivalent model of an OFDM end-to-end system is presented in **Fig. 3.9**. The data stream enters the transmitter from the left-hand side and passes through the multipath channel and a single, lumped noise source before it enters the receiver to the right-hand side.

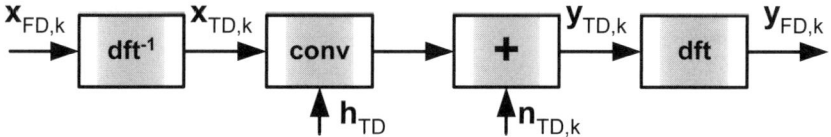

**Fig. 3.9.** Baseband-equivalent model of an end-to-end OFDM system with multipath channel and additive noise

Data are grouped into blocks of $N_{sc}$ complex-valued modulation symbols.[80] Each block represents an OFDM symbol with OFDM symbol index $k$. For each modulation symbol, standard modulation symbol alphabets such as QAM or PSK can be used together with bit-to-symbol mapping techniques such as Gray codes. It is possible to apply different modulation alphabets for each subcarrier.[81] The grouping in blocks suggests the use of a discretized vector notation as follows:

$$\mathbf{x}_{FD,k} = [x[0]_{FD,k},\ldots,x[m]_{FD,k},\ldots,x[N_c-1]_{FD,k}]^T. \tag{3.8}$$

In **Fig. 3.9**, an OFDM symbol stream $\mathbf{x}_{FD,k}$ enters the transmitter inverse discrete Fourier transform. Its output is convoluted with the stationary multipath channel impulse response $\mathbf{h}_{TD}$. A single, lumped noise source adds white Gaussian noise (AWGN) $\mathbf{n}_{TD,k}$. The receiver transforms the chain of OFDM symbols back into a stream of OFDM symbols. Mathematically, we obtain the following relationship between the receiver output $\mathbf{y}_{FD,k}$ and the transmitter input $\mathbf{x}_{FD,k}$:

---

[79] Preliminary specifications include a 500-MHz band, 128 QPSK-modulated subcarriers, 4–10 m ranges, and data rates beyond 100 Mbps at very low-power spectral density.
[80] The transmitter expects its input as constellation points in the complex phase domain.
[81] This results in an optimization problem, which is dealt with in adaptive loading [Kalet89, Thoen02].

$$
\begin{aligned}
\mathbf{y}_{\mathrm{FD},k} &= \mathrm{dft}(\mathbf{y}_{\mathrm{TD},k}) \\
&= \mathrm{dft}(\mathbf{h}_{\mathrm{TD}} \otimes \mathbf{x}_{\mathrm{TD},k} + \mathbf{n}_{\mathrm{TD},k}) \\
&= \mathrm{dft}(\mathbf{h}_{\mathrm{TD}} \otimes \mathbf{x}_{\mathrm{TD},k}) + \mathrm{dft}(\mathbf{n}_{\mathrm{TD},k}) \\
&= \mathrm{dft}(\mathbf{h}_{\mathrm{TD}}) \bullet \mathrm{dft}(\mathbf{x}_{\mathrm{TD},k}) + \mathrm{dft}(\mathbf{n}_{\mathrm{TD},k}) \\
&= \mathbf{h}_{\mathrm{FD}} \bullet \mathrm{dft}(\mathrm{dft}^{-1}(\mathbf{x}_{\mathrm{FD},k})) + \mathbf{n}_{\mathrm{FD},k} \\
&= \mathbf{h}_{\mathrm{FD}} \bullet \mathbf{x}_{\mathrm{FD},k} + \mathbf{n}_{\mathrm{FD},k},
\end{aligned}
\tag{3.9}
$$

where $\bullet$ denotes the elementwise multiplication. The discrete Fourier transform dft is defined as follows ($\mathrm{dft}^{-1}$ corresponds to its inverse operation):

$$
y_{\mathrm{FD},k}[m] = \sum_{n=0}^{N_{\mathrm{sc}}-1} y_{\mathrm{TD},k}[n] \exp\left( j2\pi \frac{mn}{N_{\mathrm{sc}}} \right).
\tag{3.10}
$$

The cyclic convolution $\otimes$ is defined as follows for each component $y_k[m]$ of the vector $\mathbf{y}_k$:

$$
y_k[m] = \sum_{n=0}^{N_{\mathrm{sc}}-1} h[(m-n) \bmod N_{\mathrm{sc}}] x_k[n].
\tag{3.11}
$$

Equation (3.9) demonstrates that a received data symbol $y_{\mathrm{FD},k}[m]$ on a particular subcarrier $m$ equals the data symbol $x_{\mathrm{FD},k}[m]$ on the same subcarrier multiplied by the corresponding frequency-domain channel coefficient $h_{\mathrm{FD}}[m]$ next to an additive noise contribution $n_{\mathrm{FD},k}[m]$. As expected from the use of an orthogonal basis function, the time dispersion of the multipath channel could not introduce any interference between different subcarriers at the receiver. The receiver can now obtain the original data bits in a two-step procedure. In a first *equalization* step, each received data symbol $y_{\mathrm{FD},k}[m]$ is divided by its channel coefficient $h_{\mathrm{FD}}[m]$ which results in a *soft* estimate $\tilde{x}_{\mathrm{FD},k}[m]$. In a second *slicing* step, the soft estimate is rounded toward the nearest symbol in the chosen modulation alphabet, providing the so-called *hard* estimate $\hat{x}_{\mathrm{FD},k}[m]$.

## Performance

Equation (3.9) has shown that a wideband OFDM system with a bandwidth of $N_{\mathrm{sc}}$ $\Delta f_{\mathrm{sc}}$, which is subject to multipath and additive white Gaussian noise, actually behaves as $N_{\mathrm{sc}}$ individual narrowband systems with bandwidth $\Delta f_{\mathrm{sc}}$ and identical spectral noise power density. The channel coefficients $h_{\mathrm{FD}}[m]$ can be approximated by a Rayleigh distribution [Medbo98, Medbo99, Hashemi93].

Hence, we can define the SNR per subcarrier as follows:

$$\text{SNR}[m] = \frac{E\{h[m]x_k[m], k\}}{E\{n_k[m], k\}} = h_{\text{FD}}[m]\left(\frac{S[m]}{N}\right). \tag{3.12}$$

In the AWGN case and for a given modulation scheme, the probability for a bit error $p_b$ (or the bit-error rate, BER) can easily be determined from the SNR [Proakis95]. Under the assumption of equal baud rate per subcarrier, the total BER $p_{b,\text{total}}$ over all subcarriers is then given by

$$p_{b,\text{total}} = 1 - \prod_{m=0}^{N_c-1}(1 - p_b[m]). \tag{3.13}$$

Equation (3.12) reveals that a subcarrier located in a spectral dip (**Fig. 3.5**) may suffer from such a high attenuation that it becomes noise dominated. Already a single affected subcarrier $m_{\text{fail}}$ with $p_b[m_{\text{fail}}] = 0.5$ biases the total BER in (3.13), even if all other subcarriers have very low BERs:

$$p_{b,\text{total}|p_b[m_{\text{fail}}]=0.5 \text{ and } \forall m \neq m_{\text{fail}}:p_b[m]\approx0} \approx 1 - \left(0.5 \prod_{\substack{m=0,\\ m \neq m_{\text{fail}}}}^{N_c-1}(1-0)\right) = 0.5. \tag{3.14}$$

The performance of an OFDM system actually depends on the performance of its *mostly attenuated* subcarrier. An OFDM system cannot work adequately without coding on an ISI channel in contrast to a single-carrier system [Sari95]. However, the loss of information on subcarriers with very low SNR can be mitigated by the introduction of redundancy on other subcarriers, either through traditional interleaving and coding techniques [Sari95] or through adaptive loading techniques [Kalet89, Thoen02a]. The latter first identify and then avoid *bad* subcarriers. In the coding case, well-known coding techniques such as convolutional, block or Turbo coding can easily be combined with OFDM.

### 3.2.4    Extension to a Practical System Model

From the evolution of OFDM (Sect. 3.2.2), we learned already that the way from the ideal, mathematical model of OFDM toward a practical system increases complexity considerably. **Fig. 3.10** shows, by means of shading, the evolution of the block diagram in three steps, starting from the ideal model. In a second step, the canonical model is built which includes the cyclic prefix and the equalization process described before. Finally, the following extensions in the transmitter and its dual operation in the receiver chain (in brackets) are applied: coding, preamble insertion (acquisition and the related tracking), transmit preprocessing (front-end compensation), and transmit front-end for the RF upconversion. Further, the equalizer is enhanced with tracking and front-end compensation facilities. Note that the frequency-domain signal processing in an OFDM system adds a fourth *signal representation* domain in contrast to the three in conventional single-carrier systems.[82] One of the key aspects in OFDM system design is the exploitation of this additional domain to reduce implementation complexity.

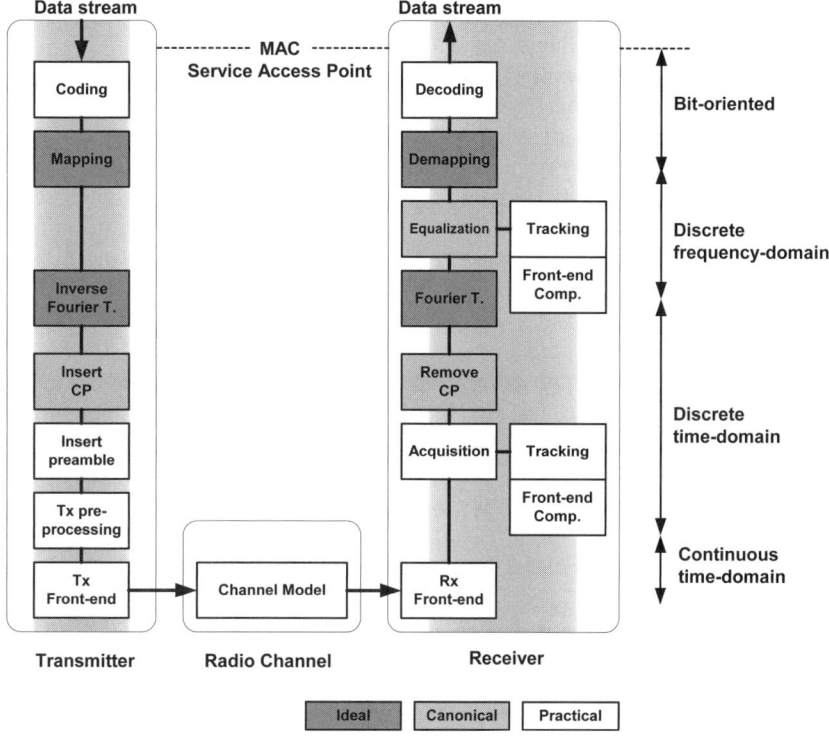

**Fig. 3.10.** Enabling the transition from an ideal to a practical OFDM system is the main application-oriented task

The transition from the *ideal* to a *practical* OFDM system is the main application-oriented task. This process is divided into three phases:

1. *Canonical OFDM*
   This includes the Fourier transform, the cyclic prefix, the modulation mapping, the demapping, and the equalization. On the one hand, we look for an efficient, low-power architecture for this core functionality. On the other hand, we may keep the scalability and flexibility of OFDM with respect to the number of subcarriers $N_{sc}$, the frequency spacing $\Delta f_{sc}$, the cyclic prefix length $N_{cp}$, the modulation alphabet, and the multiplexing process from bits to subcarriers. We address this tradeoff in Chap. 4.

---

[82] Recently, frequency-domain equalization has become popular in single-carrier system, too [Czylwik97].

2.   *Enhanced equalization, tracking, and front-end compensation*
     OFDM is very sensitive to mismatches in carrier frequency, phase, and
     timing [Saltzberg67]. Phase noise introduced by local oscillators falls in
     this category of problems too. These and other front-end imperfections
     form a receiver design problem; more particularly, extensions to the
     acquisition and the equalization process are required. These extensions
     are addressed in Chaps. 4 and 5.
3.   *Transmit preprocessing*
     OFDM is based on the summation of $N_{sc}$ waveforms, more particularly
     cosine–sine basis functions. Consequently, the amplitude varies largely
     as a function of the data stream. This *nonconstant* envelope requires
     analog components with high linearity, leading to both high-cost and
     high-power consumption. Transmit preprocessing techniques to mitigate
     this problem at the transmitter side both from a link-level and physical
     layer perspective are presented in Chap. 6.

*Transmit and receive front-end*

In this work, we do not present any specific work on analog or mixed-signal
circuit design. However, our research was performed in close cooperation with the
design of both a discrete and an integrated[83] 5-GHz front-end [Donnay00,
Côme04] that began end of 1998. These front-ends were also used for
characterization and testing of our designs.

## Protocol Layer for an OFDM Physical Layer

Our main focus is on the physical layer. However, designing the physical layer
independent of the adjacent MAC layer leads to a suboptimum solution. Hence,
we will describe briefly the assumptions made regarding the MAC layer in this
work; for a detailed motivation, we refer to the introduction on services and
networking in Sect. 3.1.2.

Wireless LANs are packet-based communication systems. The partitioning of data
streams into packets and the coordination of multiple services or users that require
wireless communication at a time is part of the MAC. The MAC tries to exploit
diversity in the radio channel to parallelize transmissions, either in the time,
frequency, code, spatial, or polarization domain. In the context of OFDM, the first
three have been addressed in [Rohling97], the spatial domain in [Vandenameele00].
Standardization in IEEE 802.11 and ETSI focused, respectively, on a CSMA
scheme with collision avoidance (CSMA/CA) and a plain TDMA scheme.
Essentially, both schemes schedule packets sequentially in time (**Fig. 3.11**).

The TDMA scheme is particularly suited for the situation where a coordinating
access point is always present. In this case, a reservation scheme with fixed
downlink (AP $\rightarrow$ terminal) and uplink (terminal $\rightarrow$ AP) transmission phases can
be used which provides less protocol overhead than the CSMA/CA-based

---

[83] With integration, we mean a system-in-a-package (SiP) approach.

approach. This is mainly due to the predictability of packet starts in the TDMA scheme. The CSMA/CA loses performance in a system with a high load (e.g., many users or many short packets) due to the backoff phase, which shall provide a fair access to the shared channel medium for all stations and reduce the number of collisions.[84] Packet lengths in HiperLAN/2 are only a fraction of the 2-ms MAC frame and can reach a maximum length of about 5.5 ms[85] in IEEE 802.11. Note that in practice packets routed from the wired Internet have either lengths in the order of 200 bytes (~0.3 ms) for interactive services and of 1,500 bytes (~2 ms) for file download. Hence, practical packet lengths are indeed much shorter than the coherence time $T_{coh} \sim 20$ ms of the indoor channel (Sect. 3.2.1). Hence, temporal variations can be neglected.

The switching between transmission and reception requires a close interaction between the physical and the MAC layer. The turnaround times between transmit and receive (Tx > Rx) and vice versa (Rx > Tx) in **Fig. 3.11** put essentially timing constraints on the combined processing chain from the MAC service access point, over the digital and analog physical layer to the antenna.

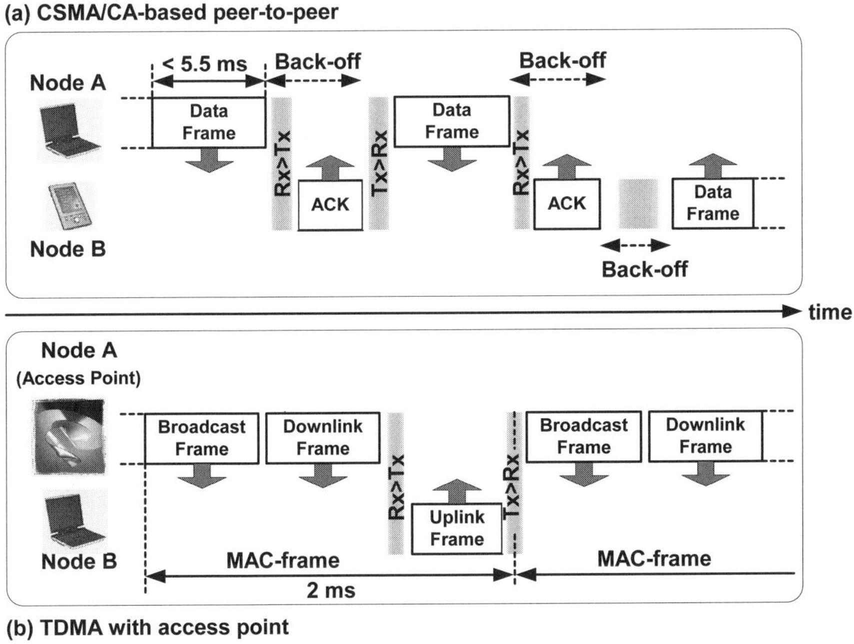

**(a) CSMA/CA-based peer-to-peer**

**(b) TDMA with access point**

**Fig. 3.11.** Both CSMA/CA (**a**) and TDMA scheme (**b**) apply scheduling in the time domain, although TDMA uses the concept of a fixed MAC frame structure and requires a coordinating access point

---

[84] For completeness, IEEE 802.11 provides also more efficient protocol flavors such as an RTS/CTS scheme.

[85] A 4,095 byte payload at the lowest possible modulation/coding rate (BPSK with half-rate coding).

In 1997, we conceived the physical layer based on a TDMA scheme [Eberle97a, Eberle97b]. Transmission and reception are separated in time, employing a time-division duplex (TDD) scheme. For the access point scenario, this allows the choice of different underlying access schemes for the downlink and uplink phase. For the uplink, we proposed plain OFDM. For the downlink, OFDM and an extension called *OFDM-A* were suggested. OFDM-A is essentially a multiuser technique and basically employs additional FDM over the $N_{sc}$ subcarriers by assigning them in groups to different receiving terminals. In [Thoen02a], this technique was extended with the adaptive loading approach.

## 3.3 Requirements Specification for a Broadband WLAN Terminal

The previous two sections form substantial parts of the requirements specification process for an OFDM-based wireless LAN transceiver. Since our aim was on prototyping R&D, we do not have a complete product specification in hands.[86] Hence, we use guidelines for the development of a system requirements specification (SyRS) such as [IEEE830] and [IEEE1233] only as an inspiration for the underlying structure. Since traceability of specification changes was not of importance, we did not employ software-based methods such as UML to capture the specifications [UML].

**Table 3.2.** Coverage of system requirements specification (SyRS) components in this document

| SyRS component | Section in this document |
|---|---|
| 1.1–1.2 system purpose and system scope | 1.1 and 1.2 for WLAN terminals |
| 1.3 overview | 1.2 and 1.3 for an outline of this work |
| 1.4 definitions | Appendix A |
| 1.5 references | Appendix B and bibliography |
| 1.6 revision history | *Not applicable* |
| 2.1 system context | 3.1.1 for WLAN |
| 2.2–2.4 major system capabilities, conditions, and constraints | 3.1.2 for WLAN, 3.2.1 for indoor channel, and 3.2.2–3.2.4 for OFDM |
| 2.5 user characteristics, assumptions and dependencies, and operational scenarios | 3.1.1–3.1.2 for WLAN, 3.2.3 for MAC and front-end context |
| 3 system capabilities, conditions, and constraints (specific requirements) | Chaps. 4–6 |

These guidelines define *requirements* as "a condition or capability that must be met or possessed by a system or system–component to satisfy a contract, standard, specification." The SyRS defines three mandatory chapters of which we cover the first two to a large extent as shown in Table 3.2, except for life cycle and

---

[86] This is fairly realistic even in industrial contexts when referring to early R&D phases. In these phases, functionality or performance is dominating metrics.

nontechnical management. SyRS in this chapter contains the specific requirements; this equals our research in Chaps. 4–6 with the same exceptions.

We can now summarize the general, functional, and nonfunctional requirements
which apply to this research in Table 3.3.

**Table 3.3.** Requirements specification

| Req. type | Description |
|---|---|
| Scope | Wireless LAN |
| | Home, office, hotspot (mainly indoor) |
| | Operation in the 5-GHz band |
| Capability | Variable data rate handling up to ~50 Mbps |
| | Flexibility with respect to OFDM parameters |
| Constraint | OFDM based |
| | Low-power operation, target laptop, *and* PDA |
| | Low manufacturing cost, i.e., high integration |
| Assumption | Alignment with standards *when available in time*[87] is |
| | desirable but not mandatory |
| | Front-end compensation techniques will be required to |
| | allow low-cost front-ends |
| | TDMA-based MAC |
| | Existing coding/decoding solutions can be reused |
| | Stationary channel and corresponding protocol |

## 3.4 Conclusions

In this chapter, we have studied the scenario context for wireless LAN
applications with a particular focus on user and service requirements and the radio
channel. Particular attention was also spent on embedding the evolution of OFDM
and wireless LAN into a historical, business, and technological context. We have
further introduced OFDM as a transmission scheme with promising properties for
these transmission conditions. The most important properties of OFDM were
reviewed from a signal processing perspective and a basic functional architecture
for OFDM was presented. Both steps are prerequisites for the development of
digital algorithms and their efficient implementation (Chap. 4). Some of the
unwanted properties such as OFDM's sensitivity to ICI and ISI will require
attention when designing the analog/RF front-end and ultimately lead to mixed
analog/digital codesign (Chap. 5).

---

[87] Note that neither IEEE 802.11a nor ETSI HiperLAN/2 were finalized when our first IC
design started. Note, however, that we designed our second IC toward the IEEE 802.11a
standard.

# 4 Efficient Digital VLSI Signal Processing for OFDM

*La perfection est atteinte non quand il ne reste rien à ajouter, mais quand il ne reste rien à enlever.*
*Perfection is achieved, not when there is nothing more to add, but when there is nothing left to take away.*
Antoine de Saint-Exupéry, 1900–1944.[88]

*To go beyond is as wrong as to fall short.*
Confucius, 551–479 BC.[88]

The technological evolution that allowed digitization of OFDM signal processing has been one of the main enablers for the long-awaited breakthrough of multicarrier modulation techniques [Bingham90]. While research into OFDM-related algorithms has resulted already in numerous publications, yet, a proof was missing that a cost-efficient integrated implementation for high-data rate wireless LANs was feasible. In particular, insufficient research results were present regarding the tradeoff between algorithmic performance and architectural complexity.

In this chapter, we propose systematic techniques for this design space exploration and describe two ASIC implementations: one with the low complexity of QPSK modulation at the expense of spectral efficiency and the other designed for high-performance and high spectral efficiency employing up to 64-QAM. Contributions

---

[88] Both Saint-Exupéry and Confucius express the need to converge to the actual goal, neither exceeding it nor falling short. In our context, the actual goal is the realization of an indoor wireless link for high-data rates; a too narrow focus, e.g., on equalization only, would lead to an underestimation of design effort and cost, as we will show. Prior to analyzing more advanced techniques such as adaptive loading, multiple-antenna schemes, etc., it is mandatory to carefully and completely analyze the plain OFDM case and develop solutions for *realistic* scenarios, especially including the acquisition process, particular transceiver nonidealities, and digital implementation effects such as quantization.

were also made in the design of low-complexity algorithms for synchronization and equalization. Next, we compare ASIC results with an equivalent FPGA implementation. Simulation and experimental results are evidence that our proposal for a distributed multiprocessor architecture yields the required low-power operation, design efficiency, and even demonstrates some of the scalability required for next-generation multistandard transceivers. Finally, we critically review our choice for an ASIC approach and give guidelines for the use of more flexible, heterogeneous architecture components such as ASIPs or coarse-grain reconfigurable arrays.

For the design of ASICs, we applied a novel design flow that integrates exploration, refinement, and test from a C++ specification to the chip-scale gate-level netlist. This flow and contributions to design technology are covered in Chap. 7.

The organization of this chapter is as follows. Section 4.1 defines the functional requirements for OFDM baseband signal processing and reviews the state of the art in wireless OFDM-based baseband design until 2001. Section 4.2 proposes a scalable, distributed multiprocessor architecture. Module design rules, on- and off-chip communications, and clocking strategy are discussed. Section 4.3 addresses the different design tradeoffs for the individual modules grouped by the most important design methodology applied. Section 4.4 presents the results of an experimental validation and a comparison with the state of the art since 2001. Section 4.5 concludes the chapter.

## 4.1  OFDM Baseband Signal Processing

Baseband signal processing is part of the complete transceiver functionality as illustrated in **Fig. 3.10**. This chapter deals with *digital* baseband signal processing only. This limitation is the result of an early partitioning into an analog and a digital portion. In our case, this early split is historically motivated and follows largely the traditional rationale for a functional split between digital baseband processing and analog front-end.

First, we will explain the rationale behind this partitioning and come up with functional requirements for the baseband processing. Using these requirements, we can compare with the state-of-the-art implementations until early 2001, which is the appearance of the first *truly* competitive design to ours.

### 4.1.1  Functional Requirements

Drawing the optimum line between analog and digital has always been a struggle between performance and cost. In the early days of radio, analog was the only option. In 1995, the other extreme was born, the *software-(defined) radio* [Mitola95]. Despite the fact that, nowadays, we face a significant shift to digitize the transceiver as much as possible for cost and time-to-market reasons [Moore65, Mehta01], the analog front-end has remained a significant component in the transceiver.

### Scope of the Digital Signal Processing and Analog–Digital Partitioning

This digital revolution was only possible with the continuous development of new analog front-end architectures (Chaps. 5 and 6). The change in architectures shifted also functionality from the analog to the digital side. Hence, when specifying functional requirements for the digital baseband processing, we have to define the front-end context. We worked out solutions that are applicable for superheterodyne, digital-IF, and zero-IF architectures (Chap. 5).

This chapter develops the *common digital functionality* for all three architectures. Specific extensions have been developed incrementally. Compensation techniques for superheterodyne or zero-IF receivers are described in Chap. 5. Similarly, for transmitters we refer to Chap. 6. The additional purely digital functionality for digital-IF transmission and reception is described in Sect. 4.4.1.

### Algorithm–Architecture Tradeoff

Exploration of the architecture space requires a requirements specification from the functional side [Kienhuis99]. However, system requirements do not necessarily describe the entire functionality that is required. Hence, the freedom in the functional (algorithmic) design space deserves an exploration [Potkonjak99, Zhang01, Verkest01a]. Even in standards such as [IEEE802.11a] or [H2-PHY], the functionality of the physical layer of a communication system is deliberately not fully described. This allows the introduction of proprietary, yet standard-compliant functional extensions through which several products can differentiate themselves. However, the degree of freedom is largely determined by the location of functionality in the inner or the outer transceiver and by the frame format [Meyr98, Speth01]:

- *Inner vs. outer transceiver*
  The *outer transceiver*, which defines the mapping from bits to a frame and signal format, and the *inner transceiver*, which defines the transformation of the signal format toward (transmission) and from (reception) the front-end. The outer transceiver is format based and hence functionally fully defined; design is limited to architectural exploration and optimization. In contrast, the freedom in the inner transceiver is large. Both functional and architectural exploration is possible under certain constraints.
- *Importance of the frame format*
  The frame format largely defines the sequence of operations in the transmitter or receiver. Importantly, standards define the frame format exactly. Hence, standardization also largely determines the sequencing of functionality.

An analysis of the outer transceiver specifications in [IEEE802.11a] or [H2-PHY] reveals that (de)framing, (de)interleaving, (de)scrambling, and (de)coding define requirements for which well-known IP solutions already exist. Hence, we *focus on the inner transceiver*, which offers a considerable joint algorithm–architecture design space. Interestingly, the freedom of functional exploration increases with the shift of front-end functionality to the digital domain. This additional freedom lies at the basis for the digital extensions described in Chaps. 5 and 6.

### Functional Flexibility and Level of Reconfiguration

With multimode and multistandard systems ahead, we were interested in designing a flexible or at least scalable baseband processor architecture. In particular, combinations of the OFDM baseband engine together with various multiple-access schemes were envisaged and taken into account in the design flexibility: OFDM-A as a downlink technique with adaptive loading and CDMA with code diversity in the frequency domain [Engels98]. A soft (multibit) equalizer output was required to enable soft-input decoding schemes such as Turbo decoding [Engels98, Deneire00a, Deneire00b]. Hence, we introduced functional flexibility at several levels (**Fig. 4.1**).

- *PHY-frame format*
  Concerning the frame format, we investigate solutions for our own, proprietary frame format [Eberle99b, Eberle00a, Eberle00c] and for IEEE 802.11a-compliant framing (**Fig. 4.1**). Contrary to the IEEE 802.11a frame, we investigate a parameterized packet frame structure that is programmable at both transmitter and receiver and allows performance–energy tradeoffs (Sect. 4.3.4). Also, contrary to most standards, we investigate a decoupled design of the preamble format from the payload format.
- *OFDM symbol format*
  We envision an architecture that allows the configuration of the OFDM symbol format through a configurable number of subcarriers from {64, 128, 256}: code spreading across multiple subcarriers, support for filter roll-off adaptation, and programmable length of the cyclic prefix.
- *Modulation symbol format*
  Modulation formats ranging from BPSK to 64-QAM with individual complex weighting coefficients to allow transmitter predistortion or adaptive loading are supported.

In contrast to classical single-carrier systems, the OFDM symbol level introduces an additional level of flexibility. The benefits of this flexibility, which come at a low implementation cost, were demonstrated in [Eberle99b, Eberle00a]. The advantage of this additional level of granularity has been exploited for diverse reasons in filter-bank transceivers [Scaglione99], transmultiplexers [Akansu98], *block-based* single-carrier frequency domain (SC-FD) [Sari94, Czylwik97], and *block-based* CDMA schemes [Zhou02].

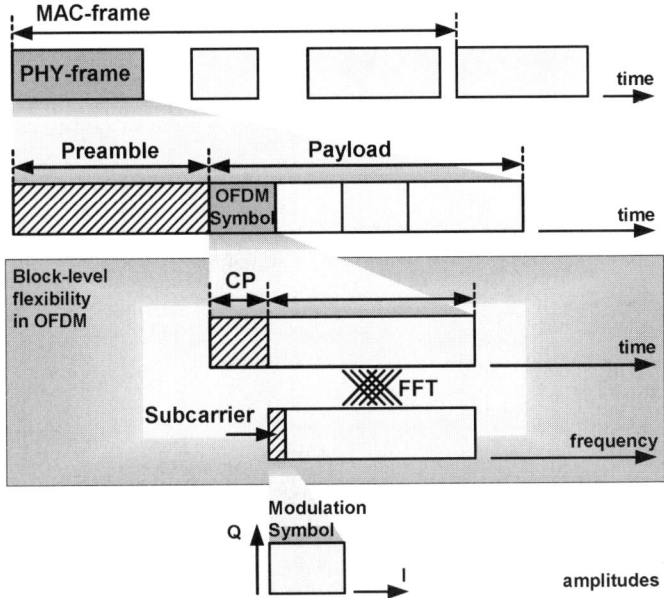

**Fig. 4.1.** OFDM introduces an additional level of granularity whose flexibility we will largely exploit

Parametric flexibility is only meaningful when associated with the frequency of parameter adaptation. Both *level of reconfiguration* [Brakensiek02] and *binding time* [Schaumont01b] refer to the same adaptation process. Four levels of reconfiguration can be distinguished:

1. *Configuration*: single-time adaptation at product shipping (commissioning)
2. *Reconfiguration with down-time*: several times during the product lifetime with the system switched off; e.g., for fundamental upgrades
3. *Reconfiguration per call or session*: dynamically, no significant down-time
4. *Dynamic reconfiguration, e.g., per time slot, MAC frame, or packet*: highly dynamic and fine granularity of parameterization

Brakensiek et al. [Brakensiek02] recommend reconfiguration per call/session as mandatory for wireless multimode devices. In fact, our VLSI implementation will go one step further in reconfiguration and show that only *dynamic reconfiguration* at the packet/frame level opens up possibilities for advanced techniques such as front-end compensation or adaptive loading.

**Table 4.1.** Comparison of the parameterization capabilities of the Festival and Carnival ASIC designs compared to the IEEE 802.11a standard requirements

| Reconfiguration capability | IEEE 802.11a | Festival ASIC | Carnival ASIC | Carnival relative to Festival |
|---|---|---|---|---|
| Number of carriers per OFDM symbol | 64 | {64, 128, 256} | 64 | Less flexible |
| Guard interval | 16 | 0:4:28 | | Same |
| Modulation | BPSK, QPSK, 16-QAM, 64-QAM | QPSK | BPSK, QPSK, 16-QAM, 64-QAM | More flexible |
| Equalization modes | Not defined | 3 | 2 | Different functionality |
| Spectral mask | Not defined | Complex, per carrier | | Same |
| FFT clipping | Not defined | 5–8b, MSB or LSB aligned | 5–10b, MSB or LSB aligned | Similar |
| Spreading | No | {1, 2, 4, 8} and programmable sequence | {1, 2, 4} and programmable sequence | Less flexible |
| Acquisition | Fixed | Programmable length, sequence, confidence factors | | Same |
| Number of zero carriers | Low: 0 High: 5, left and right | Low: 0:1:3, left and right High: 0:1:30, left and right | Low: 0:1:3, left and right High: 0:2:30, left and right | Similar |

We have designed two ASICs, called *Festival and Carnival*, that both offer a large amount of flexibility through dynamic *reconfiguration per packet* as indicated in Table 4.1. As can be seen, flexibility for both ASICs exceeds the one required for the IEEE 802.11a standard. Note that some functionality – in particular concerning reception- or front-end-related compensation – is not defined in the standard. The particular advantage of flexibility is described per parameter in the following sections.

## Design Rationale

According to our rationale (Chap. 2), we are interested in capturing all three dimensions of knowledge: the design itself (what), the design methods we apply (how), and the reasoning for all decisions we take (why). We particularly address the flexibility aspect. Yet, our goal remains designing for *just-enough flexibility* to avoid cost associated to overdesigning. Consequently, we are not interested in a detailed analysis of *isolated* algorithm or architecture aspects, but in their application in the system context. Hence, the definition of the complete, parameterized baseband architecture in Sect. 4.2 precedes the study of individual functionality (Sect. 4.3). An evaluation of the IC in a system context follows in Sect. 4.4.

## 4.1.2 State-of-the-Art Wireless OFDM Until 2001

We will briefly review the state of the art in baseband IC design for functionality similar to the previous specification. Since our own two ICs, Festival [Eberle00a] and Carnival [Eberle01a], were designed in 1999 and 2000, respectively, this section ends with the presentation of the latter of the two at ISSCC 2001. An overview and comparison with more recent designs is provided in Sect. 4.4.3.

Regarding OFDM-based WLAN implementations with an IC focus, we are only aware of the work at Macquarie University/CSIRO. This work includes the design of individual building blocks such as a 16-subcarrier fast Fourier transform (FFT) [Weste97], smaller on-chip extensions such as the cyclic prefix insertion or filtering [Skellern97], and led to a board-level demonstrator based on ASICs, FPGAs, and discrete devices [Osgood97, McDermott97]. Note that this WLAN system used only 16 subcarriers and DQPSK.

The first fully integrated 802.11a-compliant baseband ASIC was presented by Radiata in 2001 [Ryan01].

Potential similarity of OFDM implementations for different applications stimulated us to review also the state of the art in the field of DAB and DVB-T. For DAB, we find the first designs based on multiple ASICs and processors around 1989 [Alard87, LeFloch89]. In 1998, Philips presented an integrated DAB receiver (2,048 subcarriers, 0.5-$\mu$m CMOS, 127 mm$^2$, 150 mW, 640-kbit on-chip RAM) [Huisken98]. For DVB-T, an 8,192-subcarrier FFT was realized in 1995 [Bidet95]. In 2000, the first fully integrated DVB-T receiver (2,048/8,192 subcarriers, 0.35-$\mu$m CMOS, 97 mm$^2$, 2.2-Mbit embedded DRAM) was presented by Infineon [Mandl00].

Note that wired IC solutions that employ OFDM such as VDSL [Veithen99] have requirements that differ fundamentally from wireless. In the wired case, line reflections and HF interference coupling into the cable are the predominant challenges. Due to use of low frequencies, analog front-end nonidealities are significantly lower than in the 5-GHz RF case. Moreover, power consumption is less of a constraint. Also, fixed lines exhibit little temporal variation compared to the radio channel and hence allow for a long and accurate initial line calibration and reduced effort in tracking of line changes later on.

Hence, the existing state of the art until 2001 differed considerably from our goals:

- Integration of transmitter and receiver into a transceiver (as opposed to the receiver-only DAB and DVB-T)
- Development and integration of the entire inner modem including acquisition and equalization (as opposed to the partially integrated WLAN state of the art)

- Development of a complete acquisition strategy optimized for packet-based operation with low acquisition overhead and fast acquisition (as opposed to the broadcasting-based DAB/DVB-T)
- Implementation for a DSP-dominated system due to a realistic, indoor channel-based number of subcarriers between 64 and 256 (as opposed to the outdoor-oriented, memory-dominated DAB/DVB-T or the suboptimal, low-complexity 16-subcarrier WLAN)

## 4.2   Distributed Multiprocessor Architecture

Candidate architectures in digital VLSI design range from programmable but power-hungry general-purpose microprocessors to ultra-low-power inflexible, custom-designed ASIC solutions. Energy efficiency per processed equivalent instruction differs easily by a factor 100–1,000 [Claasen99], while programming and reconfigurability support ranges from nonavailable to compiler-supported software design [Schaumont01b].

An architectural exploration step allows matching functional requirements and nonfunctional constraints such as power consumption and finding an adequate architecture. Due to lack of methodologies that would cover the entire search space and allow a full quantitative comparison, we motivate our particular architecture choice through an *elimination* procedure.

First, we define a number of qualitative metrics. We use these to traverse the design space along several directions and motivate our final architecture choice. This selection resulted in a distributed multiprocessor architecture,[89] which extensively applies the interface-based design rationale along with a power-aware system state control and clocking strategy. The resulting modular concept allows a separate treatment of the functional modules in Sect. 4.3.

### 4.2.1   Directions for the Architecture Definition

Digital telecommunication systems are quite heterogeneous in terms of data throughput, processing complexity, and power constraints such that heterogeneous architectures are used and combined in their implementation. A common pattern, however, is the dominance of signal processing functionality concatenated in one or the other way [Bolsens97, Kienhuis99]. With the more involved OFDM symbol structure and our aim at a flexible solution, control complexity is added. Hence, we need a mix of dataflow and reactive paradigms which can be very well addressed using the notion of finite state machines with a local datapath (FSMD)

---

[89] Recently, the term *distributed multiprocessing* is used in the context of SoC where these processors are connected through a network on chip (NoC). We want to clarify that we assume a very heterogeneous set of processors here (where processor does not stand for a fully programmable architecture) that is connected through a rather simple data and control network compared to the proposed layer-3 NoCs.

[Catthoor88, Note91]. From the classification used in [Catthoor88], we can distinguish between lowly multiplexed datapaths and highly multiplexed datapaths; the later with a larger amount of control complexity.

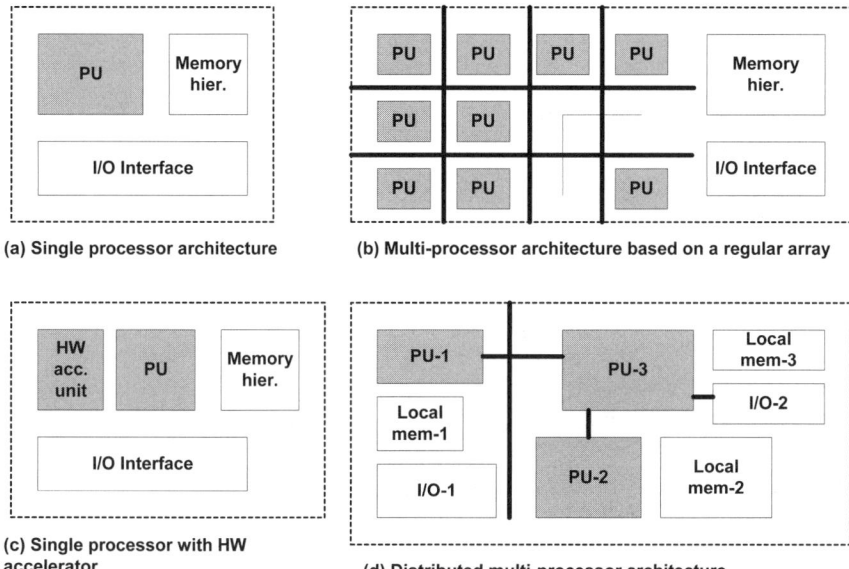

(a) Single processor architecture

(b) Multi-processor architecture based on a regular array

(c) Single processor with HW accelerator

(d) Distributed multi-processor architecture

**Fig. 4.2.** Several classical processing architectures (**a**)–(**c**) compete with the distributed multiprocessor architecture (**d**) which was used for the Festival and Carnival designs

When looking at existing OFDM implementations in 1997, we found microprocessor-based solutions for very low-data rates [Peled80], DSP-based implementations for very low-data rates [Perl87], and direct-mapped implementations [Rhett02] in commercial implementations for 1- and 2-Mbit s$^{-1}$ modems. As we show later, computational complexity of OFDM prohibits the use of single processor or DSP architectures (single processor with accelerators or SIMD) (**Fig. 4.2**). A comparison with recent designs after 2001 is provided in Sect. 4.4.3.

## 4.2.2 On-Chip Data and Control Flow Architecture

Given the performance requirements for OFDM baseband processing, we decided for a chip architecture based on communicating processing units implemented using a standard-cell ASIC design flow. Memory bandwidth and signal processing complexity are too high for a single processor. The high flexibility also makes the embedding of control into a single SIMD/VLIW solution intractable. However, our solution is actually a hybrid solution: we find both individual processing units that are partially based on highly multiplexed datapaths with intensive control and classical lowly multiplexed data processing units (e.g., FFT). The decision to limit clock frequencies[90] to more or less the maximum sampling frequencies (20 and 50 MHz, respectively) was a clear decision for a parallel architecture and a design based on standard cells.

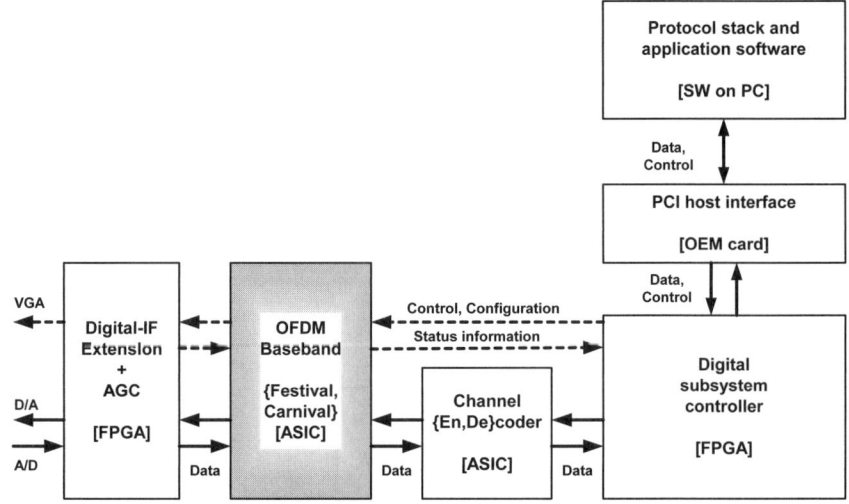

**Fig. 4.3.** The OFDM baseband design is embedded into a complete digital system. A digital-IF extension connects with the analog/RF front-end.[91] A channel encoder/decoder represents the outer baseband signal processing and connects to the higher protocol stack

Based on OFDM processing requirements, it seems that we need a hybrid solution since there is high multiplexing inside an OFDM symbol due to the high flexibility in mapping while we only face rather little control between OFDM symbols. This suggests a mix of both styles. This architecture follows to some extent the paradigm of direct-mapped architectures [Rhett02]. An important advantage that direct-mapped architectures offer is the freedom for algorithm–architecture codesign to exploit complexity reduction at both levels and to ensure that algorithmic optimizations match the available computing paradigms and operators available on the chip architecture [Zhang01].

The major focus is on the design of reusable, parameterized functional units with simple control and data communication interfaces. This approach separates communication from computation maximally and hence allows separate optimization per unit.

---

[90] Current 130- and 90-nm CMOS technologies easily allow clock speeds in the 200–300 MHz range for programmable processor designs with moderate usage of custom module design. The higher clock frequency can, however, only be used to sequentialize internal operations since the I/O interfaces operate at significantly lower symbol rates, e.g., 20 or 50 MHz in this case.

[91] The actual front-end was designed later on. First, a discrete superheterodyne front-end architecture was chosen; this front-end is further discussed in Chap. 5.

Both ASICs implement the inner transmitter and receiver datapath required for a high-speed, wireless OFDM system employing a half-duplex protocol suitable for standard-compliant time-division duplex operation. Consequently, hardware resources are shared between transmitter and receiver and various datapath reordering tasks are merged into a centralized datapath unit (symbol reordering). ASICs communicate through a FIFO-based transmit and receive interface as a master with the external data host in a slave position. Toward the front-end, they provide I/Q interfaces to dual pairs of analog-to-digital converter and digital-to-analog converters. Additional signals are provided to support analog automatic gain control (AGC) in the receiver and front-end power-up (**Fig. 4.3**).

*Data Communications*

The chip architecture relies on dataflow and token flow semantics for both data and control (**Fig. 4.4**). We first describe the dataflow in transmit and receive mode. Note that this interface-based approach supports a clean separation between communication and computation inside the design units [Rowson97, Lennard00].

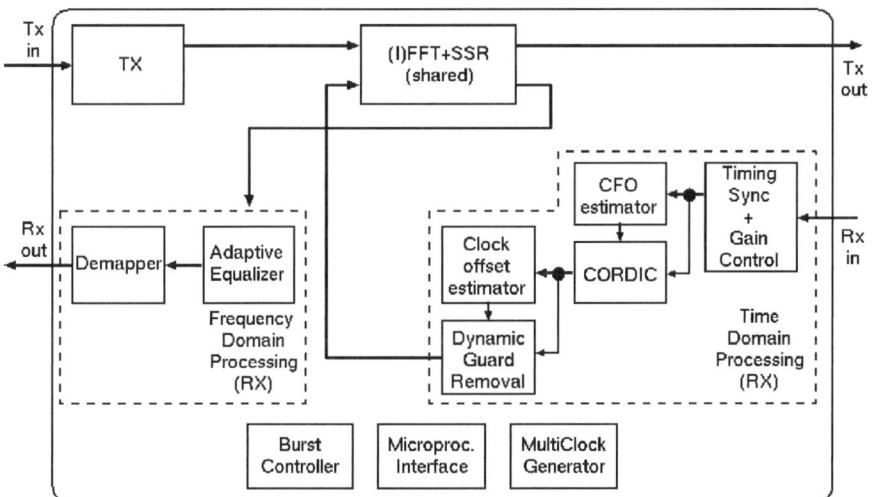

**Fig. 4.4.** The block diagram of ASICs reveals the receive processing parts (Rx) – subdivided over frequency- and time-domain processing – the transmit processing parts (Tx), and the shared modules ([Eberle01b] © IEEE 2001)

*Operation in Transmit Mode*

In transmission mode, payload data enter the ASIC through a 6 b (bits) parallel interface on request. Data enter the symbol mapper where bits are mapped onto either BPSK, QPSK, 16-QAM, or 64-QAM subcarriers. A programmable number of zero carriers is introduced near DC or Nyquist frequency to accommodate DC notch filtering and low-pass filter roll-off. A BPSK pilot sequence is inserted either on a fixed subset of four carriers or using a rotating pilot pattern with a period of 13 OFDM symbols. Each subcarrier can be individually weighted by a complex value allowing transmitter preemphasis and phase predistortion. The mapper provides a sequential series of 64 carriers, for Festival also 128 or 256 carriers, to the IFFT, denoted as an OFDM symbol. The mapper also prepends an entire BPSK OFDM symbol based on a programmable reference sequence to the payload or inserts it periodically into the stream of OFDM symbols. The inverse FFT transforms the frequency-domain constellation into a time-domain sequence. Scaling and digital hard clipping is performed at the FFT output to select a suitable crest factor and S/N ratio. OFDM symbols are then passed to the symbol reordering unit (SSR), which performs insertion of the acquisition preamble and prepends the cyclic prefix to each OFDM symbol. The SSR sends data sampled at the chip clock frequency through a $2 \times 8b$ parallel I/Q interface to, e.g., an external D/A converter pair or a digital low-IF upconversion stage. Setting the ASIC clock frequency to 20 MHz results in a standard-compliant stream of OFDM symbols.

*Operation in Receive Mode*

The coarse time synchronizer determines the received signal power, the carrier frequency offset (CFO), and the start of the FFT frames. Furthermore, it produces an AGC signal and performs a frequency correction. Importantly, it powers on the succeeding receiving circuitry upon successful detection of the frame start based on the acquisition sequence leading to power saving during sleep mode.

The adaptive equalizer (Fig. 4.3) consists of a single complex operator that sequentially processes the different carriers and removes the residual phase errors due to group delay and CFO. The equalizer performs channel estimation per carrier based on an initially or periodically transmitted BPSK-modulated OFDM training symbol. Equalization is performed in one of three ways: feedforward based on the channel estimate, based on decision feedback per carrier (SC-FB), or averaged over all carriers (AC-FB). AC-FB introduces an additional delay of one OFDM symbol that is compensated by operating the equalizer for one symbol in SC-FB mode before switching to the AC-FB mode. The calculation of the equalizer coefficients makes use of matched filtering at the expense of a gain control unit instead of dividers. Finally, the demapper despreads the modulated symbols and delivers a $2 \times 3$-bit soft output signal suitable for a convolutional decoder.

*Control Based on Token Flow Semantics*

For a modular design, a generic communication protocol is required between all design units. We implemented a scheme based on token semantics that follows the natural dataflow through transmit and receive path. A closed token-loop scheme is used between the burst controller and the datapath (**Fig. 4.5**). There is one top-level controller BC (burst controller) that only controls directly the first blocks in the transmitter chain (mapper) and receiver chain (SYN_TG and SYN_GR).

We distinguish between two different types of units: "smart" senders, "smart" receivers, and simple units. During packet reception or transmission, some tokens are only used once (single tokens) while others are used, e.g., at the periodicity of OFDM symbol starts (periodic tokens). We also distinguish between *triggering tokens* and *level-based tokens*. Level-based tokens are an elegant way to steer a receiving unit without own control; the scheduling logic of the token sender ("smart" sender) is reused. The classical triggering tokens require a "smart" receiver with its own scheduling (**Fig. 4.6**).

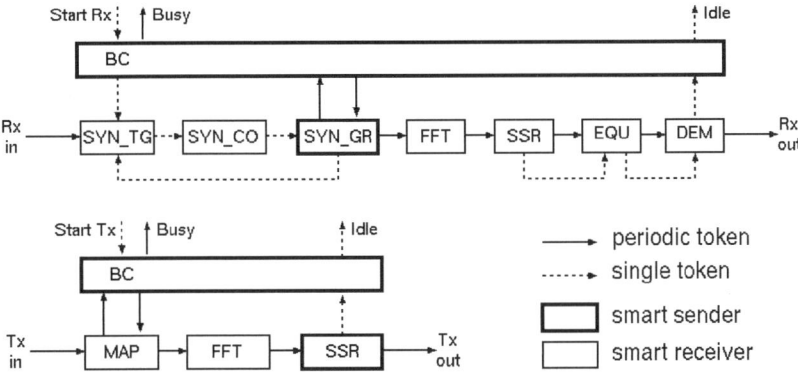

**Fig. 4.5.** Token flow concept for receive path (*above*) and transmit path (*below*) ([Eberle01b] © IEEE 2001)

An example for the detailed token flow is given in **Fig. 4.7**: the equalizer (EQU) unit is a simple unit because all control aspects with respect to the OFDM symbol structure were merged in the symbol reordering block (SSR) which appears as "smart" sender. Simple units reuse the control hardware that exists already in a "smart" sender.

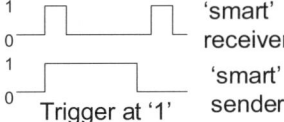

**Fig. 4.6.** "Smart" receive tokens assume a control process at the receiving unit with its own scheduling; only the beginning of a process is triggered. On the contrary, the control process is embedded into the "smart" sender when a level token is used; the receiver does not need its own scheduling

Tokens contain three sorts of information: metasymbol start, burst state information (BSI), and dynamic datapath information (DDI). Tokens are not sent at the sampling rate, but at the rate of metasymbols, i.e., at OFDM symbol rate. This token part is returned to the burst controller where it is compared against the burst length. The BSI indicates reference symbols and the last symbol of a burst and is returned by the last unit in the datapath to indicate that an entire burst has been fully processed. DDI can be added to a token by any datapath block to transfer data-dependent information synchronously with the current symbol to another unit down the processing chain. The clock offset estimator uses this to inform the equalizer in case of an FFT frame shift.

Note that this token flow semantics is compatible with distributed local clock gating which is applied here to reduce the power consumption and implement power-efficient multirate signal processing. The token scheme scales with multirate and simplifies also the design task, since a token arrival window is defined instead of a discrete point in time, keeping detailed unit latency information locally.

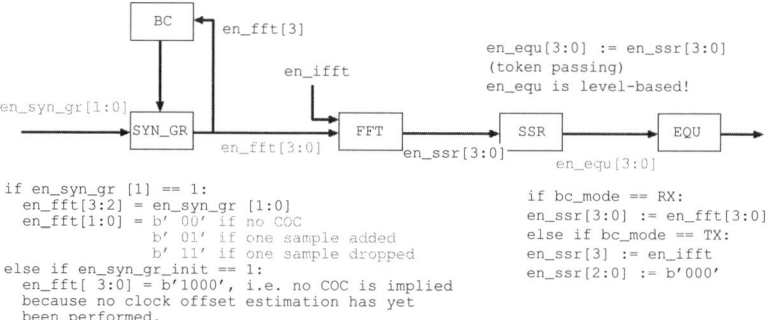


**Fig. 4.7.** Tokens can carry additional information (BSI and DDI, e.g., en_ssr[2:0]) besides the synchronization token (e.g., en_ssr[3])

*Programming Interface and Parameterization*

We designed a standard asynchronous microprocessor interface (MPI)[92] for off-chip read and write access to on-chip parameters, register content, and memory contents for programming and debugging purposes. Due to the large number of programmable parameters, an efficient configuration is important due to the large number of programmable parameters. Each building block contains a number of register files according to the local parameters needed. These are directly linked to a dedicated address value. RAMs are linearly mapped into the address space. Through the MPI, the same parameters in different components can be programmed in parallel while they can be individually read out for verification reasons (**Fig. 4.8**). This prohibits conflicting parameter programming while maintaining all debug options. Separate routing for forward and feedback paths was used to avoid tri-state buses that would require delay-sensitive arbiters.

## 4.2.3 Clocking Strategy and Low-Power Operation

Our work assumes fully synchronous CMOS design based on standard cells and hard IP macros such as on-chip memories [Rabaey96]. Hence, a clocking strategy needs to be defined that drives all registers and the memories. The choice of the clocking strategy has a large influence on area, on the achievable clock frequency due to the routing and buffering, and on power consumption due to the clock buffering.

---

[92] We took the MPI for an Analog Devices ADSP-2106x SHARC as specification.

An important input for the clocking strategy is the actual activity profile of the different modules in the design. Not all modules perform active processing at a time.

Analysis of a typical receive scenario reveals that a receiver remains a considerable amount of time in listening mode searching for a receive signal. Gaining on the average compared to the peak power consumption is thus feasible through matching activation of units with the time windows they are effectively required from the networking protocol and burst format point of view. Moreover, a transceiver operates in multiple modes. ASICs provide transmit, receive, programming, and sleep mode. All clocks are active during programming mode to allow configuration/read-out of all design units.

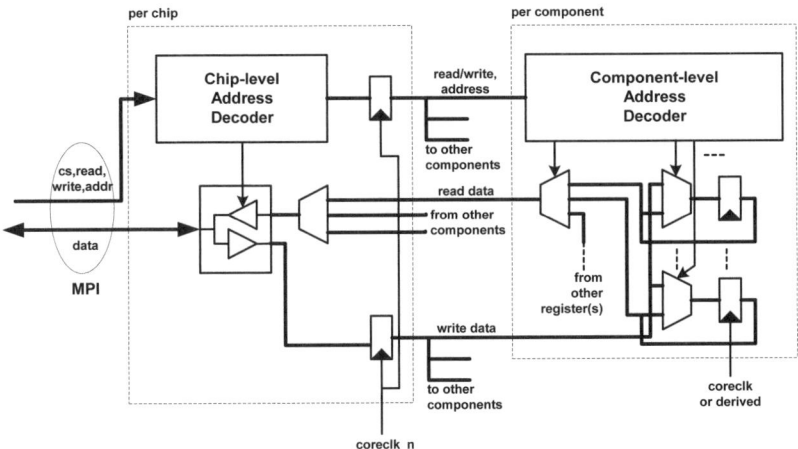

**Fig. 4.8.** Efficient configuration is important due to the large number of programmable parameters. Through the microprocessor interface (MPI), the same parameters in different components can be programmed in parallel while they can be individually read out for verification reasons

We used a clock gating scheme with a central clock generator (**Fig. 4.9**) providing several gated clocks through balanced clock subtrees. All gated clocks are directly derived from the external clock (coreclk) and not from each other. Coreclk is a single-phase edge-triggered clock. Clock enable signals are generated under the inverse of the external clock (coreclk_N) to avoid glitches. The derivation of all clocks in a single localized module guarantees a higher correlation between the delays of the generation path and hence a more predictable clock tree insertion. High testability was ensured [Favalli96].

The burst controller and decentralized smart senders (**Fig. 4.6**) control the clock generation. We also use clock gating to implement multirate interfaces between units. For example, during the guard interval period, whose length depends on programmed guard length and clock offset compensation, nonactive circuitry is disabled. Usage of several clocks can result in skew problems since clock skews can appear both positive or negative depending on routing directions and position of the clock driver. To avoid the need for verification of all clocks against each other which would largely complicate the clock tree design, we retime all transitions on a common inverted coreclk_N_out reducing the potential skew evaluation complexity from $O(n^2)$ to $O(n)$, with $n$ being the number of clocks. The reclocking imposes a tighter skew constraint on the clock tree; in particular, clock skew shall not exceed $0.5 \times (1/f_{clk})$ which results in 25 ns for a clock frequency of 20 MHz (Carnival) and 10 ns for a clock frequency of 50 MHz (Festival). This limits the amount of logic cells placeable between registers in different modules. Since all input and output stages of individual modules are retimed using registers, this did not impose a limitation. ASICs are master for all off-chip datapath interfaces and provide on-chip generated clock signals. These clocks are generated locally to the other interface I/O signals to allow joint optimization resulting in skew reduction.

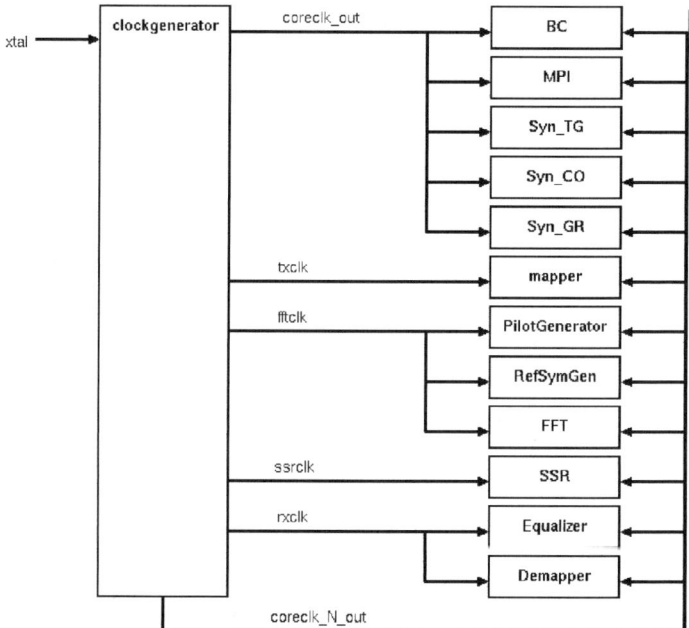

**Fig. 4.9.** A central clock generator produces the derived (gated) clocks and the commonly used inverted coreclk (coreclk_N_out) from a common reference. Derived clocks are provided through a balanced clock network to all modules([Eberle01b] © IEEE 2001)

### Activity-Driven Power Consumption Profile

The result is that we virtually do not have to insert wait states in the design. Individual design units can act as a "smart" sender and place other design units in hold mode by gating their clock (**Fig. 4.10**). Clock gating is by far more area- and timing-friendly than placing an additional register enable per flip-flop [Rhett02] and in addition, power in the clock buffer trees is also saved. As a consequence, the design follows the activity pattern of the actual data transmission and reception. Inactive phases result in reduced power consumption.

In addition to the clock gating, we consequently used operand isolation in all DSP modules to reduce dynamic power consumption in the logic [Münch00]. Note that, compared to clocking approaches such as *plausible clocking* [Yun99], we rely neither on asynchronous data FIFOs between units nor on complex handshake procedures. Scheduling is done at design time and embedded into the control logic.

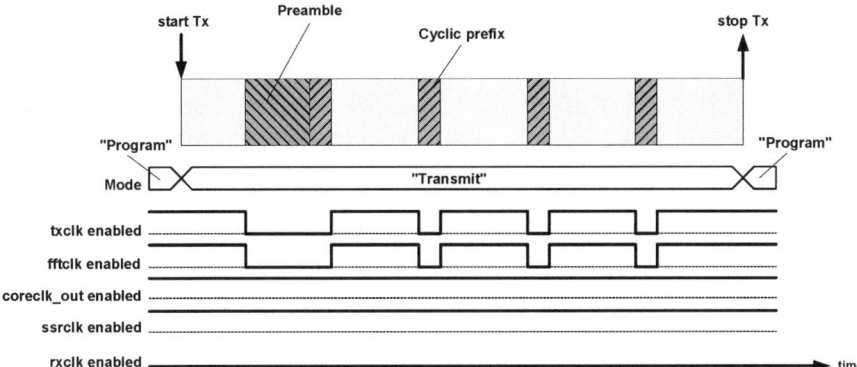

**Fig. 4.10.** Clock gating disables several modules entirely during the transmit phase. During the preamble and the cyclic prefix, mapper (txclk) and IFFT (fftclk) operation is on hold; the SSR generates preamble and cyclic prefix

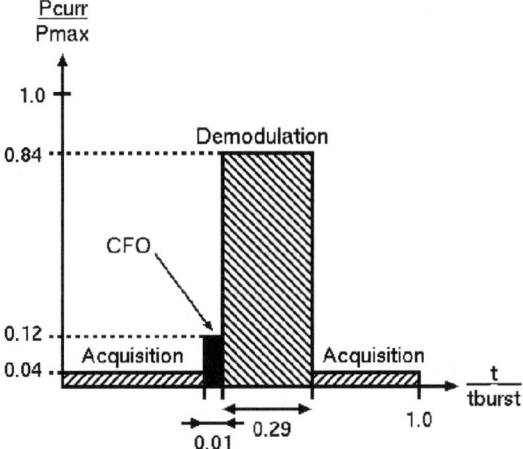

**Fig. 4.11.** The relative power consumption differs strongly between the different receive phases, e.g., more than a factor 20 between demodulation and acquisition. Since a receiver can remain rather long listening in the acquisition phase, this can lead to a substantial energy saving ([Eberle01b] © IEEE 2001)

The resulting architecture with multiple processing units and clock gating facilities follows the activity profile. In the receive scenario shown in **Fig. 4.11**, we find that the average power consumption over time for $N_{\mathrm{proc}}$ design units,

$$P_{\mathrm{curr}}(t) \sim f_{\mathrm{clk}} \sum_{k=0}^{N_{\mathrm{proc}}-1} \alpha_k(t) P_k, \tag{4.1}$$

follows the activity characteristic $\alpha_k(t)$ of the corresponding receive mode. Integrated over time, less active units, shorter activity times, and reduction of the clock rate translate into an average energy reduction.

### Outlook

On the one hand, the joint usage of a token-based communication and clock gating scheme can easily be extended toward a self-timed design in which central clock distribution could be replaced by local clock generation in synchronous islands [Chang96b, Qu01, Benini99]. On the other hand, our modular approach also allows a very easy replacement of the point-to-point communication interfaces by a packet-based NoC. Hence, both emerging trends can be supported.

Asking for the higher flexibility of instruction-set processor-based architectures calls for a tradeoff between energy and flexibility [Davis02]. We have opted for the low-power option and a largely parametrizable direct-mapped architecture.

## 4.3  Digital Signal Processing Modules

Section 4.2 described the general rationale, principles, and hence the chip architecture that we applied from a top-down perspective. This section continues with an overview of the most important signal processing modules, their architecture, and *how* algorithmic functionality has been mapped to them. Our goal here is to illustrate the *different module-dependent optimization goals* and *which* design principles and methodologies were applied to particular modules. This can be seen as motivating and describing a toolbox that enables the designer to perform all relevant nontrivial design steps. Hence, similar to [Speth01], our goal is not to describe all *functional* details.

This section is subdivided into four subsections. The sequencing is based on minimization of risk. Hence, we start with the FFT as the core functionality in Sect. 4.3.1, since it separates the signal processing into a time- and a frequency-domain partition. Main focus is on low latency. The processing latency should be smaller or in the order of the OFDM symbol duration (4 µs for 64 carriers according to IEEE 802.11a/g). In Sect. 4.3.2, we address symbol construction and deconstruction. Here, we pay special attention to translate the inherent flexibility of OFDM symbols into a parametrizable architecture. Moreover, we propose a reusable architecture for transmit and receive operation.

While FFT and symbol construction cover the main tasks of the transmitter, the receiver requires additional functionality to address equalization and synchronization. Section 4.3.3 proposes two different frequency-domain equalization schemes, trading off performance against implementation complexity. Finally, Sect. 4.3.4 describes time-domain synchronization. Here, we propose two architectures: a novel one for energy-aware acquisition and another for WLAN standard-compliant operation.

### 4.3.1 Latency-Aware Algorithm/Architecture Codesign: FFT

The Fourier transform is the central instrument of an OFDM transceiver since it provides the transition between the mathematically elegant multicarrier frequency-domain representation and the implementation-friendly time-domain representation that allows usage of sampling techniques. Efficient implementations of the Fourier transform are based on the FFT[93] in which the algorithm has been optimized for a low number of arithmetic operations [Elliott82]. Our wireless packet-based OFDM modem requires an FFT with a high throughput (1 FFT in 4 μs), low latency (less than OFDM symbols of 4 μs), low-power consumption ($<10$ nJ bit$^{-1}$ for the entire modem), and a sufficient dynamic range to accommodate the large variation present in OFDM time-domain signals. This set of constraints makes the design of the FFT rather challenging, requiring a careful tradeoff between algorithm and architecture.

*Algorithmic and Architectural Design Options and State of the Art*

The choice of an appropriate FFT algorithm in the context of VLSI design requires a careful analysis of the architectural implications such as area and power cost and performance degradation due to limited signal path accuracy [Bergland69, Thompson83, Duhamel90, Vergara98b].

Based on the initial discrete Fourier transform

$$x_{\mathrm{FD}}[m] = \sum_{k=0}^{N_{\mathrm{sc}}-1} x_{\mathrm{TD}}[k]W_{N_{\mathrm{sc}}}^{km}; \quad W_{N_{\mathrm{sc}}} = \exp\left(-j2\pi\frac{1}{N_{\mathrm{sc}}}\right), \tag{4.2}$$

the most important method to derive efficient algorithmic implementations from the original discrete Fourier transform is based on the decomposition of the length-$N_{\mathrm{sc}}$ transform into multiple transforms of smaller length and the combination of the partial results [Elliott82, Proakis96: Chap. 6, Cooley65]. The number of arithmetic operations can be drastically reduced, if $N_{\mathrm{sc}}$ can be partitioned into $n_1 \times n_2$ with radices $n_1$, $n_2 > 1$. Based on this, implementations with radix-2, radix-4, and split-radix schemes are found [Bidet95, O'Brien89, Thomson02, Melander96]. A particular interesting decomposition appears when

---

[93] Heideman et al. [Heideman84] indeed showed that Gauss' method was suitable for any composite integer. Hence, he called Gauss even the inventor of the discrete Fourier transform.

$n_1 = n_2$, since this process can be recursively continued for every partial sum until sums of length 2 are reached [Despain79]. This particular recursive radix decomposition results in a replacement of every other complex multiplication by simple rotations with integer multiples of $\pi/2$ (Table 4.2). These only need multiplexing logic to perform sign conversion and real/imaginary part swapping. This last decomposition step actually translates a classical radix-4 butterfly into a subsystem based on radix-2 butterflies and is hence also called radix-$2^2$ decomposition [He96].

**Table 4.2.** Comparison of algorithm complexity

| Algorithm | Number of complex | |
| --- | --- | --- |
| | Multiplications per sample | Additions per sample |
| Radix-2 | $\frac{1}{2} N_{sc} (\log_2 N_{sc} - 1)$ | $N_{sc} \log_2 N_{sc}$ |
| Radix-4 | $\frac{3}{4} N_{sc} (\log_4 N_{sc} - 1)$ | $2 N_{sc} \log_4 N_{sc}$ |
| Recursive radix | $\frac{3}{4} N_{sc} (\log_4 N_{sc} - 1)$ | $2 N_{sc} \log_4 N_{sc}$ |
| Split radix | $\frac{1}{4} N_{sc} (\log_2 N_{sc} - 4)$ | $N_{sc} \log_2 N_{sc}$ |

The algorithm employed has the computational complexity of the radix-4 approach

Note that we can fully reuse the FFT design to perform the inverse FFT by applying either complex conjugation or a real/imaginary part swapping on both complex input and output vectors. However, quantization needs to be revisited since the amplitude distributions differ for time- and frequency-domain signals in OFDM.

The decision for an algorithm decomposition is basically independent from the architecture to which the algorithm is mapped. Traditionally, FFT stages are in either a systolic hardwired [Swartzlander84] or pipelined form [He96, He98, Weste97, Ryan95]. For FFTs with lower throughput or special constraints such as in-place computation [Cetin97], other architectures are used, e.g., based on cached processors [Baas99a, Baas99b, Brockmeyer99, Catthoor90]. Since we focus on a high-throughput FFT, a pipelined solution appears most appealing in combination with the radix $2^2$ decomposition. Our predecessors in the WLAN area only focused on small 16-point FFTs [Ryan95, Weste97] which do neither really exhibit the control complexity nor the savings in datapath complexity of this decomposition. A drawback of this (and all other decompositions in Table 4.2) is that, mapped to pipelined architectures, they produce their output stream in bit-reversed order. This requires a postreordering step which is trivial but adds latency (Sect. 4.3.2). To guarantee low latencies, we opted for a word-serial instead of a bit-serial implementation [Melander96].

### Design of a 256-Point FFT Based on Radix-$2^2$ Decomposition

We first designed a 256-point FFT based on the radix-$2^2$ decomposition scheme. **Fig. 4.12** shows the architecture including the chain of pipelined operators. The decomposition replaces half the complex multiplier (MX) by simple rotators (RO) and reduces the complexity of butterflies to simple radix-2 butterflies. Using folding of the algorithmic flow, we achieve a 100% memory utilization for each butterfly stage when using feedback memories. This results in the absolute minimum number of memory locations $N_{sc} - 1$. Memory is implemented depending on size and availability[94] in the IP libraries as either registers or dual-port RAMs (DPRAMs). Moreover, our findings that resorting to simpler radix-2 and feedback memories results in a compacter design were confirmed in [Yeh01].

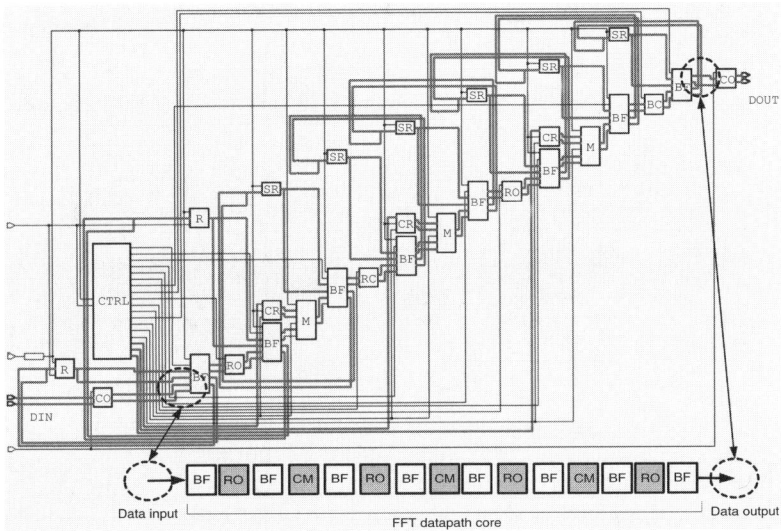

**Fig. 4.12.** The FFT datapath core is based on a radix-2 pipelined architecture. Note that, using Despain's decomposition, every second complex multiplier unit has been replaced by a simple multiplexing unit (RO) leading to a significant reduction in complexity

Importantly for packet-based OFDM, we have to analyze the latency of our implementation. The total latency $\tau$ consists of the algorithmic latency and the architectural delays introduced for pipelining reasons:

$$
\begin{aligned}
\tau \;=\; & (N_{sc}-1) && \rightarrow && \text{algorithm latency} \\
+\; & \log_2 N_{sc} && \rightarrow && \text{butterfly latency} \\
+\; & 2\log_4(N_{sc}-1) && \rightarrow && \text{complex multiplier latency}
\end{aligned}
\tag{4.3}
$$

Throughout all FFT designs, we used one register stage per butterfly and two register stages per multiplier. In total, we have designed three different FFTs based on this architecture. In all cases, the FFT-intrinsic latencies were less than

---

[94] Unfortunately, none of the technologies used offered parametrizable register files which would offer higher density and lower power consumption over the instantiation of individual registers.

20% of the OFDM symbol size and hence negligible. The design of the overall control and the approach to quantization are treated later in this section. Further details are reported in [Eberle97, Vergara98a, Vergara98b].

### Design of a Variable-Length 64-, 128-, 256-Point FFT and a 64-Point FFT

The question for the right FFT size is basically a system design question. Since standardization was not yet ready, we decided for a programmable 64-, 128-, and 256-point FFT for the Festival ASIC and, after IEEE standardization was available, we fixed the FFT size to 64 points for the Carnival ASIC (**Fig. 4.13**). Both FFTs were still based on the recursive decomposition architecture.

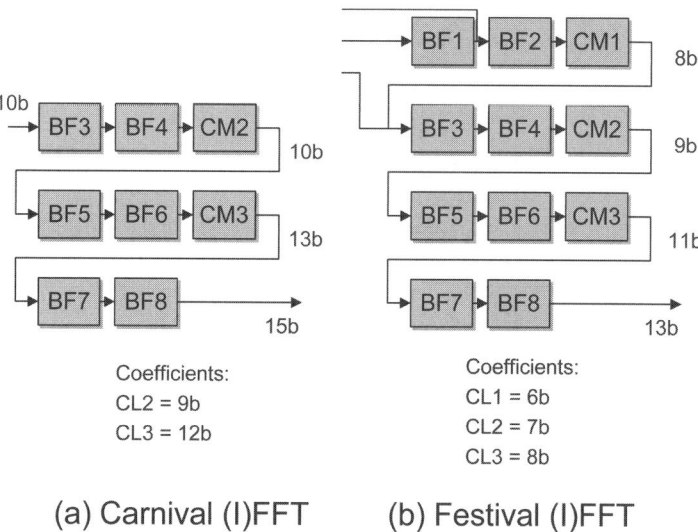

(a) Carnival (I)FFT          (b) Festival (I)FFT

**Fig. 4.13.** A block-level view of two of the three designed FFTs illustrates the key design aspects: modularization into butterfly and multiplier units and the datapath and coefficient quantization. The Festival (I)FFT (**b**) was also designed for a scalable number of carriers: 64, 128, and 256 – hence the three input paths

### Central FFT Unit Controller

A central controller (**Fig. 4.14**) modeled as a finite state machine (FSM) is used to generate enable and address signals for all FFT modules. One-bit wide enable signals generate conjugation, butterfly bypass, rotator and multiplier bypass. Because of the delay introduced by pipelining stages, the original control flow has to be rescheduled.

Each functional unit is enabled by the controller only when needed. This overall schedule is easily precomputed including register delays between the units. It structurally equals a multistage logic that is further optimized during the synthesis process resulting in a very low-complexity controller. To automate the error-prone step of rescheduling and merging, the full controller FSM is generated based on parameterization of the number and position of pipeline registers resulting in a static VHDL description. First, the flow without pipelining is generated. Individual stage synthesis reveals the required operator pipelining and hence the number of additional delay cycles. During a second step, these delays are introduced resulting in cyclically shifting the original control flow for each HW resource.

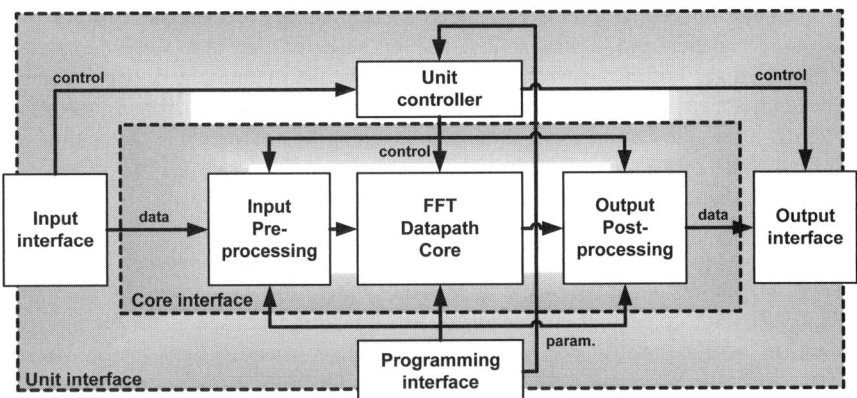

**Fig. 4.14.** The FFT architecture embeds an FFT datapath core with the required input/output pre- and postprocessing facilities, I/O interfacing, flexible control unit, and programming interface

## Quantization Aspects

Conversion to a quantized architecture is essentially an optimization problem with respect to maximum implementation loss as constraint and hardware complexity as cost. Variables to determine are the wordlengths, overflow, and rounding behavior for the datapath stages and the coefficients. Because of the rather small FFT sizes considered, we opted for a fixed-point scheme. For the larger signal dynamic ranges, e.g., in larger FFTs, block floating-point schemes may be required [Bidet95].

To limit the complexity of the optimization problem, we decided upfront for a scaling factor 0.5 at the output of each butterfly stage, such that the upper bound for the normalized SNR $\sim 1/N_{sc}$ [Trân-Thông76]. A detailed quantization analysis was carried out in [Vergara98b] from which we obtain the following approximation ($b$ is a number of magnitude bits) assuming a rounding scheme:

$$\text{SNR} \propto \frac{E\{x_{\text{TD}}^2[m]\}}{N_{sc}\sigma_{qn}^2}; \quad \sigma_{qn}^2 = \frac{1}{12}2^{-2b}. \tag{4.4}$$

Hence, for a white input signal and a radix-2-based butterfly, SNR due to round-off noise increases by 6 dB per stage. Rounding is important to avoid the introduction of a bias [Kabal86] that would be amplified in consecutive stages of the FFT.

To derive the two unknowns per multiplier, i.e., the postmultiplier datapath wordlength and the coefficient look-up table (LUT) wordlength, we performed a two-step parametric exhaustive search by simulation [Vergara98b]. This search becomes feasible since we have reduced the unknown wordlengths to only 4 in the 64/128-carrier and 6 in the 256-carrier case. In the first step, coefficient wordlengths are optimized while the datapath remains unquantized. In the second step, datapath quantization is introduced from the input stage toward the output stage. In both cases, we start with small wordlengths and increase them until the required SNR threshold is reached. For testing, we used real OFDM stimuli in time (IFFT mode) and frequency domain (FFT mode).

We have explicitly chosen for a decimation-in-frequency approach since this results in smaller memories at the later stages where the wordlength is maximum [Eberle97]. Note also that the specific decomposition chosen involves a cyclic repetition of the coefficients. As a consequence, memory for the multiplier coefficients can be reduced from $3 \times 256 = 768$ locations to $256 + 64 + 16 = 336$ locations. Further reduction is possible but involves more control complexity. This saves 25% memory in the Carnival 64-point mode compared to a decimation-in-time approach. Compared to a fixed-wordlength implementation, we even achieve a reduction of 30% in memory size from the fact that we start with 10 bit and end with 15 bit. Moreover, the choice for the radix-$2^2$ decomposition results in a reduction of rounding stages from 7 to 3 which leads to lower distortion for the same wordlength [Eberle97]. The resulting wordlengths for Festival and Carnival FFTs are provided in **Fig. 4.13**. The Carnival FFT required a requantization since it processes up to 64-QAM signals with higher SNR requirements and it works with a larger input wordlength of 10 bit.

*Results*

In total, we have designed three FFTs based on the recursive decomposition approach and following the design guidelines described above. The first was a 256-point FFT designed in 1998 in Alcatel Mietec 0.5-μm CMOS for a sampling clock of 50 MHz, a throughput of 195 kFFT s$^{-1}$ with an area of 6.25 mm$^2$ equaling 31 kGates [Vergara98a]. Synthesis and layout were performed but the design was not processed (**Fig. 4.15**). The second was a programmable 64/128/256-point FFT designed in 1999 in Alcatel Mietec 0.35-μm CMOS for a sampling clock of 50 MHz embedded into the complete Festival baseband transceiver [Eberle00a]. This FFT achieved 64-, 128-, and 256-point FFT computation in 1.28, 2.56, and 5.12 μs with latencies of $(64 + 11)$, $(128 + 13)$, and $(256 + 14)$ clock cycles, respectively. The third was a fixed 64-point FFT designed in 2000 in National Semiconductor 0.18-μm CMOS for a sampling clock of 20 MHz embedded into the complete Carnival baseband transceiver [Eberle01a]. FFT size totaled 41 kGates. This FFT achieved a 64-point FFT computation in (64 + 10) clock cycles and featured extended scaling and clipping options to shape the output signal. In all cases, the FFT-intrinsic latencies (11, 13, 14, and 10 cycles) were less than 20% of the corresponding OFDM symbol size and hence negligible.

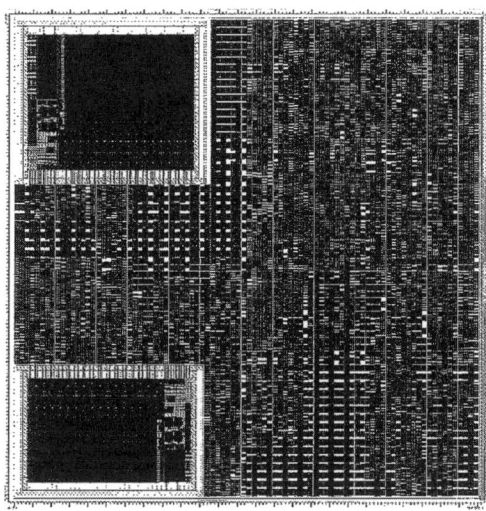

**Fig. 4.15.** Layout view of the initial 256-point FFT design in 0.5-μm CMOS

All FFTs expect input streams in natural order and deliver their output stream in bit-reversed order. Bit-reversal and other data format transformations are addressed in the next section.

### 4.3.2    Flexibility-Driven Design: Symbol (De)Construction

OFDM symbols are metasymbols compared to conventional single-carrier samples. This inherent scalability makes OFDM powerful. However, to exploit this flexibility, reconfigurable architectures supporting a discrete set of parameter choices are required. In a conventional distributed design process, the design would be first partitioned into modules and then optimized locally per module. We have, using a high-level dataflow description, first analyzed data transfer between signal processing tasks, their intraunit storage and interunit buffering requirements to handle multirate issues. The flexibility in the OFDM symbol structure leads to a large set of I/O-rate ratios. More specifically, we encountered buffering issues due to bit-reversed reordering of the FFT output, removal of pilots and zero carriers, despreading, insertion of the programmable length cyclic prefix and the preamble.

Commonly, these rate and order conversions can be described as data format conversions (DFCs) [Bae94, Bae98, Parhi92, Catthoor98a]. DFCs are a special kind of permutation architectures mainly based on storage, address generators, multiplexing, and control functionality.

In the typical functionality of an OFDM inner transceiver, we could find six major examples for such data formatting processes; three appear at the transmit side:

1. Construction of the OFDM symbol structure from data bits (mapper)
2. Reordering of the IFFT input or output stream
3. Insertion of the cyclic prefix

and the corresponding three at the receive side:

1. Removal of the cyclic prefix
2. Reordering of the FFT input or output stream
3. Deconstruction of the OFDM symbol structure into data bits (demapper)

For the FFT, we have found a distributed memory architecture to be superior to a single memory (Sect. 4.3.1). With respect to the tasks mentioned above, we found that centralizing this storage space results in a significant saving of overall memory area.

Next, we will describe two different solutions to address such data formatting issues. In the case of the symbol mapper, we face a mix of highly multiplexed small datapath operations. In the case of the symbol reordering, we found a pure data formatting task without signal processing subtasks which suited perfectly the DFC description. This subtle difference, however, leads to different architectural decisions.

*OFDM Bit-to-Symbol Mapping (Mapper)*

The first block in the transmit path, the mapper, maps the payload bits, which are provided bit parallel through an external interface, onto the subcarriers using BPSK, QPSK, 16-QAM, or 64-QAM modulation.[95]

To facilitate the filtering in the front-end, a programmable number of carriers is not modulated, both on the low and high side of the spectrum. The mapper optionally achieves frequency diversity by spreading the data over the carriers with a programmable code. Details on the programmability were provided in Sect. 4.1.1. Spreading uses a local RAM to reduce off-chip memory accesses. Fig. 4.16 reveals the architecture of the mapper which is based on a cascade of highly configurable datapath blocks with a central VLIW controller. The VLIW controller steers the individual blocks through programmable address generators.

---

[95] The mapper of Festival and Carnival ASIC differs only in details, e.g., Festival only supports QPSK modulation. They share the same architecture.

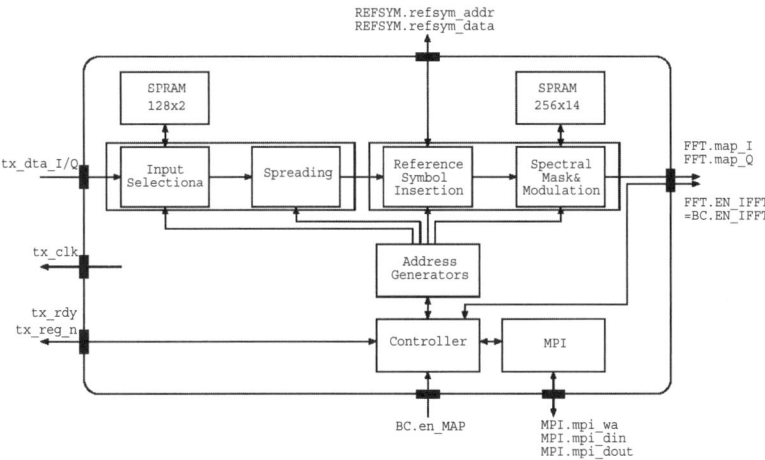

**Fig. 4.16.** The mapper generates the OFDM symbol from a set of incoming bits. Since the symbol structure is largely programmable, the mapper is a strongly programmable block with many parameters; hence, most complexity is in the controller and the address generators (numbers given are for the Festival ASIC)

([Eberle01b] © IEEE 2001)

### Centralized Symbol (Re)Ordering (SSR)

The symbol reordering consists of two single-port 64-word[96] SRAMs, memory arbiters, and a set of address generators controlled by a VLIW controller (**Fig. 4.17**). These transform the (I)FFT output to linear order. During transmission, they also perform the piecewise-linear addressing to insert the guard interval and to introduce the acquisition training sequence. During reception, they interleave the FFT output to align all data carriers in an OFDM symbol to a single continuous block of data suitable for an integrate and dump despreading operation in the demapper.

The centralization solutions efficiently use the memory transfer bandwidth while maintaining a regular access pattern. Due to this careful memory access exploration, the final on-chip datapath does not contain any caching beyond the minimum required by the signal format defined in the IEEE or ETSI standard. This caching latency is two OFDM symbols for both receive and transmit path evenly divided on FFT processing and bit-reverse reordering (SSR).[97] Since the WLAN standards require half-duplex operation only, the hardware of the (I)FFT and symbol reordering unit is shared between reception and transmission modes.

---

[96] Numbers for the Carnival chip. The Festival chip included two 256-word SRAMs.

[97] The mapper also includes storage for a half-size OFDM symbol to implement on-chip spectrum swapping. This functionality could also be merged with the interleaver for channel coding in the outer transmitter.

This leads to the fact that 98% of the datapath memory in the Festival implementation is shared.

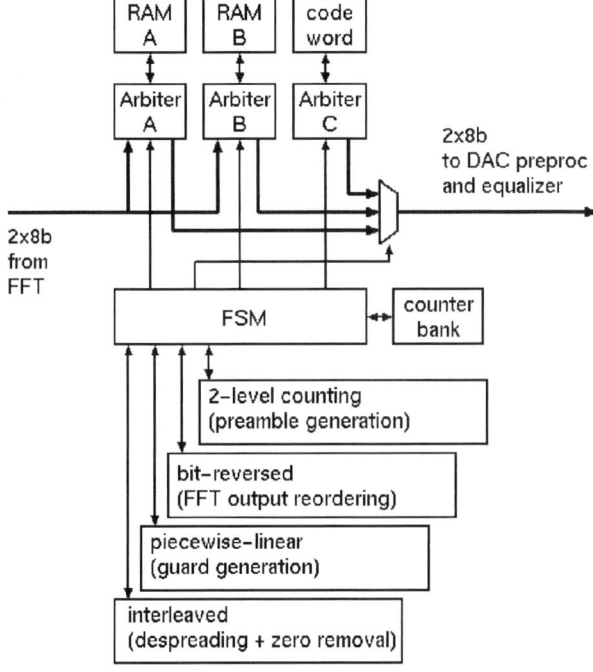

**Fig. 4.17.** The SSR combines a large number of data formatting issues such as reordering of the dataflow or generation of sequences according to a programmable repetition pattern ([Eberle01b] © IEEE 2001)

*Outlook*

The importance of data formatting tasks increases strongly with the advent of multimode and multistandard baseband transceivers, software-defined radio, and the usage of *flexible waveforms*. Increasing flexibility motivates the use of dedicated instruction-set processors or design automation for VLIW-type controllers.

### 4.3.3    Performance/Complexity-Aware Codesign: Equalization

Equalization is required in the context of demodulation if the receive signal stream is affected by frequency-selective effects. Equalization essentially restores the received, disturbed constellation pattern [Proakis95]. The availability of the received signal in frequency domain allows an elegant solution to the demodulation of the subcarrier symbol information in OFDM [Pollet00, Jones98, Kaleh95, vandeBeek95, Edfors98]. For a well-designed OFDM system, number and spacing of subcarriers are based on the channel properties, guaranteeing that each subcarrier only faces flat fading [Walzman73]. In this case, each subcarrier is treated sequentially and requires only a single multiplication with a complex correction value. In total, $N_{sc}$ different correction values are required. The complexity of an efficient FFT design together with a single sequentially applied complex multiplier is significantly lower than the one of a time-domain equalization filter with time-varying coefficients which would be needed otherwise [Bickerstaff98].

Channel equalization relies on the availability of the correction coefficients per subcarrier. For slow time-varying channels, initial channel estimation may be sufficient. For data packet lengths of several milliseconds as possible in IEEE 802.11 or ETSI HiperLAN/2, however, methods for channel tracking are mandatory.

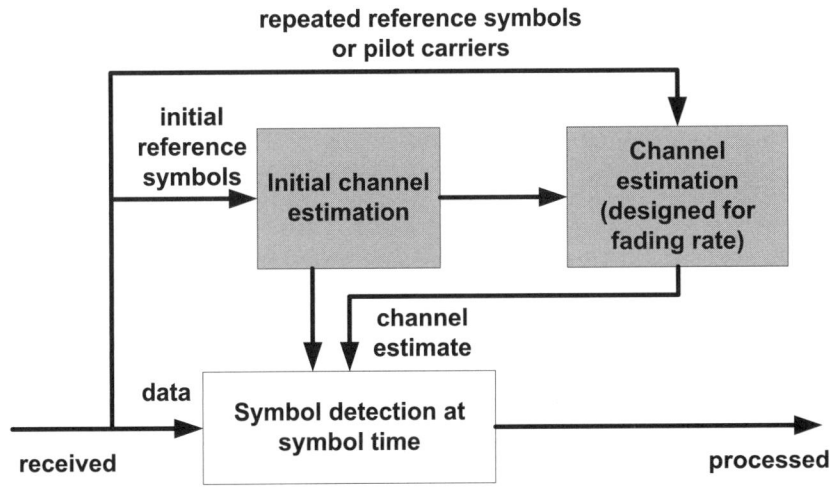

**Fig. 4.18.** The common principle of our equalizers is based on an initial channel estimation based on training symbols, symbol detection at symbol rate, and channel tracking at fading rate (or faster). For the tracking, we suggest several options – two data-aided ones using repeated training symbols or pilot subcarriers and a decision-directed one

We base our work on the architecture in **Fig. 4.18**. For channel estimation, we can rely on data-aided approaches since both our proprietary solution and the major standards foresee preambles from which the channel transfer function can be computed.

We have designed two different types of OFDM equalizers: a low-complexity equalizer for up to 256-carrier 50-Msamples s$^{-1}$ QPSK and a high-performance equalizer for 64-carrier 20-Msamples s$^{-1}$ 64-QAM. We first address the low-complexity version before we move on to the high-performance version.

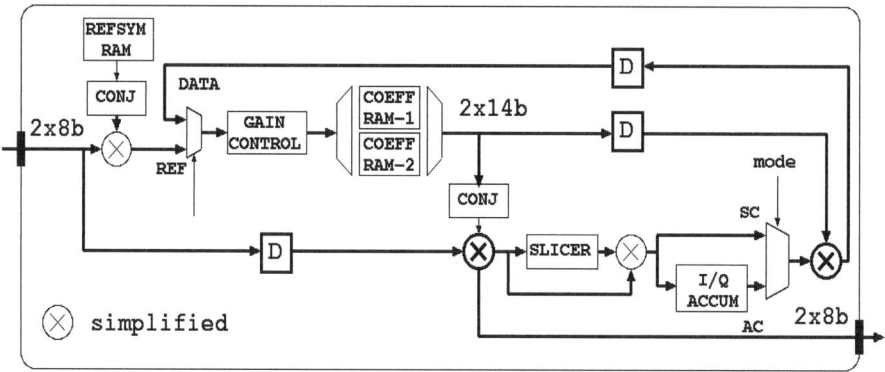

**Fig. 4.19.** The Festival equalizer reveals a low-cost solution with feedforward channel estimation and feedback decision-directed tracking. $d_i(k)$ and $d_o(k)$ are complex $2 \times 8$-bit input and output signals; $c_u(k)$ is the correction signal after the slicer/accumulator and multiplier which is translated into the $2 \times 14$-bit coefficient $G(k)$ after gain control. The latter is used for the actual multiplicate correction ([Eberle01b] © IEEE 2001)

## A Low-Complexity Equalizer (Festival)

The Festival equalizer [Eberle99a, Eberle00a] implements the basic one-tap frequency-domain equalization, consisting of a single complex multiplier with a coefficient memory to store the channel coefficients (**Fig. 4.19**).

The channel is estimated by multiplying received initial or periodic reference symbols with the known internal reference. In the equalizer, the phase difference $c_i(k)$ between the received signal $d_i(k)$ and the known reference signal $r(k)$ is measured. The division is replaced by a complex multiplication and followed by a shifter-based quantizing gain controller that maximizes the magnitude of the complex signal. $c_i(k)$ represents an initial coefficient set that is used during the following OFDM data symbol for phase compensation. The error $e(k)$ per subcarrier $k$ is computed by removal of the data information from the received complex symbol [Proakis95]. $e(k)$ can now be fed back in two different ways to obtain the updated channel coefficient $c_u(k)$:

$$c_i(k) = d_i(k)r(k),$$

$$G(k) = \frac{1}{\min_Q(\mathrm{re}\{c_u(k)\},\mathrm{im}\{c_u(k)\})} c_u(k),$$

$$d_o(k) = d_i(k)c(k),$$

$$c_u(k) = \begin{cases} c(k)e(k) & \rightarrow \text{per carrier (SC-DFE)}, \\ c(k) \sum_{m=0}^{N_{c,data}-1} e(m) & \rightarrow \text{average (AC-DFE)}. \end{cases}$$

(4.5)

In single-carrier (SC-DFE) mode, feedback per carrier is applied. In average-carrier (AC-DFE) mode, correlated phase errors on the $N_c$ data used data carriers correct the previous coefficient $c(k)$. $c_u(k)$ requires a gain controller that keeps the loop gain stable and avoids magnitude convergence toward zero. The AC-DFE mode assumes that correlated phase errors are produced by carrier offset affecting all carriers in the same way. In that case, averaging reduces the variance on the estimated sum. The variance is introduced by phase jitter and quantization effects.

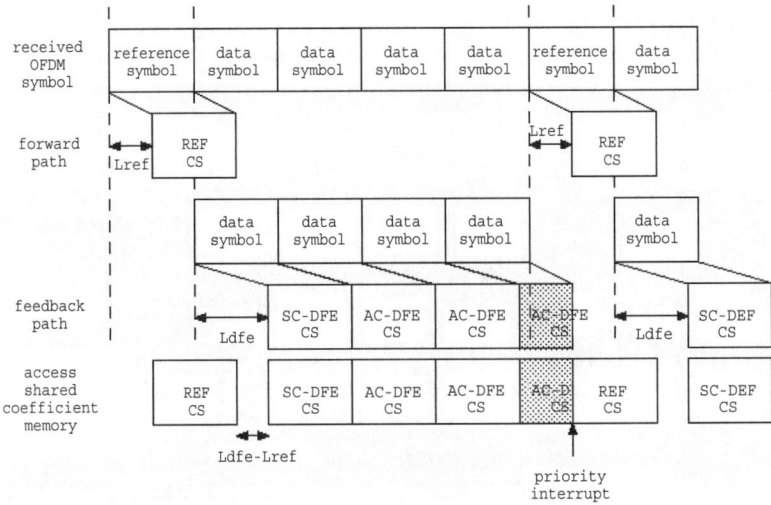

**Fig. 4.20.** Different delays of the arithmetic in forward and feedback path and the additional delay for averaging over all carriers require careful scheduling of the access to the two half-size coefficient set (CS) memories ([Eberle99a] © IEEE 1999)

Since the coefficient memory is accessed for both read and write access in every cycle, we investigated several tradeoffs with respect to RAM partitioning, including a DPRAM of 256 words, the same with 128 words, and four single-port RAMs of each 64 words. With respect to area and power, the solution with $2 \times$ SPRAMs of 128 words gains at least a factor of about 2.25 for the power $\times$ area product compared to the other solutions for the same throughput. When more but smaller RAMs are used, the fixed area overhead becomes dominant. DPRAMs have a severe initial power and area penalty. The solution with two half-size SPRAMs leads to alternate read and write per coefficient on both RAMs, putting constraints on the path delays $L_{ref}$ and $L_{dfe}$. Memory access arbitration is thus required to coordinate the alternating access between the two memory blocks and the transition between the three operation modes. Different implementation delays between forward ($L_{ref} = 3$) and feedback path ($L_{dfe} = 12$) require a priority handling during a transition from averaged to reference mode and lead to wait states during reference and single-carrier mode (**Fig. 4.20**).

Since the channel estimate is updated for phase, we can also track and reduce phase-related effects such as fine CFO and common phase noise [Muschallik95] in the averaged-feedback case. The equalizer is able to compensate for remaining CFOs up to ±3 kHz at 50-MHz sampling rate and 256 carriers. Performance results of the Festival modem with different imperfections are provided in **Fig. 4.21**. The error floor is due to the limited ADC input quantization of 8 bit. For the Carnival modem with its higher modulation schemes, 10-bit ADCs were used to lower the error floor. Measured performance with a digital-IF and 5-GHz RF front-end resulted in a bit-error rate (BER) $< 10^{-3}$ for uncoded and BER $< 2 \times 10^{-4}$ for coded transmission when using the AC-DFE mode. We did not find a significant increase in bit errors toward the end of longer bursts, which means that we face negligible error propagation in the decision-directed feedback loop.

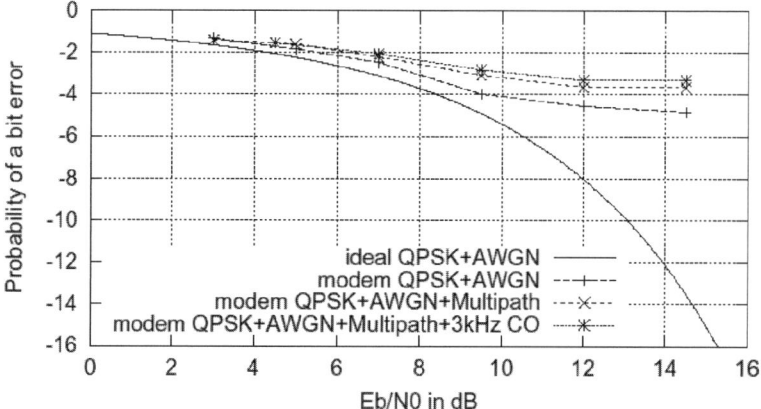

**Fig. 4.21.** Performance results of the Festival modem with different imperfections. The error floor is due to the limited ADC input quantization of 8 bit. For the Carnival modem with its higher modulation schemes, 10-bit ADCs were used to lower the error floor ([Eberle99a] © IEEE 1999)

### A High-Performance Equalizer (Carnival)

The Carnival equalizer also uses the concept of a single complex operator with coefficient memory. The 16-QAM and 64-QAM constellation schemes, however, require accurate amplitude correction, which is performed by a complex divider (**Fig. 4.22**). In addition to initial and periodic reference symbols, to update part of the channel, a pilot pattern is sent with every symbol. The equalizer provides fine CFO estimation and compensation capabilities and support for clock offset precompensation by a programmable phase rotator. This phase rotator is steered by the clock offset estimator in the time domain which is detailed in the context of acquisition (Sect. 4.3.4).

The channel estimate obtained from a single reference symbol still contains a considerable MMSE error (**Fig. 4.23**).

The received signal after the FFT is still affected by multipath fading. However, by proper choice of the subcarrier spacing relative to the coherence bandwidth, the FFT produces a highly oversampled channel response. This results in a quasi-diagonal channel matrix $H$ with insignificant contributions outside the diagonal. The equalizer can exploit this in two ways [Deneire00c, Deneire03]. First, it requires only a single complex channel coefficient per subcarrier to compensate for the channel. Second, the rank of this matrix $H$ is reduced, since high oversampling translates into correlated channel coefficients. Thus, we can apply filtering to suppress noise and interpolate a smoothed channel vector from a smaller set of coefficients.

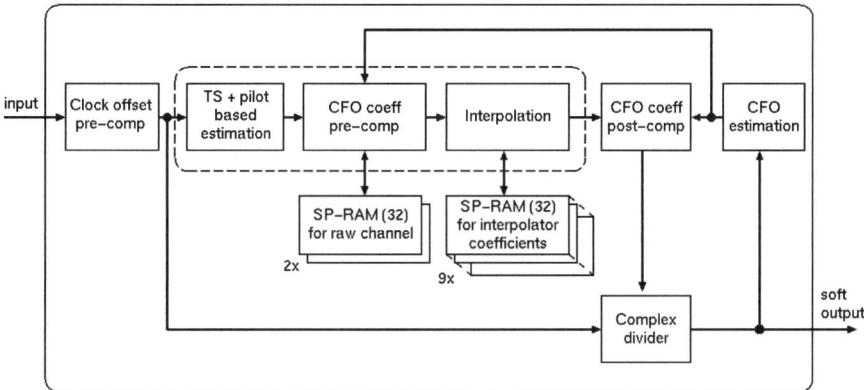

**Fig. 4.22.** The Carnival equalizer requires an interpolator and divider in addition to the Festival equalizer since modulation schemes up to 64-QAM need to be processed ([Eberle01b] © IEEE 2001)

This scheme has been implemented in the Carnival ASIC [Eberle01a]. **Fig. 4.24** provides a detailed view on the interpolator architecture. A channel interpolator, consisting of an initial "noisy" stage with the CFO phase error update, is followed by a cascade of four blocks implementing a matrix operation:

$$H_{smooth} = SS^H H_{noisy}. \tag{4.6}$$

Matrix $S$ is a $64 \times 9$ programmable complex coefficient matrix. The first two stages transform the noisy channel estimate into an impulse response vector of length 9, effectively suppressing any noise present beyond nine taps. The last two stages interpolate the full 64-tap frequency-domain channel response from this truncated impulse response vector. The first three stages employ full parallelism such that an interpolated channel tap is again available after one OFDM symbol latency. Coefficient sets are stored in nine RAMs next to a preprogrammed set in a LUT.

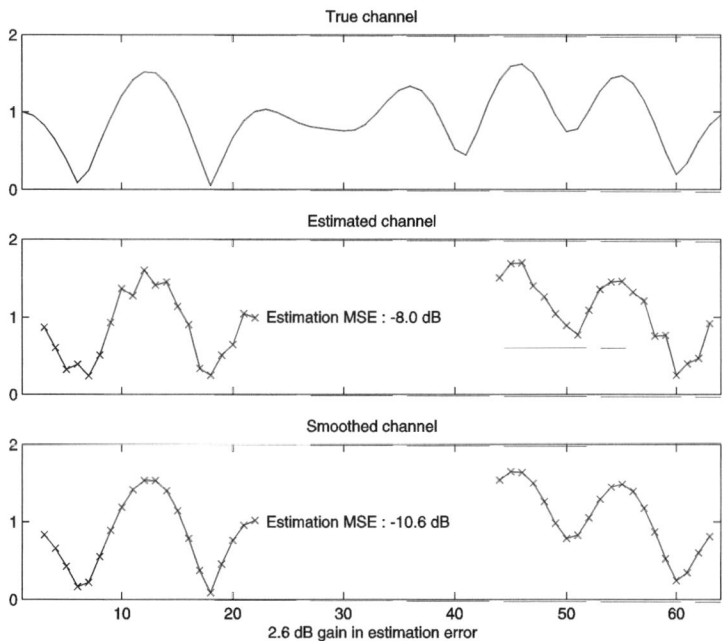

**Fig. 4.23.** The channel truncation and interpolation in frequency domain enhance the accuracy of the channel estimate by 2–3 dB on the average by interpolation (smoothing), resulting in a BER improvement for the same SNR ([Eberle01b] © IEEE 2001)

The interpolator block adds 120 kGates and hence substantial complexity to the equalizer. About 60% complexity is due to the signal processing stages Mult_A, Accum, Mult_B, and Smooth which perform the 64 × 9 matrix conversion. Seven percent is in the channel coefficient RAMs and the divider (ip_noisy). Another 7% is required for storage of the fixed coefficients in ROM and 25% for the programmable coefficient RAMs.[98] The datapath works pipelined and with maximally nine parallel MAC units requiring 54 real adders and 72 real multipliers. Per OFDM symbol of 64 samples, 864 million real additions, and 1,152 real multiplications per second are required.

The result of the interpolator is an improvement of the channel estimation accuracy by 2.5–3 dB. Together with the rotating pilot scheme, it is also able to suppress spurs, e.g., from the equalizer feedback loop, reducing error propagation. This results in a significantly reduced BER (**Fig. 4.25**).

---

[98] The latter is only required if a programmable solution is needed. The ROM can contain the same yet fixed coefficient set.

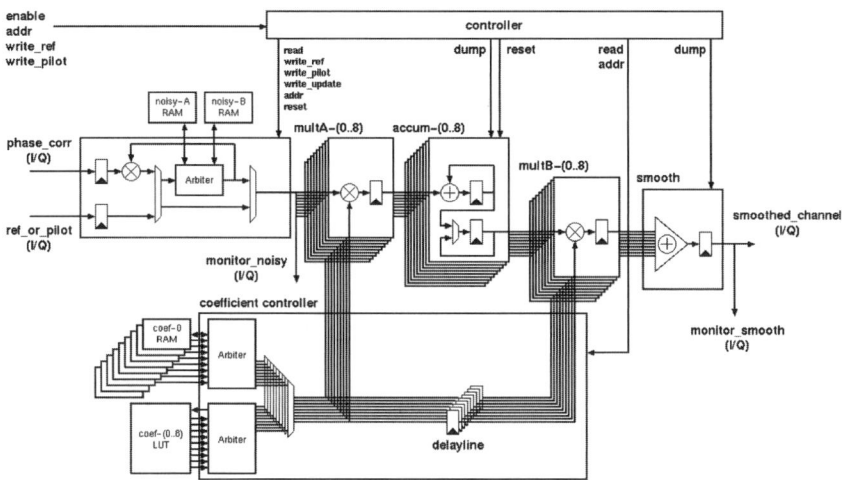

**Fig. 4.24.** The Carnival equalizer is based on the Festival equalizer but largely enhanced with an impulse response truncation and interpolation mechanism which includes a programmable 64 × 9 transformation matrix ([Eberle01b] © IEEE 2001)

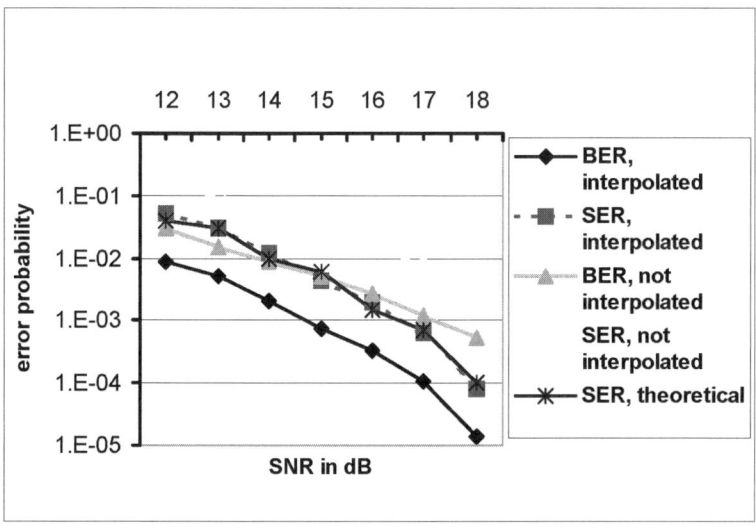

**Fig. 4.25.** Bit- and symbol-error rates are also greatly reduced when activating the interpolator in 64-QAM transmission mode

### 4.3.4 Energy-Aware Codesign: Acquisition

Acquisition is the process of achieving synchronization. For a wireless receiver, this means acquiring synchronism of the incoming signal with internal references, e.g., for sampling clock, time, and phase. In particular, receiver acquisition has to detect the incoming signal, adapt its signal power, achieve timing synchronization, and compensate for CFO introduced by local oscillator mismatches in transmit and receive front-ends. At the same time, the received signal is distorted by a number of indoor channel and front-end effects.

Due to this large number of tasks, acquisition involves fairly complex control. This is even more complicated since wireless LAN systems are packet based and depend on fast burst acquisition to minimize transmission overhead at the physical layer. Hence, we first define an acquisition strategy before we detail the individual tasks of timing synchronization, CFO compensation, and clock offset compensation. These algorithms and architectures were implemented in the Festival and Carnival ASICs. Later on, we extended our timing synchronization to a standard-compliant algorithm. Finally, we comment on the methodology we used to explore the impact of acquisition in the transceiver design context.

### Acquisition Strategy

For the acquisition strategy for wireless OFDM systems, we can hardly rely on the extensive state of the art in wired OFDM systems. Short packets prohibit the overhead of long training sequences and the wireless channel variations prohibit slow acquisition loops as typically used in ADSL, DAB [Taura96], and DVB systems. In packet-based data transmission, we cannot afford losing individual packets such as in broadcasting schemes (DAB, DVB). We also want to avoid separate signaling channels as proposed in [Stantchev99] since they require infrastructure (e.g., access point) support. Fast burst synchronization schemes were first addressed in [Speth98] but not for indoor propagation conditions. Hence, we defined a novel acquisition strategy for wireless packet-based OFDM receivers.

Correct OFDM demodulation requires precise synchronization both in time and frequency. Misestimating time leads to intersymbol interference (ISI) between subsequent OFDM symbols while a noncompensated carrier frequency error generates intercarrier interference (ICI), both reducing the signal-to-noise-and-interference ratio (SNIR) of the received signal. Moreover, the multipath propagation distorts the acquisition sequence. Eventually, both the acquisition sequences and algorithms should be designed accurately to allow fast and robust synchronization.

Our synchronization approach performs the complete timing acquisition, gain control, and coarse CFO compensation in a feedforward way before the FFT, thus avoiding the typical long reaction time of state-of-the-art DVB, DAB, and ADSL receivers. We only use data-aided techniques; blind techniques are not used due to their high complexity [Palicot03]. Moreover, standards such as IEEE 802.11a and ETSI HiperLAN/2 foresee preambles for the purpose of acquisition.

Our acquisition sequence[99] (Fig. 4.26) is similar to the sequence proposed for IEEE 802.11a but is generated in the time domain. The first part of the sequence, the relative timing sequence (RTS), is based on a repetition of a programmable code sequence (TS1) and its binary inverse (TS2), toggling between these two sequences, for a programmable number of times. COS-2 is a repetition of COS-1 and its length is also programmable.

Acquisition consists of three cascaded blocks (**Fig. 4.4**): timing synchronization and gain control (SYN_TG), CFO estimation and compensation (SYN_CO), and clock offset estimation and guard removal (SYN_GR). SYN_TG and SYN_CO exploit the RTS/ATS and COS-1/COS-2 parts of the sequence, respectively. SYN_GR estimates clock offset based on the guard removal and removes the guard removal before passing time-domain data to the FFT.

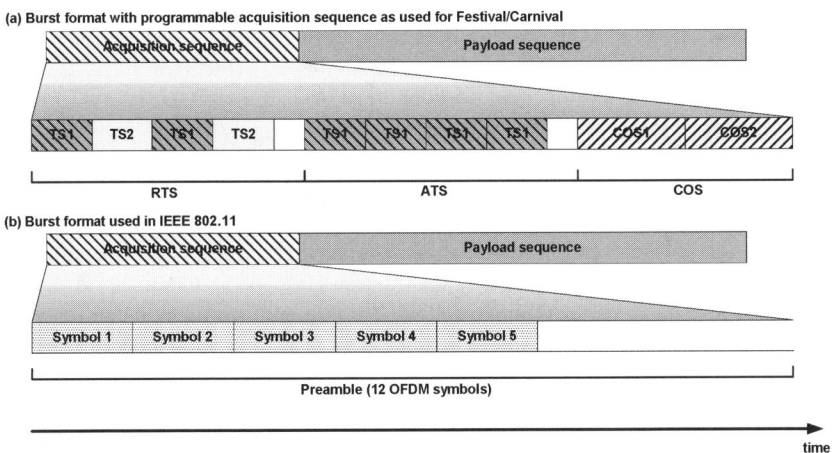

**Fig. 4.26.** We use a proprietary sequence with a more structured pattern (**a**) than the IEEE 802.11 synchronization sequence (**b**)

Importantly, we have developed a cascaded acquisition process that allows activation of the subsequent blocks on demand which is important to reduce power consumption [Williams03]. In reality, the receiver is – for a large amount of time – listening without receiving actual data. In our architecture, only the initial gain control and timing synchronization blocks will be active at that moment (Sect. 4.2.3). The acquisition strategy as a whole is detailed in [Eberle00b, Eberle02c]. We will now review its major components.

*Timing Synchronization Based on Proprietary Sequences*
The first step in the synchronization process is frame start or symbol synchronization [Speth97]. Since at this point all other properties of the receive

---

[99] The reason for a proprietary preamble was twofold. At the moment of the design (1997), standards were not yet available hence we needed our own definition of a preamble structure.

signal are still unknown,[100] we can only evaluate structural information. Typically, this means that timing synchronization is based on autocorrelation (AC) or crosscorrelation (CC) properties of the initial sequence which appears fairly robust against noise, linear, and nonlinear distortion effects.

Fortunately, OFDM with a cyclic prefix to limit ISI does not require accurate timing synchronization to within a sample such as single-carrier transmission schemes (assuming no oversampling).

In literature, several algorithms were proposed to exploit particular preamble structures for timing synchronization. Müller-Weinfurtner [MüllerW01] relies on a modified sandwich preamble. Schmidl and Cox [Schmidl97] use correlation over a two-symbol training sequence to derive timing and carrier frequency offset. [Lambrette97] exploits a single-carrier CAZAC sequence which has specific constant-envelope properties. van de Beek et al. [vandeBeek97] use correlation on the guard interval to derive timing and carrier frequency offset but did not perform well under a set of real front-end nonidealities. All approaches assumed perfect gain control as starting point which is unrealistic.

Hence, we defined a novel acquisition strategy that was tolerant against initial gain fluctuations and allowed a tradeoff between complexity, preamble overhead, and probability of correct synchronization. Our architecture for timing acquisition is based on a two-phase autocorrelation process (**Fig. 4.27**) using the RTS/ATS sequence which is a programmable BPSK time-domain code sequence repeated according to a second metalevel sequence. Since the sliding window correlator only requires complex I/Q 1-bit input, it is very robust against AGC transients and implementable with low area and power cost. A parallel sliding window signal power estimation is used to validate the correlator results. Alternating bipolar correlation peaks during phase S1 determine the relative code sequence start, while the transition to phase S2 defines the absolute frame reference (**Fig. 4.28**). Timing synchronization exploits the repetitive pattern. TS1 and TS2 are specifically chosen such that a sequence of negative and positive peaks appears after correlation. This allows detection of the sequence start. The absolute frame start is determined from the position where the alternate transmission of TS1/TS2 is replaced by a continuous transmission of TS1. Based on the known absolute timing sequence (ATS) length and correlator delay, the frame start (COS-1) is determined.

The receiver only uses information on the codeword length and the metalevel sequence; the codeword itself is not known. Probabilities of false alarm and missing detection depend on the programmed numbers of detected peaks in phases S1 and S2, respectively. Phase S3 counts until the frame start when phase S2 has obtained enough confirmations.

---

[100] This actually also included appropriate usage of the dynamic range. We focus on automatic gain control and link it to timing synchronization in Chap. 5.

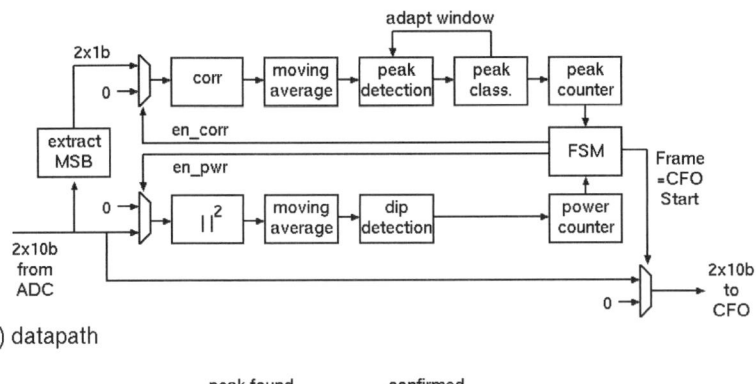

a) datapath

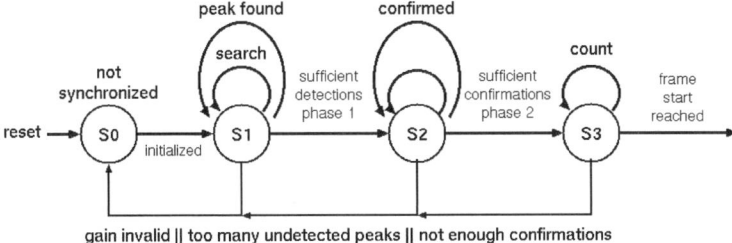

b) controller FSM

**Fig. 4.27.** Timing synchronization is based on autocorrelation followed by peak detection. Peak detection is restricted to a moving window whose position is adjusted based on previous peaks (*a – upper path*). Power estimation notices instable or weak signal strength cases (*a – lower path*). A state machine determines first the relative peak positions (*phase 1*) and then the absolute frame start (*phase 2*) ([Eberle01b] © IEEE 2001)

By repetition, the probability of false alarm and missing detection can be reduced. A periodicity switch is introduced by the ATS, which allows us to determine the absolute time, i.e., the end of the last partial code sequence TS1 in the ATS. During the ATS detection, the expected position of the correlation peaks is known from the RTS estimate, resulting in more robustness.

A typical performance result for this synchronization algorithm is shown in **Fig. 4.29** based on simulation. The synchronization algorithm finds the correct position (location = 7) within ±2 samples inaccuracy in 84% of cases and for ±3 samples inaccuracy in 93% of the cases. Results are for 5-dB SNR, 40-ns channel delay spread, and 1-kHz clock offset. Synchronization performance has also been measured on the actual Festival ASIC implementation together with a digital-IF and superheterodyne front-end. Over 1,000 packets, the sum of probabilities for failed acquisition $P_{fa}$ and missed detection $P_{md}$ was found to be below $10^{-3}$. This is negligible compared to typical error rates for the decoding if we assume effective packet retransmission schemes (e.g., ARQ) as foreseen in IEEE and ETSI standards.

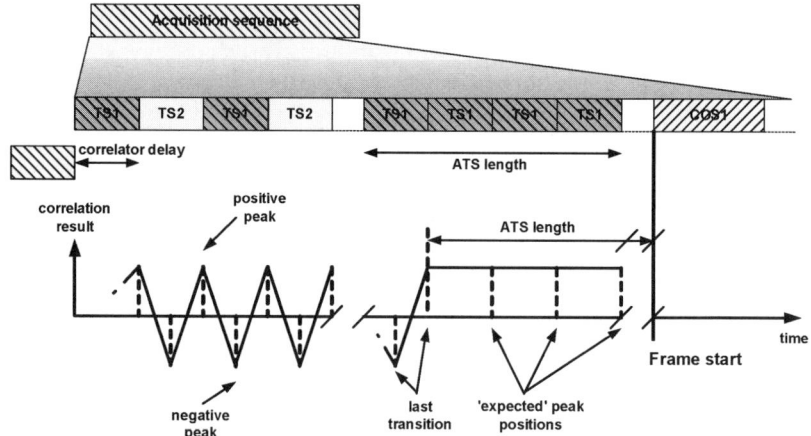

**Fig. 4.28.** Timing synchronization exploits the repetitive pattern. TS1 and TS2 are specifically chosen such that a sequence of negative and positive peaks appears after correlation. This allows detection of the sequence start. The absolute frame start is determined from the position where the alternate transmission of TS1/TS2 is replaced by a continuous transmission of TS1. Based on the known ATS length and correlator delay, the frame start (COS-1) is determined

## Carrier Frequency Offset Estimation and Precompensation

CFO appears in carrier-based communications and originates from mismatches in the local oscillators at transmit and receive side. Conventionally, analog frequency control (AFC) loops were used but for a flexible and fast compensation, we propose a digital compensation scheme [Moose94, Morelli99, Muschallik00, Luise96].

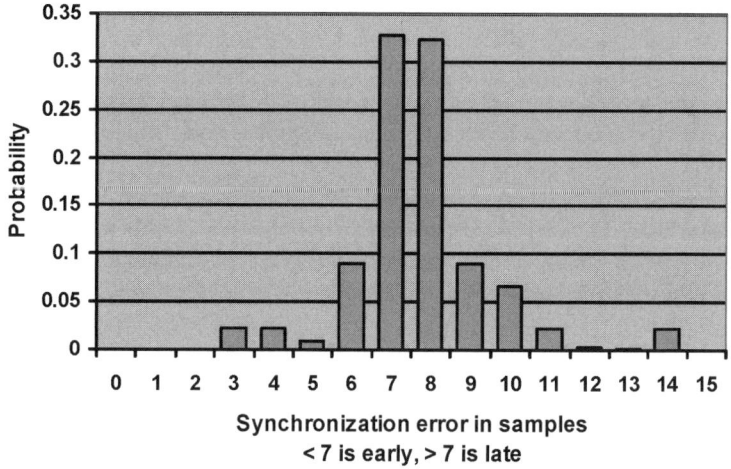

**Fig. 4.29.** The synchronization algorithm finds the correct position (location = 7) within ±2 samples inaccuracy in 84% of cases and for ±3 samples inaccuracy in 93% of the cases. Results are for 5-dB SNR, 40-ns channel delay spread, and 1-kHz clock offset

Our architecture (**Fig. 4.30**) estimates and compensates the CFO in time domain prior to decoding. We use a data-aided approach exploiting a part of the acquisition preamble. CFO is estimated using the twice repeated preamble part (COS-1, COS-2) of length 64, 128, 256, or 512, which follows the frame start, based on autocorrelation for multipath immunity reasons. A larger preamble size trades off a higher noise suppression against a lower capture range. Carrier offset must be reduced to a fraction, e.g., 1–2%, of the 312.5-kHz subcarrier spacing, to achieve negligible ICI in the FFT. A single-operator sequential CORDIC converts the Cartesian estimate into a phase difference. The evolution of the carrier offset phase is reproduced by a wrap-around phase accumulator with a second 18-stage pipelined CORDIC stage. The CORDIC uses a constant input reference $1 + j_0$ to provide a Cartesian output with a conversion accuracy independent of the highly amplitude-varying receive signal. The compensation of an entire OFDM symbol of 80 samples requires 240 real multiplications and 4,720 real additions. In the current implementation, the usage of CORDICs results in a significant reduction of hardware complexity at the expense of an additional delay (wait-state sequence, WSS).

We still need to apply fine CFO compensation to meet the performance expectations in the equalization. This second compensation has been described in the context of the equalization (Sect. 4.3.3).

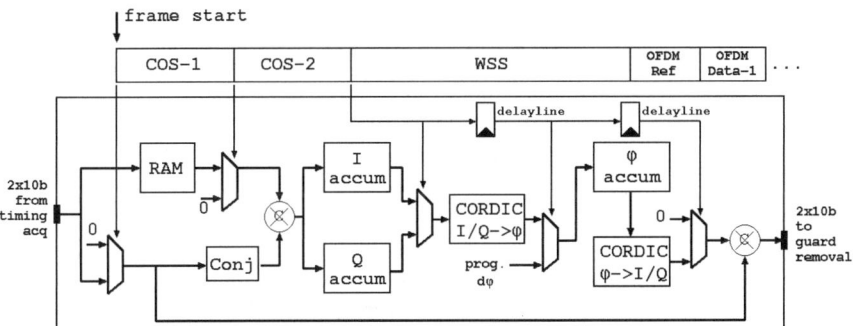

**Fig. 4.30.** Carrier frequency offset is measured by correlating the two identical COS-1 and COS-2 sequences. During the wait-state sequence (WSS), a CORDIC translates the Cartesian estimate once into its polar representation and feeds a phase accumulator. The final phase is then continuously translated from polar to Cartesian and applied to the rest of the packet ([Eberle01b] © IEEE 2001)

*Clock Offset Estimation and Compensation*

Clock or more precisely sampling clock offsets originate from different absolute clock references at transmit and receive side. Clock offset estimation and compensation are essentially a means to synchronize the two clocks. A large mismatch in sampling clocks results in sampling at bad time instances which can result in ISI [Yang00, Speth98]. In OFDM in particular, synchronization needs to take place within a particular window of the cyclic prefix to avoid ISI (Sects. 3.2 and 5.4).

OFDM symbol duration is only 4 µs over which the accumulative clock offset can be neglected. However, for the total packet lengths of 4,095 payload bytes allowed in IEEE 802.11, clock offsets based on commercial clock generators with 20-ppm frequency accuracy can result in a time shift of several samples which would move the start of the FFT outside the guard interval. Hence, clock offset appears as a rather slow effect and does not need fast compensation. We exploit both the fact that a shift results in ISI as well as the fact that only slow compensation is needed by estimating the timing shift with the help of the guard interval. **Fig. 4.31** illustrates the architecture (a) and the usage of the regular guard interval (b).

The SYN_GR unit traditionally discards the first $N_{cp}$ samples of the guard interval and provides the following $N_{sc}$ samples to the FFT. We make this process adaptive. A correlator measures the correlation between the original $N_{cp}$ samples at the end of the OFDM symbol and the copy in the guard interval. The peak position of the correlation is slightly affected by the clock offset. A single estimate is, however, too noisy such that we average over a number of OFDM symbols. The slow nature of the clock offset allows averaging over 32–64 OFDM symbols resulting in a reliable estimate. After each averaging period, we take a three-way decision: we can leave the starting point of the FFT window, we can add or drop a sample. Dropping a sample essentially means that we have to place the subsequent receive chain in a hold mode. We can easily do this using our clock gating facilities. Note that the decision needs to be passed to the equalizer too to correct for the phase shift that we introduce with adding or dropping a sample.

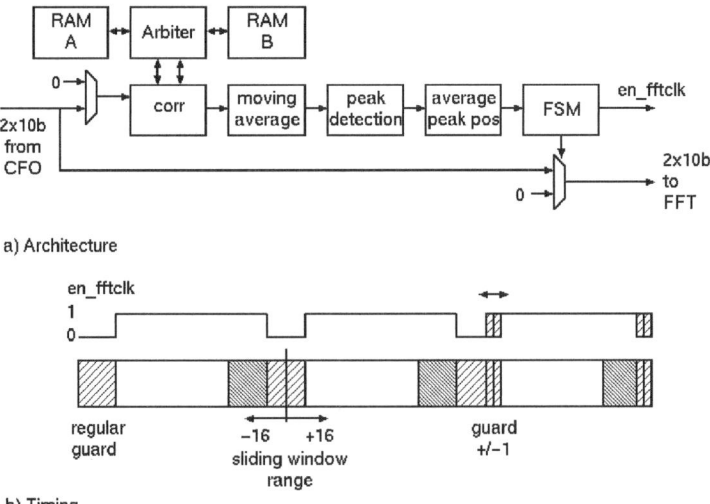

a) Architecture

b) Timing

**Fig. 4.31.** Clock offset is tracked by guard interval correlation and averaging over multiple OFDM symbols ([Eberle01b] © IEEE 2001)

The hardware complexity of the additional clock offset estimation and compensation functionality in the SYN_GR unit amounts to 64 real multiplications and 192 real additions per OFDM symbol (80 samples). The clock offset compensation can be deactitated and the update frequency is made programmable. Note that this fully digital approach is significantly more flexible and less complex than steering of the ADC clock through a PLL and VCXO [Pollet99] (**Fig. 4.32**).

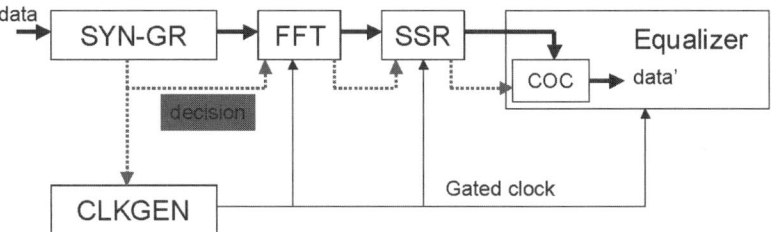

**Fig. 4.32.** Clock offset is estimated in time domain on the cyclic prefix. The decision to add or drop samples to compensate for the induced time shift needs to be communicated to the equalizer to compensate for the accompanying phase rotation in frequency domain. The clock generator uses clock gating to introduce the compensating time shift

### Improved Standard-Compliant Fine Timing Synchronization

IEEE 802.11a and HiperLAN/2 have finally defined different preambles to the ones initially selected by us (**Fig. 4.33**). Both standards provide preambles that are suitable for autocorrelation and crosscorrelation. We are interested in an improvement of the coarse timing synchronization accuracy which is in general termed *fine* timing synchronization. AC algorithms have been studied extensively in [Schmidl97, vandeBeek97, Minn00b] and were also proposed previously in this section. An OFDM CC algorithm has been proposed in [Hazy97, Tufvesson99]. This class of algorithms performs a crosscorrelation of the received signal with a training sequence known to the receiver.

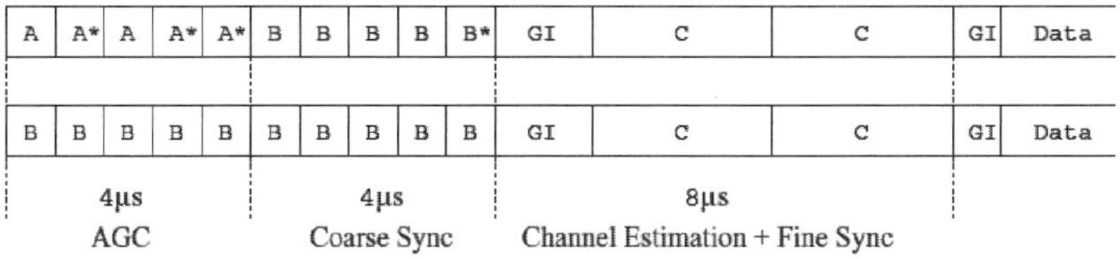

| A | A* | A | A* | A* | B | B | B | B | B* | GI | C | C | GI | Data |
|---|----|---|----|----|---|---|---|---|----|----|---|---|----|----|

| B | B | B | B | B | B | B | B | B | B | GI | C | C | GI | Data |
|---|---|---|---|---|---|---|---|---|---|----|---|---|----|----|

|        4µs        |       4µs        |          8µs            |
|        AGC        |    Coarse Sync   | Channel Estimation + Fine Sync |

**Fig. 4.33.** HiperLAN/2 broadcast preamble (*top*) and IEEE preamble (*bottom*)
([Fort03a] © IEEE 2003)

We noticed that, so far, crosscorrelation detectors suitable for both standards had not been investigated – neither from a performance nor from a complexity perspective. Hence, we developed a novel crosscorrelation algorithm for standard-compliant *fine* synchronization and compared its performance and complexity

[Fort03a]. We implemented this algorithm on an FPGA together with an AC algorithm and tested it extensively using HiperLAN/2 channel models.

In IEEE and HiperLAN/2 systems, each transmitted burst consists of a preamble followed by regular OFDM symbols. For both standards, the first 4 µs are used for AGC. The remaining B sequences can be used for packet detection and coarse timing/frequency synchronization. The C sequences are used for channel estimation and synchronization refinement. Our algorithm uses B sequences after the AGC settles. For HiperLAN/2, the final B sequence is inverted so that the autocorrelation function decreases for the last 16 samples creating a clear peak. As peak detector, we select a max function which is a good approximation of the maximum-likelihood (ML) solution. Note that the peak is less clearly visible for the IEEE preamble since it does not have an inverted B sequence. Due to this drawback, modifications are required in practice when front-end effects and AGC are considered (Sect. 5.3). The magnitude of the crosscorrelator output consists of a large peak for each multipath component, and several small peaks due to AWGN and imperfect autocorrelation properties of the preamble sequences. While the training symbols can be chosen from any part of the preamble, the C sequence is preferred so that B sequences can be used for an initial frequency correction (**Fig. 4.34**).

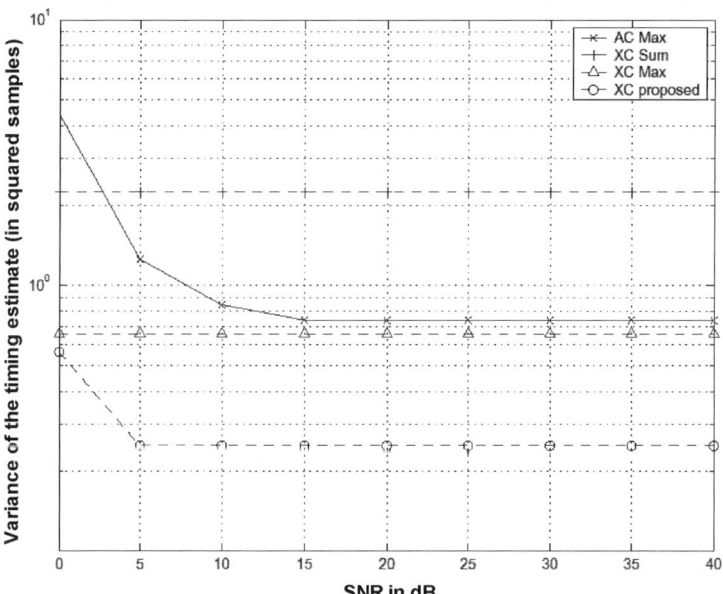

**Fig. 4.34.** The proposed crosscorrelation (XC proposed) results in a significantly better timing synchronization than the other crosscorrelation algorithms and the autocorrelation approach (multipath case with 50-ns delay spread and AWGN averaged over 5,000 channel instantiations) ([Fort03a] © IEEE 2003)

Two crosscorrelation timing detectors have been presented in literature and we propose a third: The XC sum algorithm sums consecutive correlator outputs to locate a window of length $L$ where the channel has the most energy. $L$ is ideally the channel length. The XC max algorithm selects the peak with the largest magnitude. Our proposal (XC proposed) selects the earliest peak with a magnitude greater than some percentage of the largest peak. This improvement tends to select the first multipath component rather than later reflections, thus reducing the variance of the timing estimate.

The autocorrelation algorithm (solid line) yields large timing errors at low SNRs as the noise complicates peak detection. However, this may not be a problem since IEEE and HiperLAN/2 systems typically operate above 5-dB SNR. At higher SNRs, a variance floor is reached due to the multipath channel. This floor corresponds to a symbol timing error of −2 and +2 samples 99% of the time. Under 200-kHz frequency offset and larger delay spread (150 ns), results were confirmed [Fort03a]. The crosscorrelation algorithms provide better performance at low SNRs as the averaging process did not work because many small channel reflections were indistinguishable from the noise making the energy estimate unreliable. The maximum peak detector (XC max) provides better performance similar to the autocorrelation. For our XC proposal, a 50% threshold was used for SNRs greater than 5 dB, but an 80% threshold was needed to avoid selecting noise peaks at very low SNRs. Only a single threshold is needed in practice since SNRs lower than 5 dB are not encountered. This algorithm yielded the best performance since it is more robust to channel reflections. At high SNR, the variance corresponds to a timing offset error between 0 and +2 samples during 99% of the correlation time. Hence, our newly proposed crosscorrelation algorithm outperforms the state-of-the-art detection mechanisms.

*Exploration Methodology*

A communication system acts essentially as a chain of functional units where a particular unit depends on the correct functioning of its predecessor. A particular dependency is created between the acquisition and the decoding of the actual payload in a packet. For performance investigation, it is cumbersome if we would always resort to exhaustive simulation of the entire chain as, e.g., done in [Kabulepa01]. Instead, we separated this dependency by deriving statistical models of the acquisition effects [Jeruchim92] independently of the payload decoding. This is exemplified in (4.7) for a parameter $\kappa$, for which we would derive the probability density function pdf($\kappa$) and its relevant parameter range $\kappa_{min} \dots \kappa_{max}$:

$$\overline{p}_b = \int_{\kappa_{min}}^{\kappa_{max}} p_b(\kappa)\,\text{pdf}(\kappa)\,d\kappa, \qquad (4.7)$$

where $p_b(\kappa)$ describes the effect of a particular value $k$ on, e.g., the decoding functionality. These models can be used to create a representative set of input stimuli for the simulation of the decoding part. An example is the distribution of the timing error as a function of the signal-to-noise ratio (**Fig. 4.29**). Based on this, Monte Carlo simulations can be run efficiently on a particular functionality, e.g., to derive particular statistical moments, such as the average $\overline{p}_b$. This approach has also systematically and successfully been applied for the analysis of the acquisition process at the packet level in [Huys03].

## 4.4 Evaluation

For the evaluation of the design of the Festival and Carnival ASIC, we first introduce the experimental results from chip measurements and from application demonstrators. Next, we place them in the context of the evolving state of the art after 2001. Finally, we review the major achievements for both designs.

### 4.4.1 Experimental Results

The Festival ASIC (**Fig. 4.35**) was processed in 1999 in an Alcatel Mietec 0.35-μm digital CMOS process and packaged in a 144-pin PQFP. The design complexity was 155,000 equivalent gates and the design contained ten SRAM

**Table 4.3.** Carnival outperforms Festival with respect to spectral efficiency and energy efficiency[101] at a moderate increase in silicon area despite a significantly higher complexity ([Eberle01b] © IEEE 2001)

| Same protocol and burst overhead at an uncoded rate of 76 Mbit s$^{-1}$ | Festival ASIC | Carnival ASIC | Carnival relative to Festival |
|---|---|---|---|
| Efficiency (bit s$^{-1}$ Hz$^{-1}$) | 1.5 | 3.8 | 2.5 (better) |
| Energy per transmitted bit (nJ bit$^{-1}$) | 8.8 | 2.6 | 3.4 (less) |
| Energy per received bit (nJ bit$^{-1}$) | 7.5 | 2.8 | 2.7 (less) |
| Tx DC power (mW) | 670 | 199 | 3.4 (less) |
| Rx DC power (mW) | 570 | 212 | 2.7 (less) |
| Nominal clock frequency (MHz)[102] | 50 | 20 | 2.5 (less) |
| Die size (mm$^2$) | 16.4 | 20.8 | 1.3 (worse) |
| Technology | CMOS | CMOS | Same |
| Feature size (μm) | 0.35 | 0.18 | Smaller |
| Supply voltage (V) | 3.3 | 1.8 (core)/3.3 | Lower |

---

[101] Note that the first discrete and analog OFDM-based modem in 1959 consumed 100 W for 2,400 bps achieving an energy efficiency of 41.6 mJ bit$^{-1}$. Our ASIC solution is nine magnitudes of order more energy efficient [CRC59].

[102] Nominal clock frequencies differ: Carnival was designed according to the 20-MHz channel specified in IEEE 802.11a. Festival was designed with a clock frequency limit in mind. Festival and Carnival can be configured to achieve the same throughput. Both ICs use the highest I/O clock frequency as their core clock frequency. Since Carnival uses higher constellations, it can achieve the same throughput with a lower I/O clock frequency.

cells. The chip was designed for a maximum clock frequency of 50 MHz. The Carnival ASIC was processed in 2000 in a National Semiconductor 0.18-μm digital CMOS process (**Fig. 4.35**) and packaged in a 160-pin PQFP. The design complexity was 431,000 equivalent gates.

Table 4.3 gives an overview of the most relevant measured performance and cost parameters for both ASICs. A fair comparison at the same data rate and overhead between Festival and Carnival shows the superior spectral efficiency and energy efficiency of the latter at the cost of a moderate area increase of 30%. The highly programmable equalizer occupies 63% of the area in the 64-QAM chip compared to 10% for the FFT. Fixing the coefficient set is reducing this percentage to significantly less than 50%.

Evaluation of the Festival and Carnival ASICs ranged from functional chip testing and nonfunctional chip parameter measurements over long-term scenario tests (several hours) for stability reasons to full application demonstrators such as webcam and video transmission. Both chips were functionally sufficiently operational such that a number of application demonstrators could be built upon them. First, we assess complexity and power consumption. Then, we describe the application demonstrators.

*Complexity*

Design complexity for the Carnival ASIC is now assessed in more detail in number of operations per second, number of memory accesses per second (**Fig. 4.36**), and overall memory bandwidth (**Fig. 4.37**).

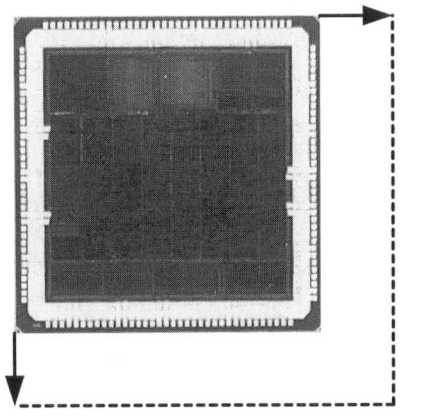

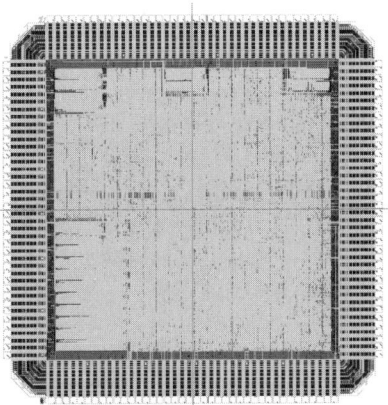

### a) FESTIVAL ASIC
### (0.35 μm CMOS)

### b) CARNIVAL ASIC
### (0.18 μm CMOS)

**Fig. 4.35.** Additional equalization and synchronization functionality increased the area of the Carnival ASIC (**b**) by about 30% compared to its predecessor Festival (**a**)

In transmission, 18% of the design is active whereas 97% of the design is active in reception. The FFT contributes 10% of the active cells. In transmit mode, the FFT contributes about 50% of the power consumption. In receive mode, the FFT is not the dominating consumer; the main power consumption goes to the interpolating equalizer with 63%.

### Measured Power Consumption

For the Festival ASIC, power consumption has been measured in 20-MHz operation in typical transmit, receive, and sleep[103] scenarios resulting in 270, 230, and 50 mW, respectively. Table 4.4 compares the power consumption and complexity.

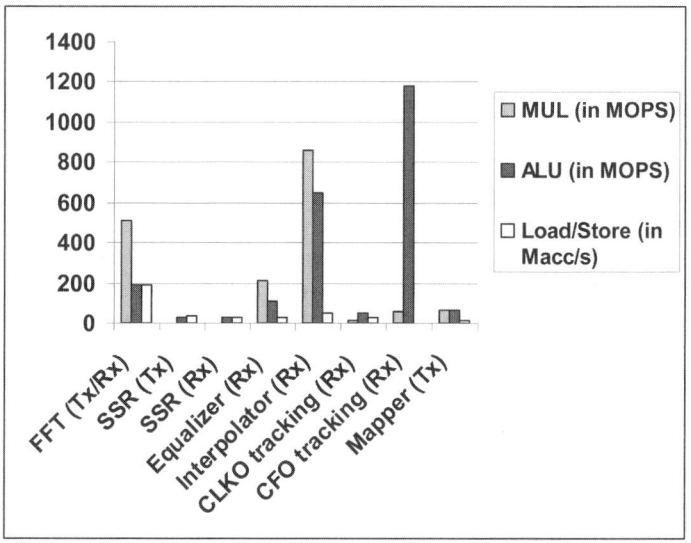

**Fig. 4.36.** Computational complexity in complex MUL and complex ALU operations and memory accesses per unit for the Carnival design

For the Carnival ASIC, power consumption has been measured separately for the 1.8-V core and 3.3-V I/O supply in typical transmit, receive, and programming scenarios. During transmission, 156-mW I/O and 43-mW core power consumption were observed. During reception, the much higher core activity dominates with 146 mW compared to a lower 66-mW I/O consumption due to less I/O switching. In programming mode, logic switching is zero but all clocks are enabled, leading to 35-mW I/O and 81-mW core consumption.

---

[103] The sleep mode has not been designed for low-standby power consumption. It only deactivates some signal processing blocks.

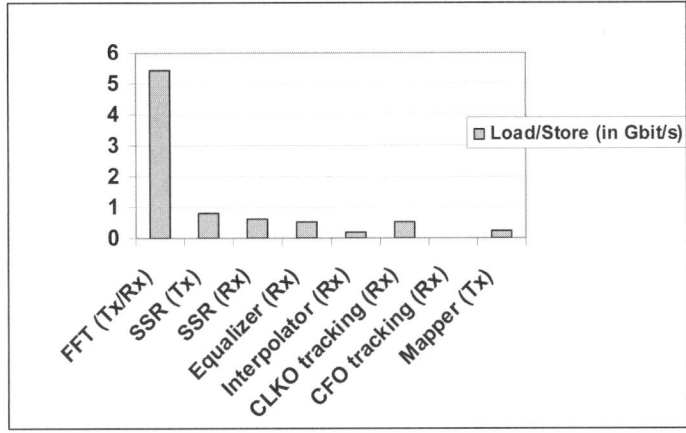

**Fig. 4.37.** Memory access bandwidth in Gbit s$^{-1}$ per unit for the Carnival design

**Table 4.4.** Comparison of power consumption and complexity for different modes of the Festival ASIC

| "IEEE" mode at 50-MHz clock frequency | Power (mW) | Operations $(10^9\,s^{-1})$ | Memory accesses $(10^9\,s^{-1})$ | Memory accesses (Gbit s$^{-1}$) |
|---|---|---|---|---|
| Transmit mode | 670 | 3.8 | 1.32 | 21.3 |
| Receive mode | 570 | 6.8 | 1.28 | 20.3 |
| Programming mode | 400 | N/A | N/A | N/A |
| Sleep mode | 150 | N/A | N/A | N/A |

*Reduction of I/O Power Consumption*

For both ASICs, chip I/O drivers contributed a large amount of the power consumption. For Carnival, 156 mW (66 mW) or 78.4% (31.1%) of the total power consumption could be attributed to external I/O, for transmit and receive mode, respectively. This amount, however, can be reduced significantly by avoiding off-chip connections, e.g., by integrating the ADC and DAC on-chip. In [Thomson02], only 25 (24) mW or 10.2 (10.5)% of the power consumption was due to external I/O.

### 4.4.2   Testing and Application Demonstrators

A wireless local area network allows a multitude of applications. If we want to prove our architecture concept, the best way is to develop a complete demonstration setup around a specific application and embed the ASIC designs in it. In this way, we will encounter also problems that originate from the interaction between different components and from real physical effects for which models can sometimes only give an approximation.

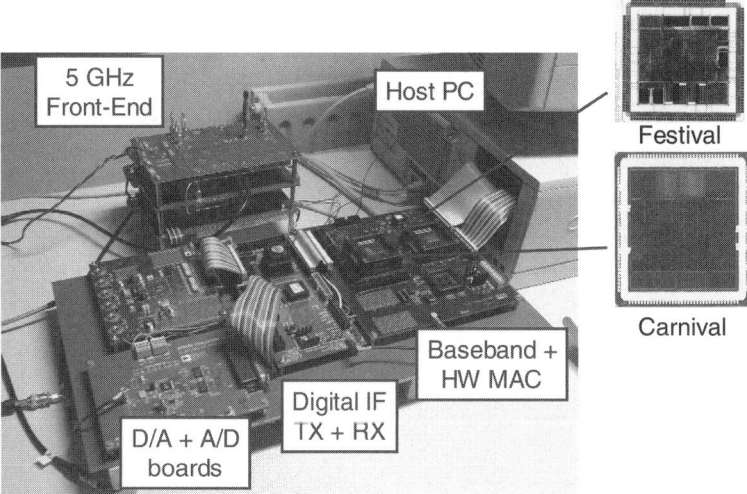

**Fig. 4.38.** A 5-GHz wireless LAN setup with socket for Festival or Carnival chip

As a consequence, both ASICs were tested in an experimental test setup consisting of a discrete 5-GHz superheterodyne front-end with a four-times oversampled digital-IF stage [Eberle00b], an FPGA-based hardware MAC, and a software MAC and API implemented on Linux host PC (**Fig. 4.38**).

### Demonstration with the Analog/RF Front-End and Digital-IF

Initial functional tests were performed by connecting two digital basebands directly with a cable. First performance tests for multipath effects were performed with a digital channel emulator. For the interfacing with the analog/RF front-end, we first opted for a digital-IF solution with four-times upsampling. Based on the 20-MHz sampling frequency at baseband, this resulted in an 80-MHz IF. A 41-tap Remez-filter was used as interpolation/decimation filter. The digital-IF block assumed 10 bit at the front-end and baseband side. It was implemented in an FPGA [Eberle00b]. The digital-IF was crucial for testing the acquisition, for the initial development of the AGC schemes and the impact of front-end nonidealities on the baseband transceiver. With the design of more advanced RF transceivers, the digital-IF was replaced by direct-conversion architectures (Chap. 5). For over the air tests, we first used the 430-MHz IF of the front-end, then replaced it later by a 5-GHz front-end. At all levels, successful tests were run.

### Application Tests

Application tests over the air were performed using webcam-based image transmission, file transfers, and video transmissions between two terminal prototypes using the baseband transceiver in transmit and receive mode. All transmissions required a full bidirectional communication link. As an example, the system architecture for the webcam-based setup is shown in **Fig. 4.39**. A more detailed description of the application setups and testing can be found in [Eberle02c].

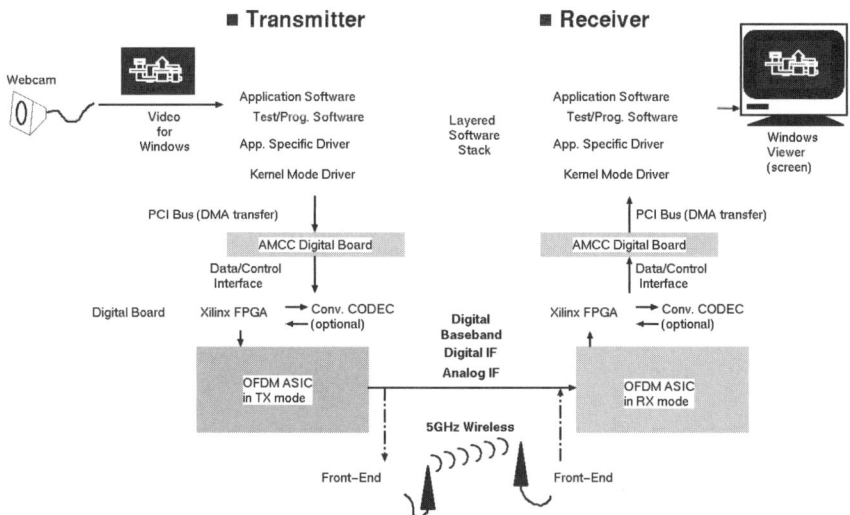

**Fig. 4.39.** As an early example, a PC-based webcam video transmission over a wireless end-to-end link was demonstrated

## Demonstration of Adaptive Loading

The Carnival architecture was already designed with support in mind for adaptive loading strategies [VanderPerre98, Thoen02a]. The OFDM frame structure makes it possible to apply diversity techniques in the frequency domain across subcarriers. The basic idea of adaptive loading is to send more bits on those subcarriers where the SNR is high and less or even no bits where the SNR is low, allowing a nearly constant BER on all subcarriers.

Adaptive loading requires a flexible symbol mapping programmable per subcarrier at the transmit side and the possibility to periodically extract channel estimation information at the receive side. Carnival provides the transmit features through providing multiple constellation schemes (BPSK, QPSK, 16-QAM, 64-QAM), a symbol interface that groups bits per subcarrier, and a spectral mask per subcarrier that allows for varying the output power per subcarrier. At the receive side, the equalizer output can be streamed out during full receive operation at OFDM symbol rate. The algorithm for the actual computation of the optimal bit loading has been mapped to an FPGA [Thoen02a]. The adaptive loading scheme was successfully tested using a combination of Carnival ASIC and bit-loading FPGA.

## Further Evolution

The code database for the ASIC was further extended later on and reused for mapping the entire transceiver to a Xilinx Virtex FPGA, termed *FlexCop* in 2002 [Wouters02b]. Many further developments such as the integration with digital compensation techniques (Chap. 5) were based on extensions on the FPGA platform. Note that the initial implementation took up about 48% of the slices and 86% of the multiplier units in a Virtex XC2V6000. In 2002, the code database of the Carnival ASIC was also reused to develop an IEEE 802.11a-compliant baseband transceiver in collaboration with Resonext Communications. Later developments toward MIMO systems required large additions of new algorithms, obviously reducing the reusable code base.

### 4.4.3  Comparison with the State of the Art After 2001

A direct comparison with the state of the art becomes rather difficult since most industry products became highly integrated and incorporate other functionality into the same SoC such as MAC, data converters, or access point functionality.

*Recent State-of-the-Art Designs*

After 2001, a multitude of integrated baseband designs has appeared on the market.

While there was a clear trend to integrate more functionality on the same die, notably the lower (hardware) and higher (software) MAC [Thomson02, Fujisawa03] as well as A/D and D/A converters [Thomson02, Kneip02], no clear trend can be seen so far with respect to the digital baseband architecture. Atheros, Cisco, Amphion, Resonext, [Thomson02], and [Fujisawa03] came up with dedicated ASIC architectures. In [Kneip02], a configurable, data- and instruction-parallel DSP core assisted by application-specific enhancements is described.

*Why Did We Design an ASIC?*

A first indication is the work of [Davis02] where the 64-point FFT at 50 Msamples s$^{-1}$ and hence comparable to our requirements is compared for different architectures. A 250 × higher energy efficiency is reported for direct-mapped hardware over a low-power DSP implementation. From a power consumption point of view, the ASIC solution remains interesting at the cost of flexibility, of course.

For a classical DSP processor such as the TI C64 series, running the FFT at 20 MHz would not be a major problem since such DSPs are optimized for FFTs and the FFT does not require special operations as such. However, mapping the entire Carnival receiver on a DSP would require more than 15 GOPS which is beyond the capabilities of a single low-power DSP [McCain].

Sandbridge Technologies propose a multithreaded multiprocessor design (SB9600) with four processors and 9.6 GMAC s$^{-1}$. Based on our complexity estimate and without additional application-specific enhancements, we estimate that a multiprocessor implementation with eight processors would be needed.

Since multiprocessor-based platforms have been only coming up recently, it was a logical conclusion that, from energy consumption and availability, ASIC design was the most promising solution. Still, given the heterogeneity of requirements for the different signal processing tasks (high-rate yet reconfigurable stream processing for the FFT, high-rate filtering for part of the equalizer or the digital-IF, and mixed control/data processing for the synchronization), a combination of several processors with possibly different processor styles appears interesting even today.

## 4.5 Conclusions

Algorithm and architecture design for OFDM baseband transceivers was at the basis of this chapter. In 1997, we started from the ambitious target to extend the throughput of WLAN solutions from the 2-Mbit $s^{-1}$ state of the art to 100 Mbit $s^{-1}$. This required decisions for a new transmission scheme, new algorithms, and new VLSI architectures for efficient implementation of these algorithms. Deciding for OFDM as the new transmission scheme also meant that we could not start from a given specification for the baseband part. Instead, significant research was needed to investigate intrinsic performance of OFDM and of algorithms in the context of OFDM. A key design goal was to keep flexibility of the design high, such that it could be configured according to throughput and power requirements. In this context, we achieved:

- An optimized yet scalable FFT design based on algorithm–architecture co-optimization and exploration of quantization and storage properties
- A novel architecture concept for OFDM transceivers with substantial resource sharing (SSR) and consequent application of time- and frequency-domain signal processing; the architecture provided also initial support for adaptive loading
- The design of a low-cost decision-directed equalizer with common phase noise cancellation for QPSK-based OFDM and the design of a high-performance novel interpolation-based equalizer
- A novel, programmable acquisition scheme with a focus on timing synchronization and a stepwise energy-consumption-friendly ramp-up of the system
- An efficient clock offset compensation scheme in time domain with simple frequency-domain postcompensation
- The complete design and successful testing of two baseband transceivers in 0.35-μm CMOS for 80-Mbit $s^{-1}$ QPSK-based OFDM (Festival) and in 0.18-μm CMOS for 72-Mbit $s^{-1}$ 64-QAM-based OFDM (Carnival)

A patent [Eberle99b] was filed in 1999 and granted in 2004. It covers the novel baseband modem architecture with in particular the split of time- and frequency-domain functionality and the usage of common resources for transmit and receive mode. Moreover, the novel acquisition process with its energy-consumption-friendly power-up and its flexible synchronization pattern is claimed.

In this chapter, we have described functionality and design of two inner baseband transceivers for wireless LAN applications that pushed throughput from the 2-Mbit s$^{-1}$ state of the art to 70 Mbit s$^{-1}$ (Festival) and 54 Mbit s$^{-1}$ (Carnival). Still, we should not forget that the overall performance of the physical layer strongly depends on the analog/RF front-end part and good codesign between digital and analog. Some analog/RF nonidealities such as imperfect gain settings and DC offsets at the receiver, transmitter nonlinearity, and I/Q mismatch in both receiver and transmitter were not yet addressed in this chapter. To obtain a complete solution, we investigate solutions for the analog/digital receive interaction in Chap. 5 and for the transmitter design in Chap. 6.

# 5 Digital Compensation Techniques for Receiver Front-Ends

*The great tragedy of science – the slaying of a beautiful hypothesis by an ugly fact.*
Thomas H. Huxley, 1825–1895.[104]

The digital signal processing aspects of an OFDM transceiver have been addressed in Chap. 4, where we achieved a low-cost, low-power, scalable solution based on digital VLSI design. This shifts the cost and performance pressure to the analog front-end.

An important cost factor in analog front-ends is a high number of heterogeneous components. For every component that designers remove from the bill of materials, a drop occurs in the overhead to maintain the supply chain, a drop occurs in manufacturing cost due to assembly, and most likely a reduction in size occurs due to increased integration [Moretti03b]. Heterogeneity and high component count originates from performance-oriented optimum selection of components. Today, a multiobjective optimization for performance, power consumption, size, and cost has come in place. This also requires changes in front-end architectures, for example a move to zero-IF techniques [Abidi97], and integration of entire front-ends in (Bi)CMOS SoCs or SiPs [Donnay00, Côme04, Shen02]. As a consequence of the multiobjective tradeoff, obtained state-of-the-art solutions were of lower performance or required significantly higher design effort

---

[104] Huxley, a contemporary of Charles Darwin, devoted most of his career defending Darwinism. Darwinism was defined in the Victorian period (and today) not only as Darwin's theory of natural selection, but also as a comprehensive network that includes a philosophical view of the ethical as well as practical significance of scientific investigation. A key idea is that the species survives which best adapts to its environment. Ideal hypotheses usually do not belong to the adaptive category. This idea motivates the introduction of the necessary flexibility at the lowest possible cost.

to achieve similar performance levels. Particularly for OFDM, there was a belief that low-cost low-power front-ends are impossible to achieve [Martone00].

In this work, we introduce the consistent application of digital signal processing capabilities to compensate for analog nonidealities either to enhance the performance, to relax the front-end requirements specification or to move analog functionality to the digital domain [Eberle02a]. Our approach requires a cospecification and codesign of analog front-end and digital compensation techniques. In [Asbeck01], this idea has been called *synergistic design*. The use of DSP techniques is advantageous for several reasons. First, DSP benefits from the availability of architectures with lower cost and lower power consumption. Second, aspects of dynamic control, reconfiguration, or adaptation[104] of function-ality can easily be integrated in digital designs in contrast with analog designs. Achieving the same in analog requires significantly more design effort and appears to be a much less systematic process [Abidi00]. Note, however, that we do not strive for a *software radio (SR)* or *software-defined radio (SDR)* as proposed in [Mitola95]. The SR and SDR approaches[105] are purely performance driven. Instead, we envisage a radio design with an optimally selected, minimum flexibility. This type of radio has been called a *software-reconfigurable radio (SRR)* in [Brakensiek02].

Since receiver performance is of first concern, in this chapter, we apply these techniques first to the receive side. Note that similar compensation techniques will be developed for the transmit side in Chap. 6. Essentially, joint treatment of analog and digital requires efficient mixed-signal codesign and modeling techniques. Development of corresponding, appropriate methodology is covered in Chap. 7. In this chapter, we focus on functional and architectural concepts.

This chapter is structured as follows. Section 5.1 introduces the receiver design with an overview of common architectures and the accompanying nonidealities. Next, we derive general requirements for WLAN receivers and review the state of the art; the identified challenges motivate our research. Section 5.2 describes solutions to the dynamic range problem in receivers through applying automatic gain control (AGC) and DC offset (DCO) compensation techniques. Section 5.3 extends the codesign to timing synchronization. Section 5.4 explains the impact of channel response and transmit/receive filter design on timing synchronization and proposes a joint optimization approach. Section 5.5 combines these and additional compensation techniques in an integrated receiver architecture. Section 5.6 concludes the chapter.

---

[105] SR minimizes the analog portion of a transceiver, assuming that RF signals can be digitized; this means that nearly all transceiver functionality appears in the digital domain and can be implemented in software. SDR assumes that a significant portion of the transceiver functionality is implemented in software; this software determines largely the operation of analog and digital hardware components. A new wireless standard shall be implementable by a software upgrade.

## 5.1   Receiver Design

Receiver design would be easy if the ideal receiver consisting of antenna and data converter (ADC) existed as postulated. The task of the receiver consists of amplification of the usually weak receive signal, downconversion from a carrier frequency to baseband, and suppression of unwanted signals such as noise and spurs. However, these components do not exhibit at the same time sufficient dynamic range, gain, bandwidth, suppression of undesired out-of-band signals, and sufficiently low-thermal noise and low-power consumption. Unfortunately, neither a linear concatenation of these basic receiver functions can be realized. Hence, actual receivers consist of a distributed chain of components that fulfill these functions. As a consequence, a range of architectures exists, trading off the different nonidealities and their cascaded effect.

### 5.1.1   Receiver Architectures and Their Nonidealities

For the understanding of the following sections, we briefly review the major sources of nonidealities in the receive path. An excellent more detailed overview on OFDM-specific nonidealities is provided, e.g., in [Muschallik00]. Next, we describe the major receiver architectures and comment on their advantages and disadvantages.

*Carrier Frequency Offset and Oscillator Phase Noise*

Carrier frequency offset (CFO) or residual frequency offset is introduced in the mixing process through absolute frequency differences between the local oscillators at the transmit and receive side of a communication link. It can be modeled as a multiplicative effect in equivalent baseband notation:

$$\underline{y}_k = \underline{x}_k e^{-jk\Delta\phi},$$
$$\Delta\phi = \frac{\Delta f_{sc}}{f_s},$$

(5.1)

where $\underline{x}_k$ is the received signal, $\underline{y}_k$ is the signal after mixing (and thus applying a CFO), $\Delta f_{sc}$ is the absolute CFO, $f_s$ is the sampling frequency at which the digital signals are processed, and $\Delta\phi$ is the ratio of CFO and sampling frequency.

Originating from the local oscillator in both transmit and receive front-ends, phase noise refers to amplitude and phase jitter in the local oscillator generation process that translates mainly into noise in the phase domain [Wojituk05, Muschallik95, Hajimiri98, Leeson66, Daffara96]. The phase noise spectrum is shaped by the VCO and PLL characteristics, especially the loop filter of the PLL. In general, it can be separated into a common phase error (CPE) which refers to the slow phase variations compared to the OFDM symbol duration, and the interchannel error (ICE) referring to the faster variations.

CFO and CPEs exhibit little variation in time such that they can be rather easily estimated and compensated for. This is not the case for the ICE induced by fast phase noise components.

## I/Q Mismatch

Mismatches in amplitude and phase between the in-phase and quadrature-phase paths, termed *I/Q mismatch*, can be introduced by unequal signal paths for I and Q components or an imperfect phase relation in the I/Q decomposition in the mixer. I/Q mismatch can be decomposed into an amplitude mismatch component $\varepsilon$ and a phase mismatch component $\Delta\varphi$:

$$\underline{y}_k = \mathrm{Re}\{\underline{x}_k\}(1-\varepsilon)\mathrm{e}^{-j\Delta\varphi} + j\,\mathrm{Im}\{\underline{x}_k\}(1+\varepsilon)\mathrm{e}^{j\Delta\varphi}, \qquad (5.2)$$

where $\underline{x}_k$ is the received signal and $\underline{y}_k$ is the signal affected by I/Q mismatch assuming the receive case. I/Q mismatch is also introduced in the transmitter.

## Nonlinearities, Imperfect Gain, and DC Offsets

All active elements in the receive chain have a limited dynamic range. Signals exceeding this range are subject to saturation or clipping. The actual saturation behavior is difficult to model; hence, saturation should be avoided as much as possible. On the other hand, all components incorporate thermal and, in case of active devices, other noise sources. A weak input signal may become unrecognizable.

Unwanted signals may occupy part of the dynamic range. These signals may be present at the antenna input in form of blockers or interferers and not perfectly filtered out. Another important source for reduction of the dynamic range in the analog baseband part is DC offset. Static DC offsets may be generated by bias mismatches in the baseband chain itself, but also generated from self-mixing of the RF signal with the LO due to imperfect LO–RF isolation on the same substrate (**Fig. 5.1**). Dynamic DC offsets can be introduced by mixing of close-in interferers with the LO signal.

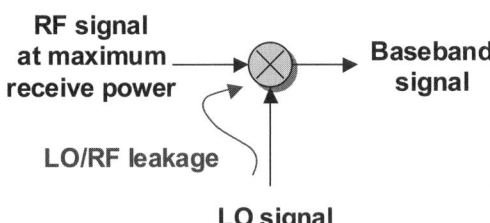

**Fig. 5.1.** A DC offset signal may be generated by self-mixing of the LO and RF signals in case of a direct-conversion receiver with LO at the carrier frequency. LO/RF leakage is subject to layout and parasitic effects and not very well controllable

*Receive Front-End Architectures*

There are several front-end architectures frequently used due to their specific properties. The traditional solution is the superheterodyne architecture where one or multiple frequency conversions to intermediate frequencies are performed (**Fig. 5.2**). This architecture allows a good tradeoff between dynamic range and filtering requirements along the receive chain. Its disadvantage is the narrow image reject filter needed which often requires large off-chip components (e.g., SAW filters) or the use of specific technologies (MEMS, FBAR). The superheterodyne front-end needs a careful frequency plan to avoid unwanted mixing products between RF, IF, and baseband frequencies. The last IF-to-baseband conversion can also be implemented in digital-IF at the cost of higher sampling frequencies and a digital downconversion. Due to the intermediate IF stages, DC offsets from self-mixing or LO leakage cannot occur.

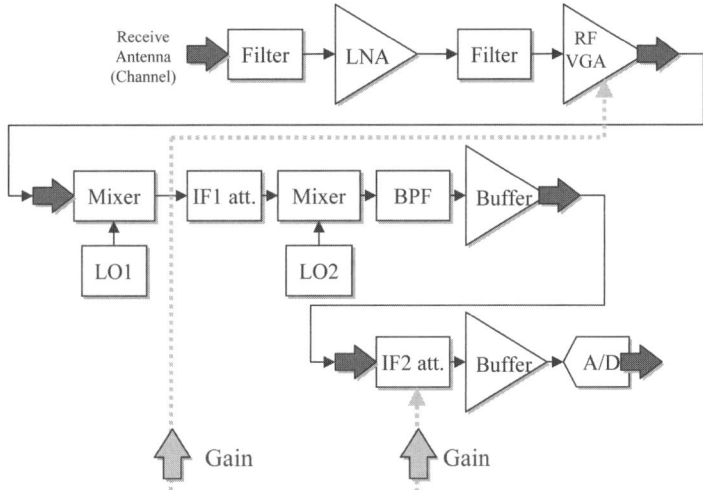

**Fig. 5.2.** Superheterodyne architecture with two IF stages

A more integrated and hence cost-efficient solution are zero-IF architectures (**Fig. 5.3**) [Razavi97a, Razavi97b, Razavi99] and low-IF architectures [Crols98]. Low-IF architectures were not considered due to the higher mirror signal suppression required which adds significant cost in the case of WLAN.[106] The RF signal is directly downconverted to the baseband frequency. A simpler frequency plan is needed, less analog building blocks need to be designed, but large DC offsets may be generated in the case of LO–RF feedthrough. With a specific harmonic mixer, a LO at half the carrier frequency, here at $f_s/2 = 2.5$ GHz, can be used in conjunction with a polyphase filter in the signal path. This increases the signal path loss but reduces the DC offset. All architectures exhibit I/Q

---

[106] In WLAN, interference in the mirror frequency bands is common and hence sufficient suppression is mandatory.

mismatches that can originate from mismatches in I and Q baseband signal paths. The digital-IF does not have this disadvantage.

Other architectures exist but have not been treated here. Most methods developed here are, however, extendible to other front-end architectures.

A careful implementation of the analog RF front-end is extremely critical to the behavior of multicarrier transmission since this scheme is used for spectrally efficient transmission schemes with many subcarriers and complex modulation schemes [Martone00]. While cellular systems using, e.g., GSM or EDGE reside at frequencies below 1 or 2 GHz with low-throughput rates in the order of tens to hundreds of kbit s$^{-1}$ and employing noncritical single-carrier phase-based modulation schemes (e.g., GMSK), wireless LAN works in the 2.4- or 5-GHz bands using a 64-carrier multicarrier scheme employing modulation schemes up to 64-QAM for throughputs up to 54 Mbit s$^{-1}$. Little can easily be reused from the large available knowledge in the cellular domain.

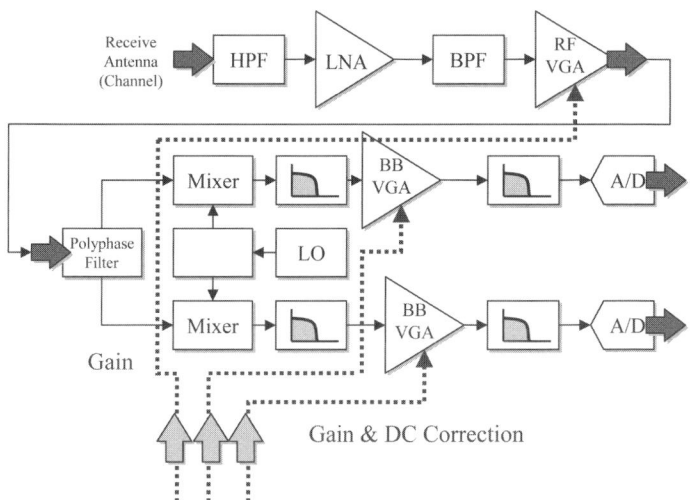

**Fig. 5.3.** Zero-IF architecture

*Achieving Higher Performance*

Analog nonidealities degrade the performance of the front-end and hence also the performance of the receive or transmit chain. For systems like Bluetooth, requirements for the analog design were reduced on purpose to allow low-cost solutions. For WLAN applications, this is not an option. The use of extremely linear components leads to high cost which is not feasible for consumer products. We advocate for more adaptation and compensation of the nonidealities in the front-end. This comes at the cost of more design effort for partitioning between front-end and baseband and more design effort for the actual compensation techniques.

Analog-only compensation is possible but usually very specific for a particular type of transceiver design. It is not very scalable if we want scalability toward multimode receivers, very large parameter ranges, fast settling times in packet-based schemes, and high accuracy at the same time. However, a mixed-signal solution with digital support improves reuse, reduces design effort, and allows for more intelligent control algorithms [Sakurai03]. Heterogeneous digital techniques have been proposed in [Bouras03] with mixed-signal loop-backs for compensation and digital estimation and in [Lanschützer03] for the transmit side. Advantages of digital techniques are the high precision and repeatability, the flexibility, and the low implementation cost [Asbeck01].

The focus in this chapter is entirely on performance improvement. The performance gains that our digital compensation techniques achieve can also be translated into relaxation of analog design constraints. However, it is difficult to quantify cost reductions or reductions in power consumption[107] due to this.

### 5.1.2   Our Contributions

In this chapter, we describe digital and mixed-signal techniques for the mitigation of analog imperfections. In particular, we derive two AGC schemes: a simple scheme applicable for constant-envelope preambles and a flexible scheme jointly addressing AGC and DC offset compensation using design-time information of the front-end architecture. We suggest a solution for robust timing synchronization in conjunction with AGC. Next, we come up with a technique to minimize the distortion effect of filters in the receive path and show its relation to timing synchronization performance. Finally, we have integrated all these techniques together with I/Q mismatch compensation [Tubbax04] in an overall architecture concept.

The underlying concept of digital compensation is illustrated in **Fig. 5.4**. Digital techniques can be applied as postprocessing or active feedback control of front-end resources [Eberle02e]. Our goal is to fit all required compensation techniques for analog impairments in this framework of digital estimation and digital or mixed-signal compensation.[108]

AGC and DCO compensation techniques have been evaluated for four different front-end architectures in simulation and partially on real front-end implementations[109]:

---

[107] A power reduction could be obtained if the corresponding circuitry was designed, such that its I/Q accuracy increases with power consumption or its noise figure decreases with higher power consumption. Since late 2005, inspired by performance–power scalable design (see Chap. 6), such circuits have been designed at IMEC, e.g., for a scalable baseband filter and a scalable transmit preamplifier chain.

[108] Quoting Y. Tsividis: "Analog has a big friend, [...] and his name is 'Digital'." [Ohr04]

[109] All of these designs were implemented in IMEC but not all were published.

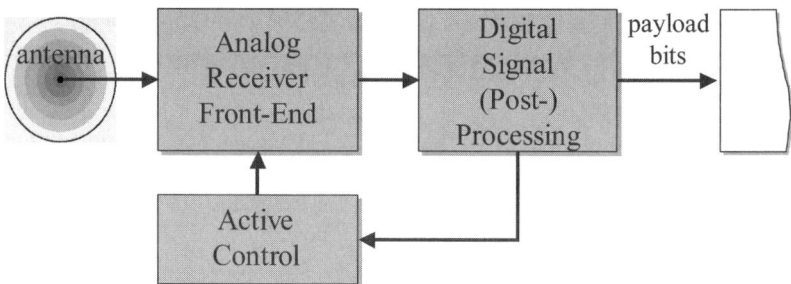

**Fig. 5.4.** Instead of embedding all compensation directly and in a rather inflexible and specific way into the analog/RF receive front-end, we suggest an embedding with digital functionality. A hybrid approach employing postprocessing and active feedback control is most appropriate

1. A 5-GHz superheterodyne receiver with two IF stages at 880 and 140 MHz implemented as a discrete board design
2. A similar 5-GHz superheterodyne receiver implemented in a system-in-a-package (SiP) approach and a digital-IF stage
3. A zero-IF receiver with a 5-GHz local oscillator
4. A zero-IF receiver with a subharmonic mixer operating at 2.5 GHz [Côme04, Shimozawa96]

All receivers had the same design target, meeting minimum SNR requirements for each modulation scheme up to 64-QAM at the minimum sensitivity level and coverage of a receive range from −85 to −30 dBm.

The overall digital compensation architecture has been validated on an integrated zero-IF receiver front-end [Eberle02a, Côme04].

## 5.2 Automatic Gain Control and DC Offset Compensation

AGC is an estimation and compensation process to arrange optimum dynamic range usage in a receiver front-end to guarantee optimum signal quality at the input of the decoding stage.[110]

Both estimation and compensation can be performed in the analog or digital domain. We find the following topologies:

---

[110] A not fully accurate but more traditional statement is that AGC maintains constant output level independently of the input level [Bertran91].

- Digital-only (both estimation and compensation in the digital domain); *toward software radio*
- Digitally controlled (estimation in digital domain, compensation partially analog); *toward software-controlled radio*
- Mixed-signal (including analog decision/control facilities)
- Analog-only (both estimation and compensation in the analog domain; not considered here)

Conventional designs often employ analog estimation and compensation; however, this results in customized designs with little reuse and low flexibility. Pure digital gain control is often not possible since dynamic range in the analog front-end must already be controlled. A hybrid solution with mixed analog/digital compensation and low-cost and flexible digital estimation appears interesting due to the low analog/RF complexity and the low-cost yet flexible digital design portion; hence, we opt for this.

AGC is a control technique and can rely on open- or closed-loop control. Note that open-loop or feedforward techniques in a receiver context would require signal strength estimation prior to compensation and hence estimation in the analog domain. Consequently, we go for a closed-loop technique. Note also that, due to the large number of nonidealities and hence unknown characteristics of the receive path, open-loop techniques for high-performance wireless receivers are not practical. This is also the reason why we cannot treat AGC as a gain-adjustment technique only. Dynamic range optimization is hampered if unwanted signal components dominate the usage of the dynamic range. This is, for example, the case for DC offsets created due to device mismatches, self-mixing, or mixing with interference signals. Hence, we need a joint optimization of AGC and DCO compensation.

We apply AGC to packet-based wireless LAN schemes. Packet lengths are in the order of several milliseconds. This is short compared to the slow time variation typically seen in circuit components.[111] Hence, we can restrict ourselves to developing AGC techniques for initial acquisition and can ignore the tracking of gain and DC offset variation over time.

In this section, we first review the state of the art in AGC techniques (Sect. 5.2.1). Next, we propose a simple AGC approach that works for particular acquisition sequences (Sect. 5.2.2). We extend this approach to a generic AGC/DCO compensation approach that is compliant with IEEE and HiperLAN/2 standard requirements (Sect. 5.2.3). Finally, we extend our studies to an exploration framework in which different front-end architectures can be compared with respect to their dynamic range and signal quality properties (Sect. 5.2.4).

---

[111] This applies both to front-ends based on discrete and integrated designs.

## 5.2.1  A Survey of Existing Techniques

A multitude of solutions for AGC and DCO compensation is found in literature and patent databases. Notably, most of the approaches are ad hoc and very specific to a particular front-end architecture or application. Moreover, most of these approaches did not fit the requirements for a flexible and fast technique applicable to a variety of front-ends.

Most of the general theory on AGC assumes continuous analog control loops [Ohlson74]. From our experience, analog-only receive signal strength information and gain control are not accurate enough which are also confirmed in [Fujisawa03] in the context of WLAN.

In this context, we need convergence of the AGC within the first 8 μs of the short training sequence (STS). The digital control techniques presented in [Wang89, Nicolay02, Perl87, Shan88] are too slow and cannot be easily scaled. The open-loop feedforward AGC in [Walker94] relies on a fairly well-characterized front-end and constant-envelope signals. The AGC for DAB in [Bolle98] relies on the availability of null symbols in a continuous stream and is also too slow. Jia and Mathew [Jia00] and Weber [Weber75] involve the digital decoding which introduces an unacceptably large delay. Shiue et al. [Shiue98] apply limiting techniques which are only applicable for constant-envelope signals. Morgan [Morgan75] and Lovrich et al. [Lovrich88] describe discrete-time techniques which are, however, only applicable to a single gain stage. Minnis and Moore [Minnis03] abandon the AGC at the expense of a high oversampling ratio and hence increased power consumption. Prodanov et al. [Prodanov01] define an AGC-less architecture for Bluetooth which, however, has relaxed specifications compared to WLAN.

Interesting details were found and taken up into our implementation. Victor and Brockman [Victor60] and Elwan et al. [Elwan98] introduce the log-domain process-sing to reduce digital dynamic range. Schwanenberger et al. [Schwanenberger03] and Fertner and Sölve [Fertner97] suggest AGC with variable or switchable step sizes. In [Hulbert96], we encounter a bandwidth-adjustable loop filter.

For the removal of the DC offset, solutions can be subdivided into two main categories. First, the insertion of a high-pass filter which eliminates signals close to DC; this solution is only applicable if there is no modulation information close to DC and may induce distortion due to its delay (Sect. 5.4). For OFDM, this technique limits effectively the distortion-free synchronization range too much. Second, subtraction of the estimated DC offset. Note that, to prevent saturation, DC offset must be removed before baseband amplification can be applied. Digital techniques are reported in [Yee01, Sampei92, Lindoff00].

In the end, none of the reviewed techniques for AGC and DCO compensation meets our constraints.

## 5.2.2 A Simple AGC Approach and Analysis of Preamble Properties

One of our first system designs involved the coupling of the Carnival ASIC design with a 5-GHz superheterodyne front-end[112] to arrive at a complete physical layer implementation. The architecture of the front-end is described in **Fig. 5.5**.

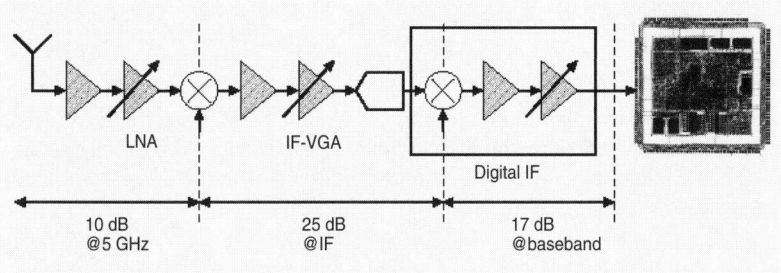

**Fig. 5.5.** Dual-IF receive chain with distributed digitally controllable gain stages. The challenge for AGC is to find the optimum gain configuration for each receive signal strength, such that the digital baseband receives a signal with minimum noise and maximum resolution

We first review the requirements for AGC from which the dimensioning of the variable gains in LNA, IF VGA, and digital-IF is derived. As a control loop, AGC is sensitive to delays in the loop. Hence, we analyze the delay properties in all elements in the front-end receive path and the variable-gain elements. We also show that the envelope properties of the receive preamble on which the signal strength estimation is based have an influence on AGC performance. We propose a proportional-integral (PI) controller for AGC and evaluate its performance for the entire specified receive range and several preamble types.

*Requirements*

The dynamic range specified in the standards requires a receiver sensitivity of at least −82 dBm for an SNR and of 4.8 dB for BPSK and the receiver must be capable of handling input power levels up to −30 dBm. Thus, the goal of the AGC is to achieve a gain control range of $\Delta G = 52$ dB. There is, however, a gain difference of 17 dB between the minimum sensitivities depending on the bit rate. We cope with these 17 dB entirely in the digital domain; thus all modes require the same maximum analog $\Delta G$ of 35 dB. The limited dynamic range of analog components leads the designer to distribute the variable gain along the front-end. For 64-QAM with 4σ-clipping, a maximum of 8-bit precision is required [Côme00]. Thus, there is a tradeoff between noise contribution appearing early in the front-end and quantization noise at the A/D converter [Clawin98]. An accuracy of ±3 dB around the desired power level ensures negligible performance loss in the digital baseband processor. A backoff of 6 dB between the maximum signal power and the mean signal power accommodates a crest factor based

---

[112] The discrete superheterodyne front-end design has not been published.

on 4σ-clipping. This accuracy complies with the minimum tolerance of ±5 dB required in ETSI for the measurement of the received signal strength level (RSSI).

The time constant of the AGC loop must be short enough to allow convergence of the gain within the first 8 μs to allow undisturbed channel estimation and fine carrier offset tuning.

### Detailed Architecture Analysis

Dynamic range considerations and noise figure require a variable LNA gain of 10 dB. A single bypass switch allowing 0 or 10 dB is sufficient. Propagation delay and settling time depend mainly on the decoupling capacitance at the control input and the drive capability. With a 20-mA drive at 3.3 V and a low 1-nF load, we obtain about 400-ns delay or eight samples at 20 Msamples s$^{-1}$. For the IF VGA, we have obtained a sum of settling and propagation time of maximum 100 ns for a gain variation up to 25 dB with a programmable attenuator (measured). The 41-tap FIR filter running at 80 Msamples s$^{-1}$ represents with 11 samples delay at 20 Msamples s$^{-1}$, the major delay component. An additional budget of two samples delay is added to all gain control paths, thus also for the digital multiplier, to account for DSP pipelining delays.

### Preamble Properties

AGC relies on the estimation of the average signal strength based on the first part of a receive packet. For this purpose, acquisition sequences are foreseen partially or entirely dedicated to the AGC process. It is obvious that the estimation of the average signal strength depends on shape and properties of such a preamble. Therefore, we analyzed and compared three different preambles – our proprietary preamble used in Festival and Carnival designs, the IEEE 802.11a-based PLCP standard preamble, and the ETSI HiperLAN/2 BCH standard preamble (**Fig. 5.6**).

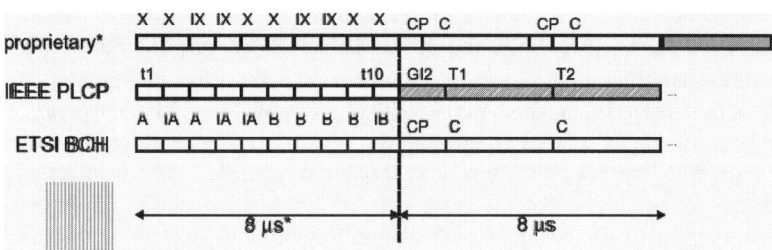

**Fig. 5.6.** Comparison of the structure of the IEEE 802.11 PLCP preamble, the ETSI BCH preamble, and our proprietary preamble. All preambles use repetition patterns with original (X, A, B) and inverted sequences (IX, IA, IB). Our proprietary sequence has a programmable sequence pattern allowing also longer or shorter sequences

It is clear that any variation of the envelope of these sequences causes deviations from the average, resulting in contributions for higher-order moments. This results in an estimation inaccuracy for the average value. We assume that the first 8 μs (160 samples) are used for AGC in all three cases. A suitable criterion is the minimum-to-maximum power ratio $\min_{k \in [0,159]}\{S(k)\} / \max_{k \in [0,159]}\{S(k)\}$ per sample. This results in 21 dB for IEEE and 31 dB for BCH. Our proprietary sequence achieves 0 dB. Hence, we expect a significantly better AGC performance due to the lower disturbance.

### A PI-Based Automatic Gain Control Loop

The AGC loop actually contains three loops controlling the LNA, VGA, and digital baseband gain (**Fig. 5.7**). All amplifiers are linear in dB. We estimate the power of the digitized and downconverted signal, optionally filter the power estimate, and compare it with the reference level. The error enters a PI controller that produces the final gain which is distributed over the three amplifiers based on a distribution rule (5.4).

An AGC loop is in general a nonlinear system producing an output signal with a gain acquisition settling time depending on the input signal level [Khoury98]. However, our AGC loop should be designed to settle for all operating points within $\Delta G = 52$ dB in a fixed amount of time. A linear control system cannot accomplish this; instead, a nonlinear dependency of the gain $g$ in dB is required. An exponential relationship exhibits a suitable I/O behavior:

$$a_{\text{out}} = k_1 \exp(k_2 g) a_{\text{in}}. \tag{5.3}$$

For a lumped amplifier model, the PI controller achieves a constant settling time [Khoury98] with proportional and integral gain as parameters. We used this topology slightly modified with an upper and lower bound of 0 and 52 dB for the integrator. An extension to a gain distribution rule was required to address a distribution of gains over several amplifier stages. On an experimental basis,[113] a simple rule was chosen with nonoverlapping gain ranges (all numbers in dB):

$$g_{\text{LNA}} = \begin{cases} 10 & g_{\text{sum}} \geq 10, \\ 0 & \text{else,} \end{cases}$$

$$g_{\text{VGA}} = \max\{0, \min\{25, g - g_{\text{LNA}}\}\},$$

$$g_{\text{dig}} = \max\{0, \{17, g - g_{\text{LNA}} - g_{\text{VGA}}\}\}. \tag{5.4}$$

---

[113] Specification ranges originated from an initial cascade analysis and design specification of the front-end design team.

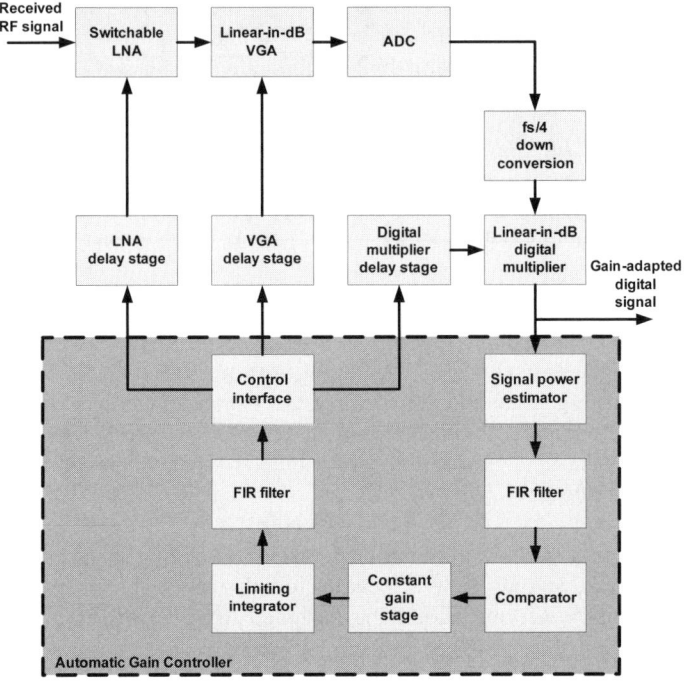

**Fig. 5.7.** The AGC employs three gain control loops with a control interface distributing the total gain over the different gain stages. We proposed a PI controller with integration limits

## Performance Evaluation

To evaluate the controller performance, we define a hard criterion: gain has to converge within ±3 dB after 160 samples. Since the AGC has to work over the whole specified input level range, we also request an uninterrupted range of input power levels for which the AGC converges. Therefore, the lower and upper boundary of the gain control range is extracted from simulations. The gain control range is a continuous input power range over which the signal gain is adjusted with the required accuracy.

Simulations were performed using the following parameter sweeps:

- Initial gain from 0 to 52 dB in steps of 6 dB
- Set of 121 indoor multipath channel models
- Input signal level range from −90 to +60 dB

The target to meet is specified as follows:

- Signal within ±3 dB of reference level

- Convergence in less than 8 μs
- Take the maximum continuous input level range for which this holds

From these simulations, we derived the upper and lower boundary of ∆G. We aim at achieving the full 52-dB gain control range under all circumstances. Simulations, however, show that we cannot achieve this in all cases: success rates are 97, 84, and 75% for our proprietary sequence, the IEEE preamble, and the ETSI preamble, respectively (**Fig. 5.8**). We conclude that a sequence with low crest factor introduces less disturbance allowing for a larger loop bandwidth and thus a faster settling time. While meeting the specification for our own proprietary sequence in 97% of all cases, the simple linear-in-dB PI-based controller does not reach similar performance for IEEE and ETSI preambles. For the latter two, we need a different AGC approach taking into account in more detail the specific distributed nonidealities of the front-end.

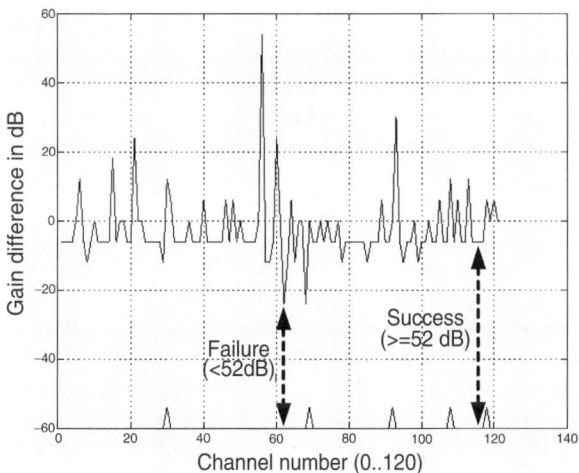

**Fig. 5.8.** Based on simulations across 121 multipath channel instantiations, the gain difference between minimum and maximum input signal strengths is computed. Correct gain control over the complete range of 52 dB is required for success

### 5.2.3 AGC/DCO Using Design-Time Information

The goal of our joint AGC/DCO approach [Eberle02c] is to reach the optimal front-end configuration for a desired signal quality within the required receive signal strength range using only very limited a priori knowledge of the preamble. We present here a systematic approach that exploits design-time information on the front-end during the run-time process.

First, we classify the effects of saturation and DC offset on the receive signal at the analog–digital interface. We propose a three-phase estimation/compensation process to iterate toward appropriate gain/DC offset correction settings. Next, we derive the signal strength and DC offset estimators. We continue with the actual gain mapping and run-time configuration and the overall AGC/DCO architecture. Before we conclude with a performance evaluation of our approach, we describe the design-time methodology and tool flow associated with our approach.

## Classification of Receive Scenarios

The effect of inadequate gain settings and remaining DC offsets can be classified into three cases (**Fig. 5.9**):

- Case I: large DC offset is present, hence perform DCO compensation first
- Case II: large gain and a small-to-moderate DC offset, hence adapt gain
- Case III: moderate gain and DC offset mismatches, compensate both

Case I denotes saturation due to a high DC offset; the DC offset is clipped to either the positive or the negative limit $A_{sat}$ and any signal content becomes invisible. Case II is the result of a negligible DC offset but too much gain; the signal is typically clipped at both the positive and negative limit since even small signal values are amplified too much. In both cases, nonlinear effects appear. Case III can be corrected since no nonlinear distortion is present; moderate gain and DC offset correction within the dynamic range are required. The case to the right denotes optimum usage of the dynamic range.

Note that the nonlinear cases I and II introduce a severe bias for linear signal strength estimators. We address this problem by appropriate estimator design.

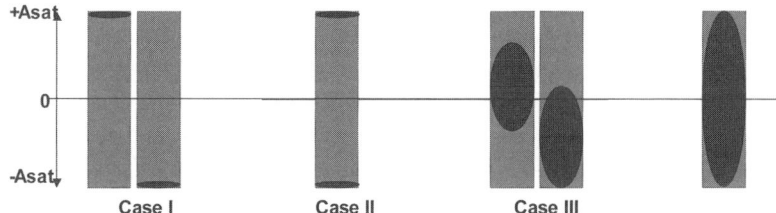

**Fig. 5.9.** Case I denotes saturation due to a high DC offset. Case II is the result of a negligible DC offset but too much gain. In both cases, nonlinear effects appear. Case III can be corrected since no nonlinear distortion is present; moderate gain and DC offset correction within the dynamic range are required. The case to the right denotes optimum usage of the dynamic range

## Three-Phase Run-Time Compensation Approach with Joint Signal Strength and DC Offset Estimation

Our AGC/DCO strategy uses up to 80 samples of the STS part of the preamble and partitions them into three pairs of estimation and compensation phases (ECPs) (**Fig. 5.10**). The length of the compensation phase should be minimized but depends in practice on the achievable settling times of the variable-gain elements. We measured settling times in the order of 25–100 ns for a 1-dB programmable digital 50-dB attenuator, 60–100 ns for an IF VGA, and up to 250 ns for an LNA. Delays introduced by filters in the signal path should also be taken into account. In total, we assume delays in the order of 300 ns (six samples) per adjustment. This prevents transients and spurs from the adjustment process influencing the following estimation.

We reserve the other half of the STS for coarse timing and CFO. The first estimation will thus start with maximum front-end gain to detect weakest possible signals (highest sensitivity level) and thus face, with a high probability, a signal and DC offset saturated signal at the ADC (case I).

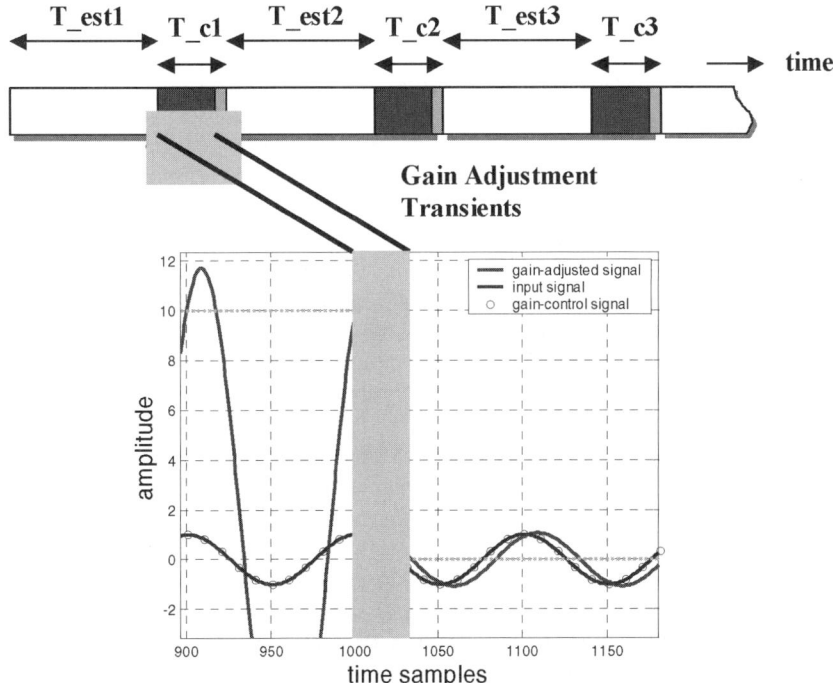

**Fig. 5.10.** Subdivision of the preamble into three estimation phases ($T\_est$) and three compensation phases ($T\_c$). Note that gain adjustments typically result in transients during which no new estimation can be performed reliably ([Eberle03] © IEEE/ACM 2003)

The first ECP mainly targets DC offset removal. Analysis of typical zero-IF architectures shows that the worst-case DC offset does not exceed the dynamic range of a 1-V ADC by more than ±1/2 MSB for maximum gain. Using a binary search tree to converge to the optimum DC removal level, we only need to investigate two input regions. Thus, a single decision is sufficient to leave the DCO saturation region.

If the result appears within specifications, we reach case II, where a second ECP tries to bring the input signal from nonlinear saturation to the linear range. The third and final ECP can then use linear operations to directly set the front-end configuration to a reliable optimum maximizing SNR. Such a partitioning is required to allow a treatment adapted to the specific signal properties in each phase. A first aspect is the shortened estimation time for phase I since saturation will lead inevitably to a biased gain estimate. Second, we can apply different transition strategies for the reconfiguration of the front-end resources, which we will describe next.

The possible decision trajectories of this process are shown in **Fig. 5.11**.

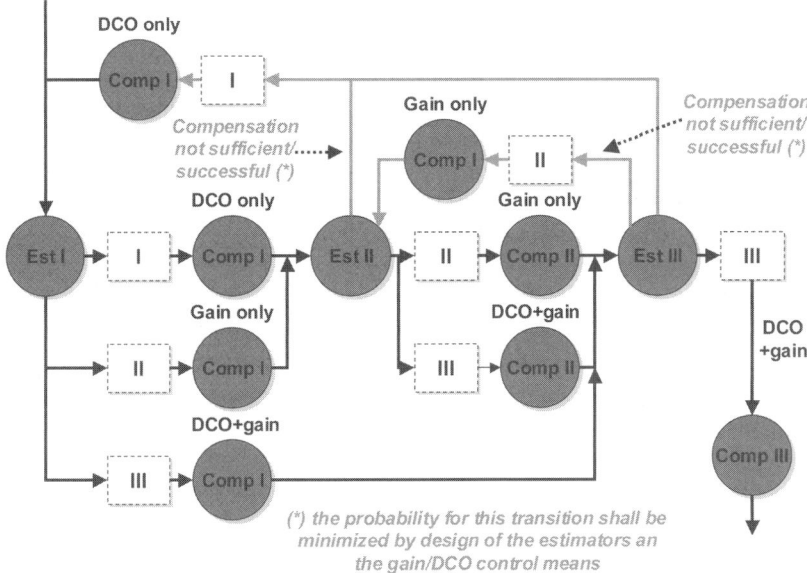

**Fig. 5.11.** Decision trajectories for the three-phase approach. Cases are indicated in *rectangular boxes*. Estimation and compensation phases appear in *round boxes*

*Derivation of a Joint Signal Strength and DC Offset Estimator*

Using only the signal assumptions from the preamble discussion, we derive the minimum mean square error (MMSE) estimators with an estimation length of $L$ samples for a DC offset-compensated signal strength $P_{m,i+q,\text{corr}}$

$$P_{m,i+q,\text{corr}} = -D_{m,i}D_{m,q} + \left( \frac{1}{L}\sum_{k=0}^{L-1} si_{m-k}^2 + sq_{m-k}^2 \right) \quad (5.5)$$

and DC offsets $D_{m,\{i,q\}}$

$$D_{m,\{i,q\}} = \frac{1}{L}\sum_{k=0}^{L-1} s\{i,q\}_{m-k},$$ (5.6)

where $m$ indicates the index of the first estimation sample. The estimators work on ADC input samples $s_{i,m-k}$ and $s_{q,m-k}$. The sampling rate can be flexible. In our practical examples, it is equivalent to the minimum sampling rate, i.e., 20 MHz. DC offsets are estimated separately on I and Q paths. By subtracting the DC offset powers from the signal strength estimate, we can eliminate the DC-related bias on the average.

*Performance Analysis of the Joint Estimator*

To investigate the performance of the estimators in the linear region, we consider an additional lumped additive white Gaussian noise source with power $N$ integrated over the signal bandwidth producing noise samples $n_{m-k}$. Preamble signal power is denoted as $S$ and DC offset power as $d_i^2$ and $d_q^2$, respectively. From

$$D_{m,\{i,q\}} = \frac{1}{L}\sum_{k=0}^{L-1} s\{i,q\}_{m-k} + d_{\{i,q\},m-k} + n_{m-k}$$ (5.7)

and the DC-free signal properties of signal and noise follows that the DC estimator is unbiased,

$$E\{D_{m,\{i,q\}}\} = d_{\{i,q\}}.$$ (5.8)

Both signal and noise power increase the variance

$$\mathrm{var}\{D_{m,\{i,q\}}\} = \frac{1}{L}(S+N)$$ (5.9)

of the estimator, since we cannot exploit any further knowledge on the preamble characteristics. The signal strength estimator

$$P_{m,i+q,\mathrm{corr}} = -D_{m,i}D_{m,q}$$
$$+\frac{1}{L}\sum_{k=0}^{L-1}(s_{i,m-k}+d_{i,m-k}+n_{i,m-k})(s_{q,m-k}+d_{i,m-k}+n_{i,m-k})$$ (5.10)

requires compensation based on the DC estimates to remain unbiased:

$$E\{P_{m,i+q,\text{corr}}\} = -d_i d_q + \frac{1}{L}(LS + Ld_i d_q) = S. \tag{5.11}$$

The variance of the estimate reaches it maximum for equal DC offsets $d = d_i = d_q$. The variance of the estimator

$$\text{var}\{P_{m,i+q,\text{corr}}\} = \frac{1}{L}\left(N(S+N) + 2d^2(S+N)\right) \tag{5.12}$$

indicates that both DC offset, noise, and the signal power reduce the performance of the estimator. As expected for the MMSE approach, the SNR for both estimators increases linearly with the estimator length $L$. Estimator lengths of $L = 12$ (initial estimates) to 24 (fine estimate) provide sufficient SNR to obtain estimation accuracies[114] in the order of ±1.5 to 3 dB.

### Implementation Complexity

Our estimator architecture requires only two multiply-accumulate (MAC) stages, two accumulators, and a single multiplier for the DC path. The estimators can be implemented as moving averages or block-based accumulators. For the targeted estimation accuracies, multiplication in the signal strength path can also be replaced by less accurate, but multiplier-free min/max power estimation.

### Gain Mapping to Distributed Front-End Resources

The estimators derived before provide us with the estimates of signal strength and DC offset. At this moment, our control approach still lacks the run-time compensation path to adjust gain and DC offset using the configurable front-end elements, derived at design time. This run-time architecture consists of the state controller (**Fig. 5.11**) and a gain mapping algorithm that traverses the front-end configuration space from an initial configuration at startup to the global optimum based on the state information and the signal strength and DC offset estimation.

We need to distinguish between the linear case III and the saturation cases I and II. This depends on the actual signal power at the A/D converter in the $k$th adaptation step, $P_{\text{adc},k}$, the maximum nonsaturating power level at the A/D converter, $P_{\text{adc,max}}$, and the desired *backoff*.[115] If $P_{\text{adc},k} < P_{\text{desired,adc}} = P_{\text{adc,max}} - \text{backoff}$, we are in the linear case, where

---

[114] This accuracy has been found sufficient in practice based on tests with a practical front-end. Higher accuracies may not necessarily lead to improved results since variations in actual transmit power and channel can easily come close to these values. Hence, a better estimation accuracy would not lead to lower margins for the usable ADC range.

[115] A backoff is required to avoid saturation for signals with a nonnegligible crest factor.

$$\left(\sum_{n=0}^{N-1} g_n\right)_{k+1} =$$

$$\max\left\{\begin{array}{l}\min_{\text{range}}\left(\sum_{n=0}^{N-1} g_n\right), \\[2ex] \min\left\{\max_{\text{range}}\left(\sum_{n=0}^{N-1} g_n\right), P_{\text{desired,adc}} - P_{\text{adc},k} + \left(\sum_{n=0}^{N-1} g_n\right)_k\right\}\end{array}\right\}. \qquad (5.13)$$

Otherwise, we are in the saturation case, where

$$\left(\sum_{n=0}^{N-1} g_n\right)_{k+1} = \frac{1}{w+1}\left(w\left(\sum_{n=0}^{N-1} g_n\right)_k + \min_{\text{range}}\left(\sum_{n=0}^{N-1} g_n\right)\right). \qquad (5.14)$$

The system offers $N$ independent gain elements[116] with adjustable gain settings $g_n$. $w$ acts as a forgetting factor. This algorithm trades off linear computation for binary-tree search, controlled by a forgetting factor $w$. A higher $w$ reduces the impact of the previous gain setting and increases the RF input power convergence range. We experienced values of 0.5 for $w$ as useful (**Fig. 5.16**). The next gain configuration $(g_{0,\ldots,N-1})_{k+1}$ is computed based on the estimated ADC signal strength $P_{\text{adc},k}$, the estimated DC offsets $D_{i,k}$ and $D_{q,k}$, and the previous gain configuration $(g_{0,\ldots,N-1})_k$:

$$(g_{0,\ldots,N-1})_{k+1} = f\left(S_k, D_{i,k}, D_{q,k}, (g_{0,\ldots,N-1})_k\right). \qquad (5.15)$$

The gain configuration unit computes the new gain step. $f(\ )$ is not an analytical function but a computational procedure where first actual signal strength is corrected for the DC offset. Next, with the current gain configuration of the front-end, a virtual RF received power is computed. Based on the desired ADC input level and this virtual RF received power estimate, the new configuration settings are selected from a look-up table. For each compensation step, the gain control unit reads the new configuration from the look-up table and stores the configuration of the previous step for comparison. The look-up table contains one configuration word per RF input power step. We used an accuracy of 1 dB. For a worst-case dynamic range of −85 to −30 dBm, we need 56 words. A configuration word is a tuple, providing the configuration bits for all gain elements. We used an 8-bit word as follows: RFVGA_ctrl<0> controls the RF VGA (two options), IF2att_ctrl<0> controls the second IF attenuator (two options), IF1att_ctrl<0>

---

[116] For example, gain elements can be implemented as attenuators or variable-gain amplifiers.

controls the first IF attenuator (two options), and IFVGA_ctrl<3:0> controls the IF VGA (16 options); the last bit is unused (**Fig. 5.12**).

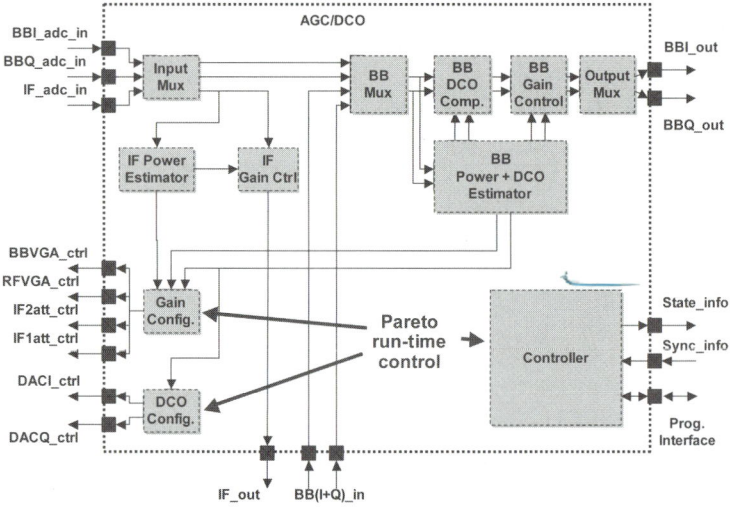

**Fig. 5.12.** Overall architecture for a baseband and digital-IF input. Design-time information is stored in the gain and DCO configuration. The controller simply selects the appropriate configurations from the database based on state information and estimated signal strength/DC offset ([Eberle03] © IEEE/ACM 2003)

Clearly, several combinations of gain settings exist that can achieve similar performance. Hence, the solution in the table may not be unique. To achieve a unique solution for each RF input power level, we apply a Pareto-based design-time optimization which eliminates this redundancy.

### Pareto-Based Design-Time Optimization

Approaching the optimum gain and DC offset setting, for particular RF input power requires a complete exploration of the configuration space offered by the front-end. We can easily add additional optimization goals such as meeting a minimum SNR requirement if we know a priori which type of modulation is used.[117]

Hence, the design-time phase starts with an extended cascade analysis. The analysis considers noise figure, second- and third-order intercept point, frequency selectivity, and includes saturation analysis from which we can conclude about the valid receive range for a given SNR (and thus modulation scheme). An exhaustive design-time search approach has been described in [Sinyanskiy98]; however, important nonidealities of the front-end were neglected.

---

[117] This information is available, e.g., in ETSI HiperLAN/2.

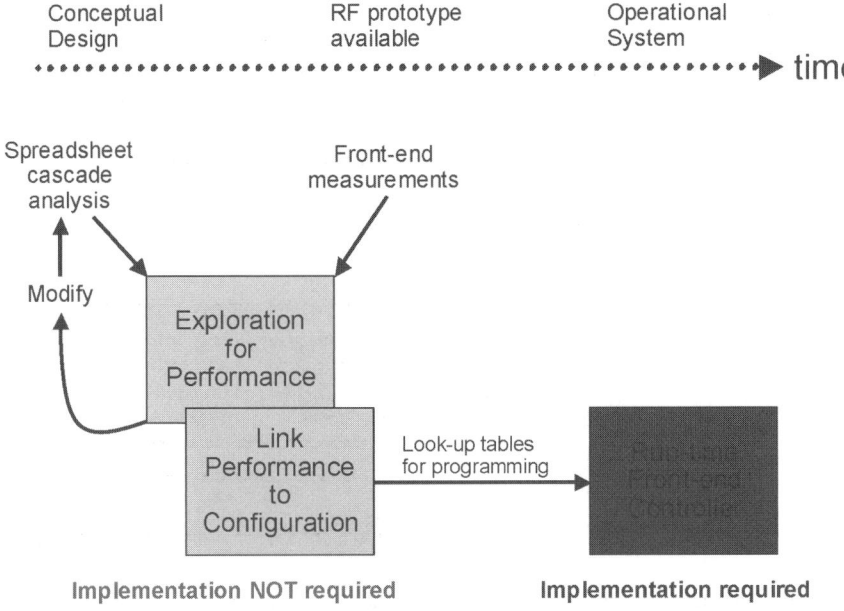

**Fig. 5.13.** Design-time run-time partitioning in our flow

The design-time part of our flow (**Fig. 5.13**) has been implemented in a MATLAB-based software tool. The program takes cascade analysis information as usually expressed in a spreadsheet way as well as optimization goals such as the minimum required SNR to achieve and the constraints such as the desired RF input power range. A second option is to provide a full measurement set to the tool. As a result in both cases, the optimum configuration per RF input power is provided as a look-up table. This design-time information can be downloaded into the embedded memory of our run-time architecture.

The optimization process iterates on an extended cascade analysis which has been implemented in MATLAB. Criteria are the cascaded noise figure and linearity requirements. Clipping and quantization in the ADC sampling process introduce noise and nonlinearity. The gain settings of each variable-gain element are the tuning parameters. They are specified through tuning ranges or discrete settings, e.g., for the switchable LNA. From the optimization, we obtain the switching points, as a function of the input signal power, between different gain configurations (**Fig. 5.14**). The process is successful if we can find a valid gain setting for the entire specified input signal strength range for a specific set of front-end constraints. Given an initial front-end configuration parameter range, it is not guaranteed that this process is successful; in this case, the results of the process will indicate that, for particular input powers, no satisfying solutions exist. As a consequence, the parameter range for one or more components needs to be adapted, new architecture topologies need to be considered, or performance improvement measurements in the digital domain must be added.

The effect of saturation and DC offset in a surface plot as in **Fig. 5.14** can be easily visually understood since they reduce the region where acceptable receive conditions are available (**Fig. 5.15**). Low VGA gains at low RF input powers result in unacceptably low signal-to-noise ratios (lower-left corner); high RF input powers with high VGA gains drive the receiver into saturation either because parasitic DC offsets or the wanted signal is amplified too much (upper-right corner); even for low RF input power, high VGA gains can lead to saturation when parasitic DC offsets are present (lower-right corner).

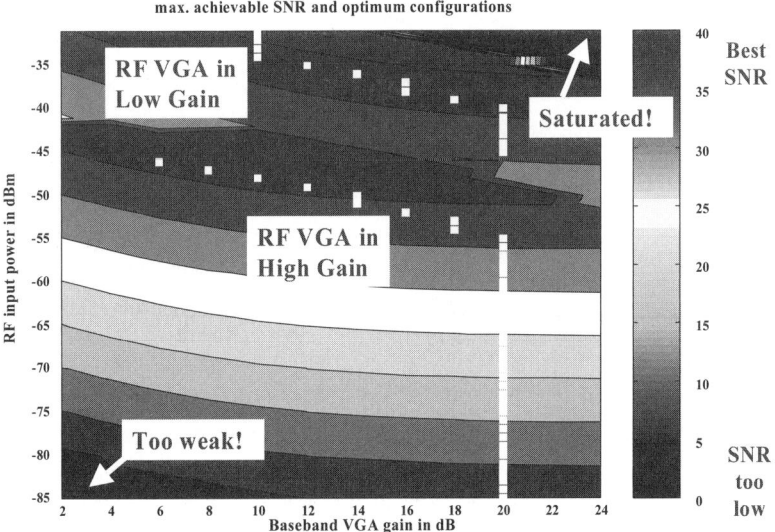

**Fig. 5.14.** During design time, the optimum gain configuration (*white boxes*) for each RF input power is determined based on an extended cascade analysis. The gain configuration defines RF VGA gain (*high/low*) and the baseband VGA gain

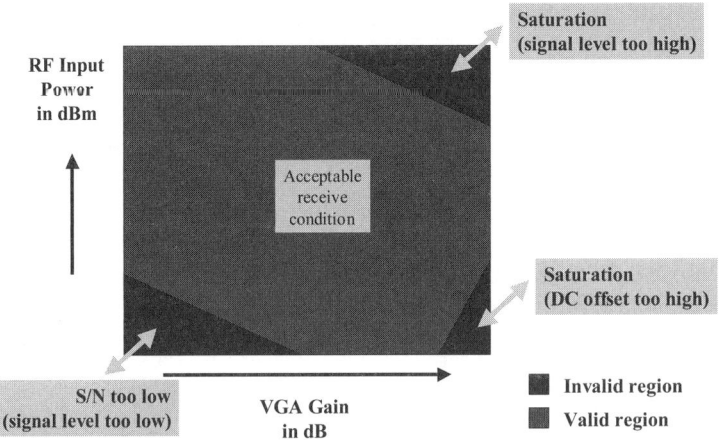

**Fig. 5.15.** Saturation and noise limit the acceptable receive region

## Performance Evaluation

Results are based on simulations in MATLAB using front-end specifications from actual IC designs, e.g., [Côme04]. **Fig. 5.16** shows that our three-phase discrete estimation/compensation approach guarantees a stable signal strength at the ADC input from −70 to −30 dBm.[118] Each ECP phase increases the flat region. After three iterations, the desired range is covered.

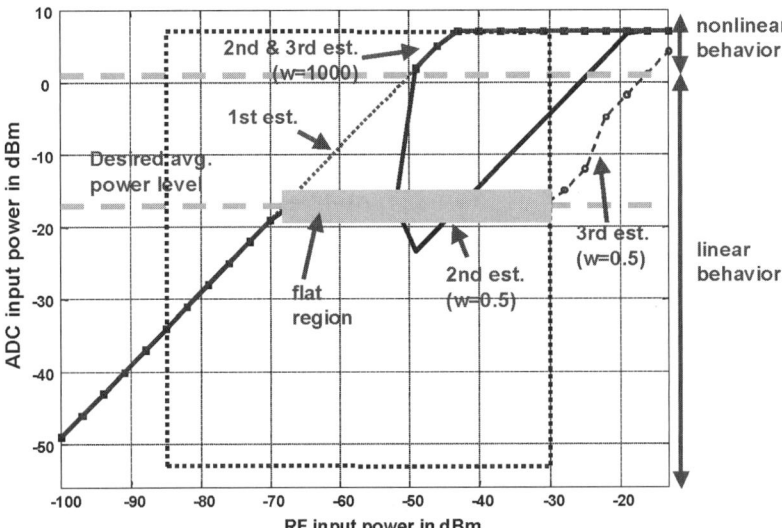

**Fig. 5.16.** ADC input power as a function of the RF input power after first, second, and third estimation and compensation step. Three steps and a forgetting factor $w = 0.5$ are required to achieve a flat gain region from −70 to −30 dBm. For lower RF input powers, the front-end does not offer enough gain

**Fig. 5.17** verifies that our strategy achieves an acceptable SNR level for all modulation schemes. Tolerances for the gain elements and the estimation were introduced during the determination of the optimum configurations to avoid accidental saturation to reflect analog component tolerances and estimator variance.

**Fig. 5.18** shows the timing behavior of the FPGA implementation[119] of our AGC/DCO approach together with timing synchronization (Sect. 5.3). First, we

---

[118] This range was derived based on the IEEE 802.11a standard specification. Actual product designs may extend this specification toward increased sensitivity (below −70 dBm) and increased robustness (above −30 dBm) to differentiate from other designs. Our techniques are intrinsically scalable with an extension of this range at the expense of an eventual additional decision step.

[119] This FPGA implementation has been effectively used together with actual front-end designs. A separate chip design has not been considered since the chosen architecture is dominated by its control complexity and can be translated in a straightforward way from the FPGA implementation to a chip implementation.

face a saturated signal (ADC data). The first gain reduction is too large, the second step results in a too weak signal, and the third step produces a signal with the right average signal strength.

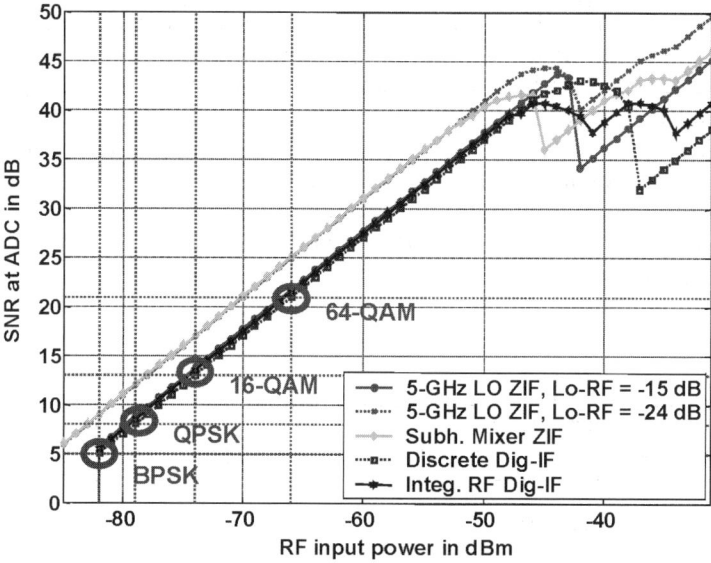

**Fig. 5.17.** Achievable SNR at the ADC input after AGC and DCO for five front-end architectures. The minimum SNR requirements for several modulation schemes are indicated and were met by the design

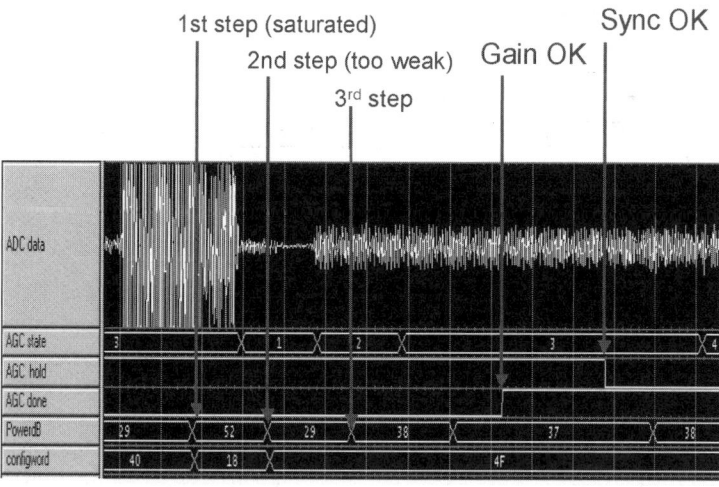

**Fig. 5.18.** Timing behavior of the combined AGC/DCO and timing synchronization as implemented on FPGA. The right ADC input signal level is reached after three iterations

Gain settles already earlier before the AGC_done signal becomes active and signals the timing synchronization to start. The synchronization lowers the AGC_hold signal after successful synchronization to prevent the AGC from further gain adaptations.

### 5.2.4    Exploration of Gain Selection and LO–RF Isolation

During the development of the AGC/DCO concept, it became visible that our methodology and software tools were also useful for an exploration of the performance properties of a front-end architecture [Eberle02b]. This is not surprising since an extended cascade analysis is at the basis of our approach. Consequently, we used this tool to compare different front-end architectures with respect to their performance and design cost. Of particular interest were the maximum gain, gain control range, and DC offset margins.

*Exploration Across Four Different Front-End Architectures*

Four front-end architectures – involving both discrete and integrated, super-heterodyne and zero-IF, 5-GHz LO and subharmonic mixing – have been analyzed regarding performance and optimum AGC/DCO configurations over the entire receiver dynamic range. So far, design decisions at such a high architectural level are often taken in a semiquantitative way in industry.

We applied our design-time optimization technique with joint AGC and DCO compensation to the four front-end architectures. The input to this process was standard cascade analysis information. Simulation, postprocessing, and optimization for the complete test suite took about 30 or 7 min per architecture. This allowed the designer to use this tool in an interactive way.

*Results*

Exploration runs for all four front-end architectures result in estimates for the maximum available gain, the required gain range, and the margin for DC offset at the ADC (Table 5.1).

We see that the superheterodyne/digital-IF architecture requires a significantly larger gain range due to its distributed nature. For the typically achievable 15-dB LO–RF isolation, the 5-GHz LO zero-IF solution has a severe DC offset problem unless a very good LO–RF isolation of 24 dB is achieved. The subharmonic mixing-based zero-IF receiver appears the most favorable solution with moderate gain range and maximum gain requirements and basically no DC offset problems. For the 5-GHz LO zero-IF, the SNR depends nonlinearly on the LO–RF isolation of the mixer (Table 5.2). Increasing the isolation improves the SNR only in a moderate way. Still, applying a DC offset correction resulting in equivalent 24-dB isolation would result in 3.2-dB gain in SNR at the specified minimum sensitivity levels.

**Table 5.1.** The superheterodyne/digital-IF architectures require a large gain range in the last receiver section

|  | LO–RF isolation (dB) | Maximum available gain (dB) | Gain range (dB) | Margin between maximum DC offset and ADC limit (dB) |
|---|---|---|---|---|
| Discrete superheterodyne/ digital-IF | N/A | 28 | 32 | N/A |
| SiP superheterodyne/ digital-IF | N/A | 21 | 31 | N/A |
| Subharmonic mixing zero-IF | N/A | 28 | 23 | N/A |
| 5-GHz LO | 15 | 37 | 24 | 2.1 |
| zero-IF | 24 | 37 | 24 | 11.1 |

The standard zero-IF requires high maximum gain, which leads to DC offsets close to the ADC limit for low LO–RF isolation in the mixer. The subharmonic mixing offers the best tradeoff

**Table 5.2.** If the DC offset based on self-mixing is not compensated, the worst-case SNR is unacceptably low and increases only slowly with better LO–RF isolation for a 5-GHz LO zero-IF receiver

| LO–RF isolation (dB) | Worst-case SNR (dB) | |
|---|---|---|
|  | QPSK at RFin = –79 dBm | 64-QAM at RFin = –66 dBm |
| 15 | 8.8 | 21.8 |
| 18 | 10.3 | 23.3 |
| 21 | 11.5 | 24.5 |
| 24 | 12.0 | 25.0 |
| 30 | 12.5 | 25.5 |

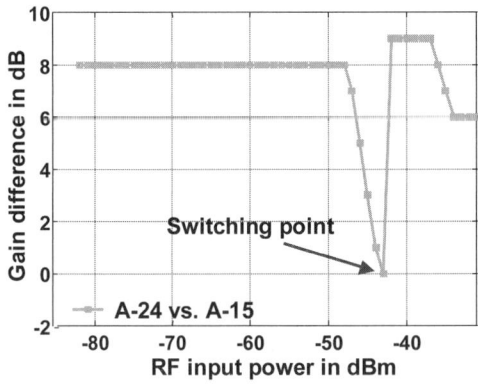

**Fig. 5.19.** When DC offset correction is applied on the 5-GHz LO zero-IF receiver (equivalent to the 24-dB LO–RF isolation case), 6–9 dB additional gain at baseband can be applied as compared to the 15-dB LO–RF isolation case

**Fig. 5.19** takes a closer look at the 5-GHz LO zero-IF receiver and compares two options for LO–RF isolation – 15 vs. 24 dB. Except for the switching point between high and low RF gain, we can add between 6 and 9 dB more gain in the baseband section without saturation.

Our design-time exploration strategy has been proven useful and efficient for comparing four largely different receiver front-end architectures. The benefit of an early architectural comparison is that realistic specifications can be developed for the actual front-end design in which overdesigning is avoided. For future multimode and multiband terminals, the number of architectural choices and parameter ranges increases even more. Hence, the design cost due to a too tough specification potentially increases. Our methodology can reduce such risks and will result in significant design time and hence potentially design cost savings.

## 5.3 Codesign of Automatic Gain Control and Timing Synchronization

When assessing methods for the digital acquisition process (Sect. 4.3.4), we mentioned already the importance of codesigning the digital baseband with the analog/RF front-end to prevent performance loss. This is particularly important for the earliest stages in the acquisition process, namely automatic gain control and timing synchronization. At this early moment of reception, barely any information on the incoming signal sequence is available and the two algorithms largely depend on each other. For this reason, we investigated the codesign of the AGC with the timing synchronization, particularly addressing IEEE 802.11a-compliant WLAN systems [Fort03b]. Several IEEE-compliant synchronization algorithms have been proposed in [Schmidl97, Sony99, Minn00a]. However, these proposals do not take into account the rapid gain fluctuations introduced by the AGC until a convergence has been reached. This prevents them from working under practical conditions. We have codesigned a novel autocorrelation algorithm together with an AGC interface that meets the IEEE 802.11a expectations.

### 5.3.1 Preamble Structure and Improved Synchronization Algorithm

The preamble is divided into two parts: the short training sequence (STS) and the long training sequence (LTS). The first 4 µs of the STS are used for AGC. The remaining B sequences can be used for packet detection and coarse timing/ frequency synchronization. The LTS has a length of 8 µs and is used for channel estimation and fine synchronization. We use the second part of the STS for timing synchronization based on a moving autocorrelation (AC) scheme (Sect. 4.3.4). We advocate for a first improvement of this algorithm for the IEEE 802.11a preamble since the traditional AC detector rather faces a ramp-up toward a plateau instead of a clear peak, making synchronization more difficult. We add a second autocorrelation and sum up the results: the first correlates over the conventional

16-sample repeating pattern while the second correlates over a 32-sample repeating pattern.

## 5.3.2    Codesign of AGC and Timing Synchronization

In general, synchronization algorithms are assessed, assuming that the entire preamble is received intact and with perfect gain control. In practice, AGC alters the front-end gains in discrete steps until the ADC operates in its linear range and a proper power estimate can be obtained. This process lasts typically 48–80 samples of the STS (2.4–4 μs). The large gain fluctuations and saturation effects that can occur during this convergence process often cause the conventional autocorrelation detectors based on maximization to fail.[120] Hence, we need to synchronize the synchronization and the AGC process (**Fig. 5.20**).

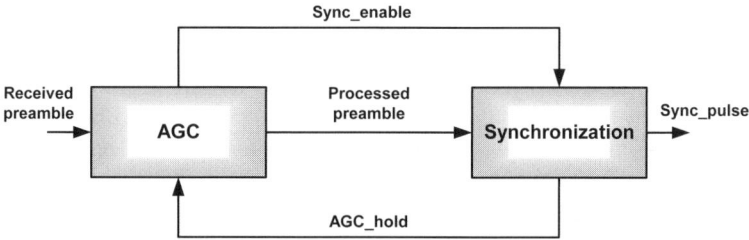

**Fig. 5.20.** The AGC triggers the synchronization when its gain adaptation settles (*Sync_enable*). The synchronization can prevent the AGC from gain adaptation during estimation and after successful detection of a frame (*AGC_hold*)

However, we still face the problem that it is unknown where in the preamble the AGC settles. Conventional algorithms assume that the length of the autocorrelation process is known and hence matched to the length of the remaining preamble. The conventional detector creates a plateau instead of a clear peak in the detector output due to this mismatch. Hence, we need a normalization with respect to the actual correlation length. This is introduced by weighting the proposed double-autocorrelation function with the power of the received signal $r(m)$, obtaining a novel detector function $S(n)$:

$$S(n) = \frac{\sum_{m=k}^{n} |r(m-16) + r(m-32)| r^*(m)}{\sum_{m=k}^{n} |r(m)|^2}. \tag{5.16}$$

The new algorithm only considers samples after AGC settling and works for all lengths $n$ of the remaining preamble. Hence, it does not rely on AGC convergence

---

[120] Conventional autocorrelation is sensitive to amplitude perturbations and only works satisfactory under small absolute amplitude variations.

at a particular time. This new algorithm creates a significantly sharper peak which is easy to detect with a max detector (**Fig. 5.21**).

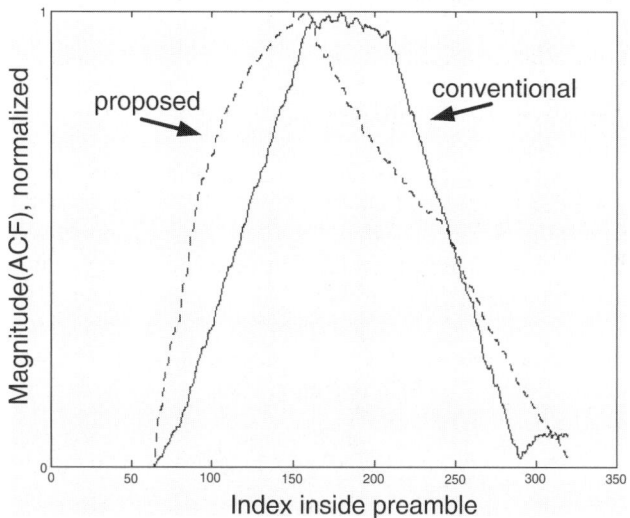

**Fig. 5.21.** The proposed double-autocorrelation algorithm performs better than the conventional single autocorrelation algorithm in that it produces a clearer and hence better detectable peak ([Fort03b] © IEEE 2003)

### 5.3.3    Complexity Assessment

The proposed solution reduces hardware requirements compared to the conventional autocorrelation approach. Due to the synchronization with the AGC, the timing synchronization is only active when needed and hence the moving sum along the entire preamble is replaced by a simple accumulator. The division operation in (5.4) is only needed for relative comparison and can be implemented as a single scaling multiplication. Hence, hardware complexity remains mainly in the delay line which in our case has a depth of 32 [Johansson99].

### 5.3.4    Performance Evaluation and Results

Evaluation was performed through simulation and emulation on an FPGA implementation[121] using AWGN and HiperLAN/2 channel models [Medbo98]. Results were compared against the conventional autocorrelation system with perfect gain control. Our proposal achieves a synchronization accuracy which is

---

[121] Complexity estimates and measurements obtained from the FPGA implementation were sufficient to prove the performance/complexity benefit of this approach. The proposed architecture allows a straightforward chip implementation.

sufficient over the expected operating range from 5 dB (for BPSK) up to more than 20 dB (for 64-QAM). Simulated timing variance corresponds to an accuracy within 2–6 samples 99% of the time. Even tighter bounds can be achieved by refining the timing synchronization in a second step using crosscorrelation based on the LTS [Fort03a].

## 5.4   Codesign of Filtering and Timing Synchronization

During the testing of the application demonstrators (Sect. 4.4) and the design of the AGC scheme (Sect. 5.2), we found that filters in the signal path had a large effect on the performance of the OFDM system. Analog anti-aliasing filters as well as the digital-IF interpolation/decimation filters were originally designed in the classical way focusing on meeting a particular frequency mask specification. This appears suitable for a multicarrier system that allows per-carrier equalization in the frequency domain. However, OFDM uses a block-based transmission with symbol durations of 4 µs. The FFT operation (Sect. 4.3.1) essentially assumes a periodic signal which is not the case.

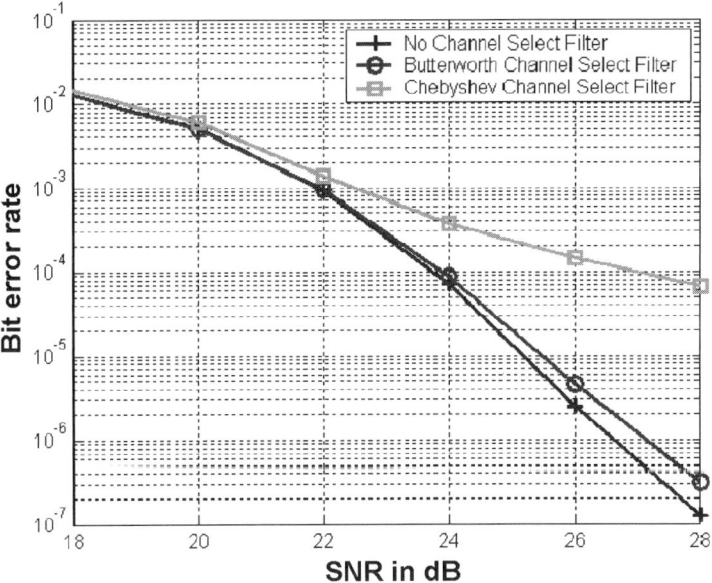

**Fig. 5.22.** While designed for the same frequency mask, the Butterworth characteristic has a dramatic impact on the bit-error rate as compared to the Chebyshev characteristic. In both cases, tenth-order analog IIR filters were used ([Debaillie01a] © IEEE 2001)

In [Debaillie01a], we found that different filter design techniques, while using the same frequency mask, resulted in different degradation of the overall performance (**Fig. 5.22**). While designed for the same frequency mask, the Butterworth characteristic has a dramatic impact on the bit-error rate as compared to the Chebyshev characteristic. In both cases, tenth-order analog IIR filters were used. So, we are interested in a systematic assessment of this effect. We investigated the sources for the performance degradation and devised a methodology and software tool implementation that can be used for appropriate filter design and performance evaluation [Debaillie01b, Debaillie02].

### 5.4.1   Reasons for Performance Degradation

The block-based nature of OFDM was identified as the source for the performance degradation. Normally, the cyclic prefix (CP) prevents the creation of intersymbol interference (ISI) and intercarrier interference (ICI) (Sect. 3.2). The length of the CP is minimized to save overhead and designed based on the length of the equivalent channel impulse response. Most theoretical studies only focus on the multipath channel as source for a scattered impulse response. However, we identified the transmit and receive filters as the main source of group delay in the case of wireless LAN systems which use a CP of only 16 samples (0.8 μs).

### 5.4.2   Mitigation

In systems with a large group delay, an additional time-domain impulse shortening filter is used at the receive side at the expense of considerable additional system complexity [Chow91b, Melsa96]. In [Armour00], a time-domain filter is introduced with frequency-domain estimation of the time-varying coefficients which also adds substantial complexity. We looked for an alternative approach that could be performed at design time and hence would not result in additional complexity at execution time.

### 5.4.3   Synchronization Range and Filter Impulse Response

An analysis of the impact of ISI/ICI due to impulse response patterns revealed the following interesting dependency between the optimum synchronization location as starting point for the FFT and the ISI (**Fig. 5.23**). A higher filter order resulting in a longer impulse response reduces the synchronization range which is identified by the open horizontal width for a reasonably high signal-to-interference (SIR) level (e.g., 50 dB). We specify the synchronization range as the range over which the SIR is negligible to the other noise sources (thermal noise and ADC quantization noise).

Hence, the amount of ISI/ICI perceived by the OFDM equalizer depends on the impulse response of the system but can be controlled by the timing synchronization location. In [MüllerW98, Pollet99, Malmgren96], the optimal synchronization location for minimum ISI/ICI is theoretically computed. However, for practical design, these approaches are not suitable. Müller-Weinfurtner et al. [MüllerW98] do not take the total impulse response into account. Pollet and Peeters [Pollet99] and Malmgren [Malmgren96] require time-consuming sample-based simulations.

### 5.4.4 Analysis and Optimization Methodology

We devised a practical analysis and fast simulation methodology to investigate the impact of ISI for a given filter response which also identifies the useful synchronization range (**Fig. 5.24**). Our technique is based on the comparison of OFDM symbols concatenated in a normal frame to the transmission of isolated OFDM symbols. ISI is created in the frame-based convolution while avoided in the isolated case. We evaluate the ISI for each possible synchronization location over the length of the cyclic prefix and derive from this the SIR. Based on a threshold, we can identify the synchronization range for acceptable performance.

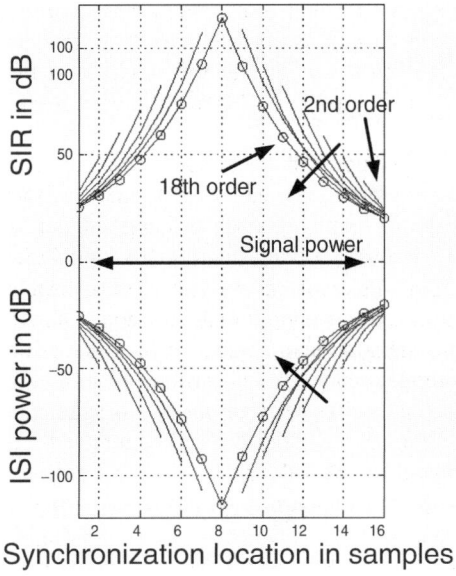

**Fig. 5.23.** Causal and noncausal ISI due to FIR filter group delay as a function of the synchronization location. The synchronization range reduces with the filter order. Signal power is normalized to 0 dB

This methodology was implemented in a software tool in MATLAB. The fact that we split the identification of the SIR and synchronization range from the actual time-consuming BER performance simulation results in a considerable time saving. While the full BER simulation based on the MATLAB tool proposed in [Côme00] requires about 5 h, a complete analysis of the synchronization range with our technique requires only about 6 s, resulting in a speedup factor of about 3,000. The short simulation time allows this approach to be interactively used to tune and optimize filter characteristics. Note that our proposed methodology still implements the actual signal processing per OFDM symbol as in the Monte Carlo BER simulation. Hence, differences in the computational accuracy can only origin from a difference in the randomly selected set of OFDM symbols for both cases and from computational inaccuracies. Since we only need an accuracy of about ±0.1 dB for taking design decisions, this is easily met.

Importantly for the system and filter designer, designing potential filter characteristics based on their spectral behavior is not sufficient but only a first step. Particularly in OFDM systems, we suggest the use of a time-domain analysis to estimate the impact on ISI and synchronization performance as a second step. This two-step approach prevents unnecessary lengthy BER simulations. These have to be performed now only after both frequency- and time-domain behavior checks were passed successfully.

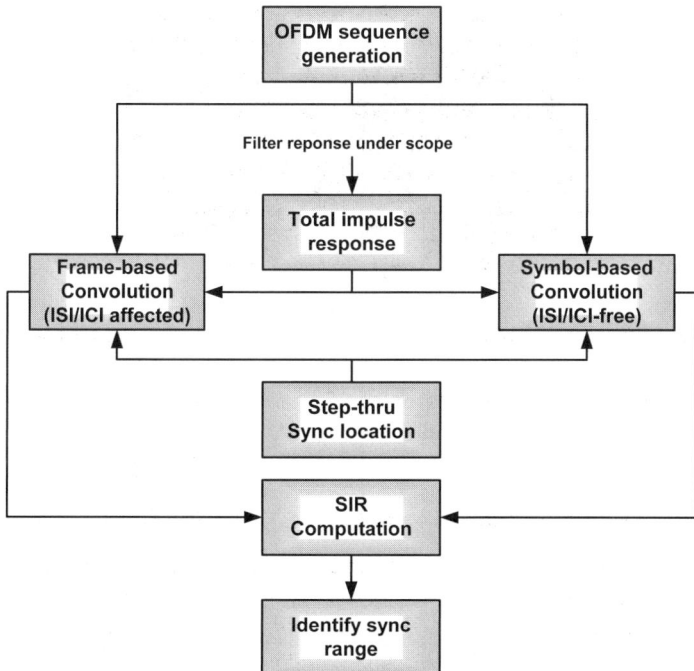

**Fig. 5.24.** Our ISI/ICI analysis flow can be used for filter design. Based on a given filter response, ISI/ICI effects are compared to the ideal case and the synchronization range is derived from the signal-to-interference (SIR) results

### 5.4.5   Results

Simulations were performed to investigate the impact of the tenth-order analog IIR filters for a Butterworth and a Chebyshev characteristic. We observe that the insertion of the filters results in a significant drop of the synchronization range and the achievable SIR (**Fig. 5.25**). Still, the Butterworth filter only drops to 53-dB

SIR which is negligible to the system noise of 30 dB while the 29 dB achieved by the Chebyshev filter creates an additional system loss. Corresponding but time-consuming BER simulations (**Fig. 5.22**) verify these results, showing a significant 3-dB performance penalty for the Chebyshev filter at a BER of $10^{-4}$.

The results obtained in this section do not only apply to the receiver design. They actually influence any part of the transceiver chain in which filters occur. The filter also plays an important role with respect to group delay in the transmitter. In general, the filter characteristic can be optimized depending on the sensitivity of the modulation scheme. In a multimode design with, for example, configurable analog baseband filtering, our methodology can be applied to derive the optimum filter configuration settings for each transmission mode.

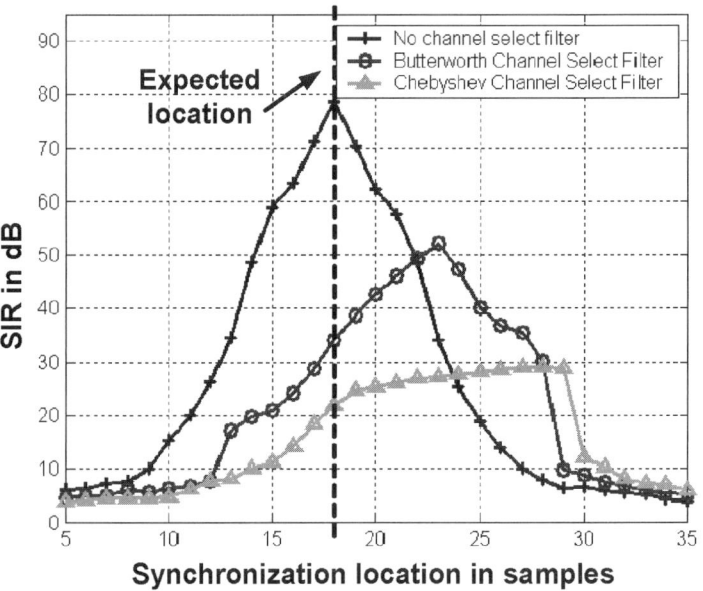

**Fig. 5.25.** Different filter shapes affect the SIR depending on the actual synchronization point ([Debaillie01a] © IEEE 2001)

## 5.5   An Integrated Digitally Compensated Receiver

The development of digital compensation techniques in this chapter was inspired by the fact that we need a low-cost solution to mitigate the imperfections in the analog/RF front-end which, otherwise, severely degrades the performance of the baseband receiver. We have investigated several techniques, in particular AGC and DCO compensation, the link to timing synchronization and filtering. In Chap. 4, we have already proposed solutions for CFO compensation, clock offset compensation, and common phase noise compensation. The important aspect of

I/Q mismatch was not addressed in this work but parallelly developed in [Tubbax04]. To arrive at a complete system solution, these works were combined resulting in an integrated digitally compensated receiver based on a 5-GHz SiP front-end for OFDM [Côme04].

First, we describe the approach for overall integration of all techniques. Then, we conclude with a review of the contributions in this chapter and an outlook.

### 5.5.1   RF Single-Package Receiver with Digital Compensation

*Direct-Conversion Receiver Architecture*

Our digital compensation techniques were applied to a single-package radio receiver with a 3-V core-IC in 0.35-μm SiGe BiCMOS [Côme04]. The receiver was designed for 5–6 GHz OFDM signal reception. The package has no RF input since the antenna was integrated in the form of a patch antenna in the cover of the ball-grid array (BGA) package. The outputs are analog baseband I/Q signals which will interface with a baseband CMOS chip. Many solutions today are heterogeneous systems that require many external components such as RF filters, GaAs power amplifiers, and antenna switches. Instead, this SiP receiver front-end combined a BiCMOS IC, high-quality integrated RF filters, and a balun on a thin-film technology on glass substrate (MCM-D), with laminates for the BGA package and the integrated patch antenna.

The front-end uses a special mixer architecture for harmonic direct down-conversion. Its main benefit is that the high-power local oscillator signal is never at the signal frequency, avoiding self-mixing in the mixers and high DC offsets at baseband. While achieving high integration, the impairments of this front-end are still too high as such to allow high-quality reception of 64-QAM constellations. I/Q mismatch and phase noise have significant impact on the BER, CFO originates from LO mismatches, and an optimum AGC scheme is required to steer the distributed gain stages. Hence, this chip is evaluated with a complete digital calibration, compensation, and control (C3) engine implemented on FPGA and tightly integrated with the OFDM core modem.

*Digital Compensation Approach*

The concept of the C3 engine was described in [Eberle02a, Côme04]. It relies only on digital estimation techniques and minimizes the amount of mixed-signal interactions with the front-end to reduce design dependencies and integration risks (**Fig. 5.26**). All estimation and compensation steps were designed for compliance to the IEEE 802.11a preamble. Only the residual DC offset due to baseband I/Q path mismatches is coarsely precalibrated offline.[122] Remaining slow DC offset variations are taken up by the AGC/DCO compensation.

---

[122] Offline DC calibration requires an estimation period that exceeds the available time from the STS but remains comparable to the overall acquisition sequence length (STS + LTS) for sufficient accuracy.

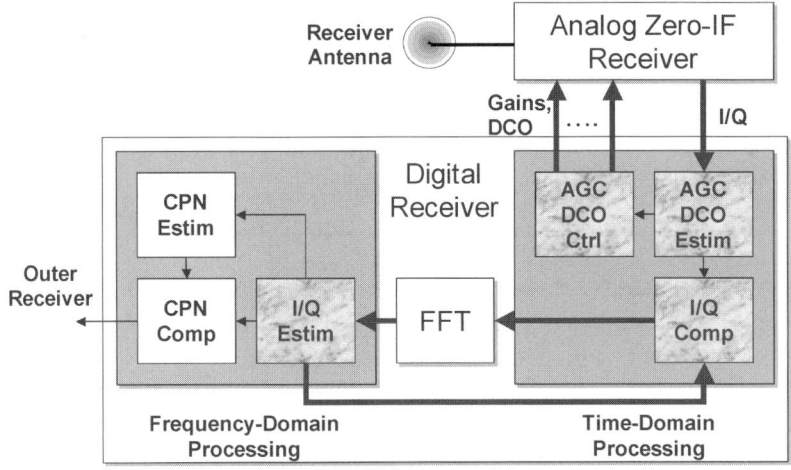

**Fig. 5.26.** Our concept of compensation minimizes the design complexity in the front-end by exploiting digital techniques both in the time and frequency domain ([Eberle02a] © IEEE 2002)

## *Resolution of Interdependencies During Integration*

Integration of all individual compensation techniques requires a careful check of dependencies. While this check had been carried out already for fine and coarse timing synchronization, filter effects, CFO, clock offset, AGC, and DCOs in our previous work (Sects. 4.3 and 5.2–5.4), this proof was still missing for the interaction with the I/Q mismatch compensation scheme proposed in [Tubbax04]. Since Simoens et al. [Simoens02] describe the impact of the two effects on each other, we verified the cascade of CFO and I/Q mismatch estimation and compensation stages [Eberle02a].

CFO estimation and compensation have to take place in the time domain to limit the leakage in the FFT processing. For I/Q mismatch estimation and compensation, all four combinations of time- and frequency domain processing are available. We suggest the following ordering of the processes:

1. CFO estimation in time domain based on the LTS
2. CFO compensation in time domain (both LTS and payload)
3. I/Q compensation in time domain (not for the preamble, only for the payload part since I/Q estimation will be performed in frequency domain)
4. I/Q estimation in frequency domain based on the LTS

Since the CFO estimation is performed first, it has to be robust against I/Q mismatch. We found[123] that CFO estimation performance is acceptable up to I/Q mismatches of 10%/10°. A combination of phase noise and I/Q mismatches higher than 10% and 10° at once results in not acceptable performance of the CFO estimation algorithm. The remaining length of the LTS available for CFO estimation is sufficient to reduce the remaining CFO to values below 500 Hz. I/Q mismatch estimation performance is practically unaffected by carrier frequency offsets up to 4 kHz. The proposed cascade architecture with CFO estimation and compensation in time domain, I/Q mismatch compensation in time domain (once I/Q coefficients have been estimated), and I/Q mismatch estimation in frequency domain exhibits adequate performance for all reception situations for IEEE 802.11a and HiperLAN/2.

### Acquisition and Digital Compensation Flow

The overall acquisition and compensation flow is illustrated along the time axis in **Fig. 5.27**. We can decompose the flow along the timeline in four phases: offline calibration, AGC/DCO and synchronization, I/Q mismatch and channel estimation, and data reception. Different parts of the preamble and payload are used in the four phases as follows:

*Phase 1.* Coarse DCO compensation is performed during idle time (offline) and does not rely on a valid receive signal.

*Phase 2.* AGC/DCO and synchronization: When a signal rises from the noise floor, a discrete three-step joint AGC/DCO algorithm adjusts the front-end gains and performs DC removal. This algorithm uses design-time information (extended, worst-case cascade analysis-like) and/or lab/fab precharacterization of the front-end, to converge in three steps to front-end settings where DC offsets do not result in saturation at any stage in the receive chain and where the signal-to-noise-and-distortion (SINAD) ratio at the FFT input of the OFDM demodulator is maximized. This and timing synchronization are performed in 8 μs during the first half of the preamble (STS).

*Phase 3.* I/Q mismatch and channel estimation: During the second half of the preamble (LTS), CFO is estimated and compensated in the time domain. After conversion to the frequency domain, I/Q mismatch is jointly estimated together with the channel response, the latter being immediately compensated for the I/Q mismatch effect. The applied algorithm is insensitive to CFO: measured on a hardware baseband channel model with 10%, 10° I/Q mismatch, 250-kHz CFO, and 30-dB SNR, the EVM increases from 9.4 dB (barely enough for QPSK) to 27.4 dB (more than sufficient for 64-QAM) with I/Q compensation turned on.

---

[123] Internal report by W. Eberle and J. Tubbax, these individual findings have not been published.

*Phase 4.* Data reception: After this phase, all estimates are acquired and the payload data can be treated: CFO in time domain and I/Q mismatch in frequency domain. Constellation rotations due to slow-varying oscillator phases (CPN, from DC to 20 kHz) and channel/CFO variations are continuously tracked and corrected for. As a measure of complexity, the I/Q estimation and compensation are 20% larger than the complete timing and frequency synchronization.

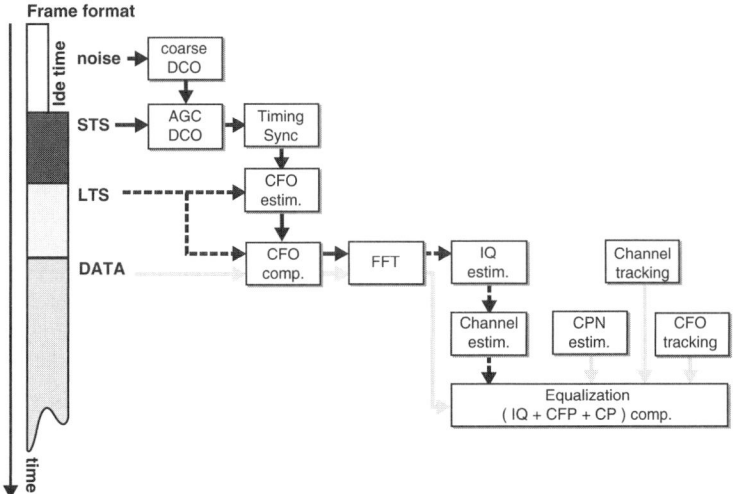

**Fig. 5.27.** Digital compensation techniques and classical receive functions like timing synchronization and channel estimation/equalization are interweaved and exploit the structure of the frame ([Côme04] © IEEE 2004)

## 5.6 Conclusions

In this chapter, we extended the design of an OFDM baseband receiver toward practical usability in the context of an analog/RF front-end with its large number of impairments. We briefly reviewed the range of analog imperfections that affect the working of the baseband receiver and proposed:

- A novel algorithmic and architectural solution for fast and flexible joint AGC and DCO compensation based on design-time information
- An architectural solution to elegantly link AGC and timing synchronization prohibiting mutual disturbance
- An investigation of the effect of transmit and receive filters on timing synchronization and a fast methodology to design filters in an interactive way

- A complete architectural solution embedding all digital compensation techniques resulting in a high-performance solution around a low-cost analog/RF receiver design

Our solutions were verified through simulations and implemented in FPGA-based extensions that were used in system tests together with different instantiations of front-ends ranging from a superheterodyne discrete to an integrated direct-conversion receiver front-end in the 5-GHz band.

The results presented in this chapter have clearly shown that digital compensation techniques are mandatory for a cost-effective design of analog/RF front-ends to meet performance expectations.

Moving toward multiantenna, multistandard, and energy-aware front-ends, requirements become even more extreme or cover a larger range of options to meet, such that the need for adaptive and hence digital solutions even increases. We stressed that design-time exploration is a fundamental means to explore design alternatives early in the design cycle and distribute functionality adequately across the analog/digital boundary.

Unfortunately, we noticed during the actual design that today's EDA tool support for such mixed-signal exploration at an early stage in the design is rather weak. Most design tools focus on analog/digital interaction at the transistor and gate level, resulting in rather long simulation times. There is clearly a need for higher abstraction to keep system exploration tractable in time.

As a consequence, we investigated modeling and simulation techniques, addressing this problem in Chap. 7. The concept of the AGC/DCO controller is extended to the transmit part in Chap. 6 and generalized in Chap. 7 under the term *resource controller*.

Importantly, our generic approach exemplified on AGC, where run-time operation is assisted and simplified by design-time information, can also be extended to cover other calibration aspects in analog or mixed-signal circuits. It basically exemplifies the digital and hence low-cost calibration and compensation of imperfect devices. Since imperfections can be reduced at run time, this allows lower design margins and hence lower design costs and higher yields specifically for analog/RF components and subsystems. Note also that the same approach may be applied to mitigate process variation effects, both in analog and digital circuits [Sakurai03].

# 6 Design Space Exploration for Transmitters

*Strength is born of constraint and dies in freedom.*
Leonardo da Vinci, 1452–1519.

Power dissipation has become *the* major constraint for portable devices. An example from the cell phone world shows that battery capacity has effectively only doubled over an 8-year period.[124] Current Li-Ion batteries can supply an average power consumption of 2 W to the electronics. This is not sufficient for portable multimedia devices such as the Nokia 7700, where a power consumption of 3 W is stated. From this budget, about 40% is reserved for the analog front-end including the power amplifier (PA).

Performance improvements through low-cost digital compensation techniques at the receiver side in Chap. 5 have shifted the key problems to the transmit side. **Fig. 6.1** indicates that power consumption is one of the rather weak factors of WLAN protocols when compared to protocols designed for low- or ultra-low-power consumption such as IEEE 802.15.4 (ZigBee) or IEEE 802.15.1 (Bluetooth). The trend toward increasing complexity is similar for the evolution of cellular systems from 2G/2.5G over 3G (e.g., UMTS) toward beyond 3G/4G but beyond our application driver.

The major remaining challenge appears the high-power consumption in the transmit front-end (**Fig. 6.2**). Also in a typical wireless LAN transceiver, it is the PA that accounts for about half the power consumption in transmit mode.

The power amplifier represents a major design challenge, particularly for OFDM. The OFDM signal is formed through a superposition of waveforms, which results in a large dynamic range for the instantaneous amplitude. Common practice is to design the PA for the worst-case fluctuation and the maximum output power. This

---

[124] Y. Neuvo, CTO Nokia mobile phones during ISSCC 2004 plenary presentation.

worst case determines power consumption even when less average output power or less dynamic range is required.

In fact, the usage of battery supplies exhibits two constraints to the user: maximum peak current and lifetime. While the maximum peak current limits the instantaneous availability of power and hence concurrent operation, battery lifetime depends on the average power consumption. Higher performance requirements increase power or energy demands. Hence, the problem is dependent on application scenarios and service choices.

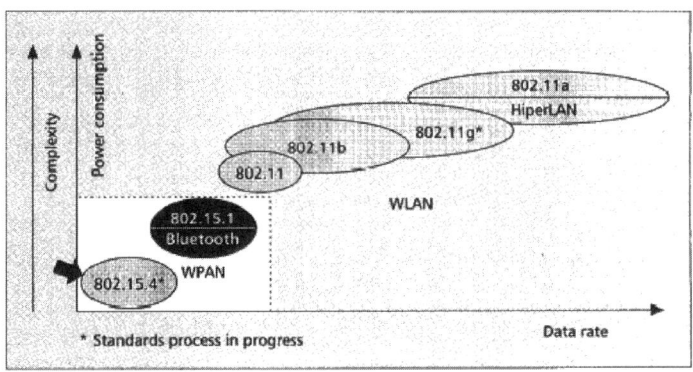

**Fig. 6.1.** Power consumption in general is a rather weak factor for standard WLAN implementations and requires a serious improvement, such that battery lifetime and hence convenience improves or at least remains with increasing application complexity and new services ([Gutierrez01] © IEEE 2001)

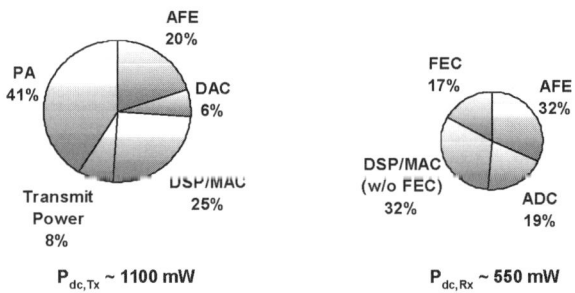

**Fig. 6.2.** Power consumers are differently distributed in transmitter (*left*) and receiver (*right*) of SISO WLAN [Bougard04]

Consequently, we have to first evaluate for each relevant application scenario, whether our system is energy or power limited. Since WLAN systems are supposed to operate in various scenarios and offer a wide range of performance through several communication modes, the system may appear limited in specific cases only. Solutions to overcome power and/or energy limitation depend on the probability of occurrence. Higher probability increases the relevance of finding an adequate countermeasure. Hence, besides a functional and architectural exploration, the solution space includes an application space exploration. The necessity of this extension was already described in Chap. 2.

Apparently, the design space for transmitters can be divided into three layers of abstraction, in which we have to search for solutions, given the optimization criteria power and energy dissipation:

1. *Design space exploration at the link level*
   A performance/power optimization at the link level takes into account properties and likelihood of application scenarios. The goal is to specifically replace the traditional worst-case-based PA specification through a probability-based specification. Behavioral modeling enables quick exploration, which allows, for example, an early coverage or cost analysis coupled to the PA technology selection.
2. *Signal-specific digital-only compensation techniques*
   Digital compensation techniques take influence on the shape of the transmit signal through functional modifications. For OFDM, the dynamic range of the signal can be adapted to the nonideal properties of the analog front-end.
3. *Digital control of the power amplifier*
   Digital techniques can also adapt performance/power properties of the PA, for example through a supply voltage adaptation.

Particularly, interesting is the fact that these three techniques are intrinsically orthogonal to each other and can be combined. However, we will show that, in particular application scenarios, a combination of less than all three techniques is sufficient.

This chapter is organized as follows. Section 6.1 proposes a systematic flow for performance/power optimization at the link level. First, we describe the modeling and computational flow. Next, we illustrate the approach for various optimization goals in the physical layer and discuss an extension to the crosslayer case. Section 6.2 deals with the power/performance optimization around the RF power amplifier directly. We propose a run-time adaptation strategy and experimentally prove its viability. Finally, Sect. 6.3 concludes the chapter with a discussion of results.

## 6.1 Power/Performance Optimization at the Link Level

In this section, we describe two techniques for power–performance optimization. The first explores the power–performance design space using a probabilistic approach and aims at fast and early performance–power investigation of architecture alternatives in various environment and usage contexts. The second extends this view to a true crosslayer perspective crossing the boundary to the data link layer while at the same time proposing a run-time control technique.

### 6.1.1    Use Case-Driven Power/Performance Optimization

The last decennia have faced the growth of wireless voice, data, and multimedia communications as independent tracks with their own roadmaps and innovations; excellent examples are 2G/2.5G/3G voice and wireless LAN-based data/multimedia communications. A major challenge for the future is the integration of different services, networks, and transmission schemes into a transparent scheme for the end customer. Besides performance, limited battery energy in terminals will be a main design driver, requiring optimum control over energy usage over time. Multistandard multimode capability will require a flexible terminal design, known as *software-defined radio (SDR)* or reconfigurable radio [Pereira01]. However, flexibility usually comes at the price of power consumption. Consequently, an optimum tradeoff between the required flexibility for a range of applications and services, and the power consumption in each mode is required.

Exploration effort using traditional design methods such as link budget or cascade analysis[125] for the physical layer does not scale reasonably for future systems for which a necessity exists to support multiple protocols, modes, and standards. The rich variety of modes in today's WLANs already represents an exploration challenge [Doufexi02]. On the other hand, even a limited exploration of the design space employing ray-tracing simulation or Monte Carlo simulation appears clearly too slow. Simulation time becomes even worse if power consumption is included as an additional optimization parameter.

We propose a probabilistic, strongly model-based approach that leads to quick yet quantitative design decisions [Gielen02, Donnay94] based on performance/power metrics across the digital/analog and protocol/physical layer domains. Importantly, we start from actual user scenarios and derive high-level performance and power figures; this supersedes the classical $BER = f(SNR)$ approach at the digital communications level. Our approach relies on adequate correlated performance and power modeling. However, it does not require the same accuracy as expected from actual sample-based simulations; hence, effort in modeling remains reasonable, yielding mainly equation-based or statistical models with good relative and reasonable absolute accuracy. Also, exploration time, particularly simulation time, is drastically reduced. With more refined models, the next step toward a framework for actual multiobjective optimization at design time is possible, as recently proposed in the context of *energy-aware* systems [Bougard03a].

Our approach has been successfully evaluated in a 5-GHz wireless LAN. Still of significantly lower complexity than multistandard radios, the design space for a WLAN terminal is already huge and hence adequate as a test case [Bougard03a]. Results relate quality-of-service (QoS) parameters such as goodput in Mbit s$^{-1}$

---

[125] Classical link budget and cascade analysis focus on performance only (e.g., neglecting power or energy cost) and do not take into account the statistical relevance of a particular configuration. These are two important extensions we introduced to establish a performance-cost tradeoff and to prune irrelevant configurations.

(effective error-free throughput across a radio link toward the network layer) and coverage to design parameters (PA technology, maximum output power), run-time configuration parameters (modulation schemes, transmit power control levels), or cost (system power consumption).

### Design Space Exploration

The process of design space exploration requires a definition of the design space and its analysis criteria. With this information, an efficient methodology for its exploration will be developed. But first, we situate the challenge of design space exploration for the system designer with the WLAN transmitter example.

*Challenging example.* System power efficiency, transmit power, and technology choice for the power amplifier, for example, are strongly dependent (**Fig. 6.3**). Preferable low-cost system-on-chip (SoC) solutions in CMOS are only feasible up to 12-dBm average transmit power in a WLAN context [Zargari02]. We will show that this kind of cost boundaries has a large impact on flexibility and applicability of the complete system. Complexity and dependencies make it impossible for the system designer to derive easily quantitative decisions. A methodology is needed to efficiently traverse this rich design space.

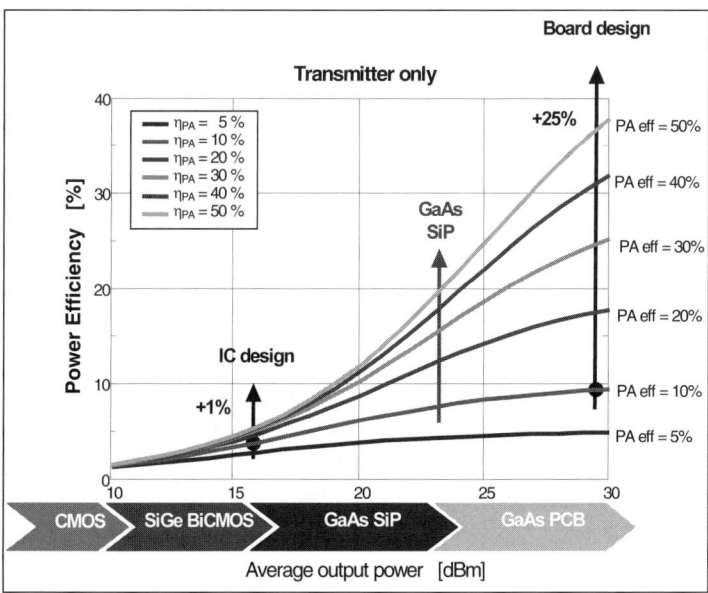

**Fig. 6.3.** The design space spanned by system efficiency vs. PA technology, maximum power, and intrinsic efficiency[126]: where is the performance/cost optimum for your particular use case? ([VanDriessche03] © IEEE 2003)

---

[126] Based on IMEC internal study (B. Côme, W. Eberle, and J. Van Driessche) derived from commercially available state-of-the-art PA specifications.

## Design Space

Performance and cost metrics[127] span the design space for (wireless) communication systems. Actual implementation of protocol and physical layers introduces a link between performance/function and implementation cost metrics describing the subspace of feasible, i.e., implementable solutions. Use cases and physical implementation bounds constrain this search space.

Today and even more in the future, the multitude of usage scenarios on the one hand and power and cost constraints on the other hand require a systematic exploration driven by metrics. First, results on performance/cost reduction in RF/digital codesign are promising [Asbeck01], but definitively deserve the wider system-level scope as explained here.

Exploration of the design space for performance/cost optimization requires a clear view on the bounding quality and implementation cost constraints. In a wireless communication system, a representative subdivision can be made into service requirements, actual network environment and terminal placement constraints, and implementation specifications for transmitter and receiver (**Fig. 6.4**). The first two define quality demands from a user perspective such as effective data throughput (goodput) in Mbit s$^{-1}$ and spatial network coverage; the latter two link performance to implementation cost, particularly to power consumption.

We consider a single transmit–receive WLAN link. Physical and MAC layer are modeled. Scenarios (use cases) span the combinational space of service and environment constraints, sometimes combined with implementation constraints (e.g., maximum transmit power). Goodput, power efficiency $\eta$ (transmit vs. dissipated system power), and total dissipated power consumption are determined for each scenario.

Clearly, a brute-force method is not applicable due to the combinatorial complexity and the mix of discrete and continuous variables. In contrast to recent system- or crosslayer optimization methods [Zhao03], our primary goal is not so much to find the optimum configuration for each scenario but to weigh its probability of occurrence, i.e., its relevance. A typical example is the probability for a particular distance between transmitter and receiver given a certain room topology and access point (AP) placement; we see already a significant impact on transmit power if we only move from optimum to suboptimum AP placement

---

[127] Besides applying the right metrics, a challenge is also obtaining performance/cost models at a fairly high abstraction level. However, modeling is here also often misunderstood: we do not need a model that accurately describes the operation of a single transistor or gate. We need a model that captures a particular transaction or operation, often on the average or for a best and a worst case. The goal is not absolute accuracy but an improvement compared to the fact where we have no upfront model at all to take design decisions.

(**Fig. 6.5**). Hence, a methodology for efficient exploration is required, involving statistical information.

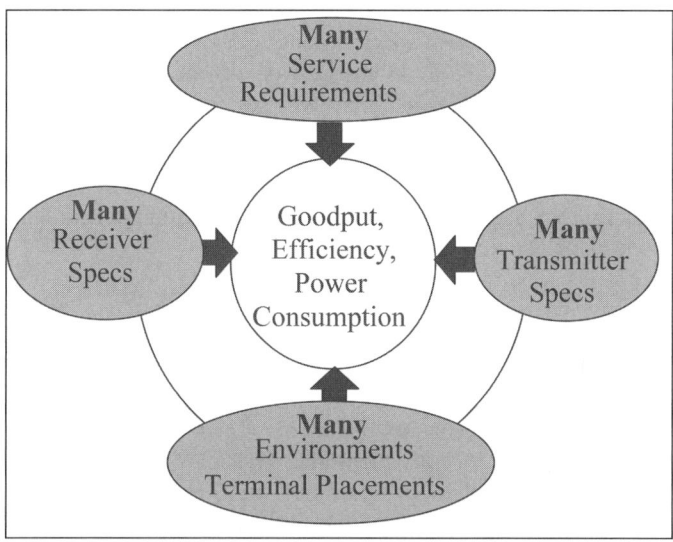

**Fig. 6.4.** Relevant performance and cost parameters (goodput, power consumption) depend in a complex way both on user-centric parameters and implementation properties (specifications) ([VanDriessche03] © IEEE 2003)

*Exploration Methodology*

Our methodology aims at fast, initial exploration [Donnay94, Gielen02] of the design space in three steps. First, we have to acquire power/performance models for each subsystem. Second, we combine them in an efficient way to speedup evaluation. Finally, the resulting system model is evaluated across the design space parameters.

In the model combination process, dependencies become visible. The ordering in the model computation is not generic; a sort of dependency graph for each application needs to be derived. This, however, is easy for the required accuracy during initial exploration, where the main goal is to reduce the design space by ruling out extreme design choices.

When comparing to classical simulation methods, our methodology proposes an application-driven weighting between Monte Carlo (MC) methods [Jeruchim92] and importance sampling (IS) techniques [Smith97]. The subdivision into three phases, however, decouples the complexity. Hence, our approach is significantly faster than MC runs at system level. Still, it would be slower than full IS; but the required detailed knowledge on probabilities is not yet available during initial exploration. More accurate models based on actual designs, which will make the integration of IS models more economical, will be usually available only in a later stage of the design process.

## Energy–Performance Modeling

The energy–performance models must cover all relevant aspects of the design space in an adequate accuracy yet remaining as computationally simple as possible to yield fast evaluation. As our design space spans from RF to the data link control (DLC) layer, each with its particular model types, we end up with a mix of heterogeneous models to be combined to yield a system model. Modeling is described as follows in a top-down manner, starting from the data link layer and reaching down to the RF components.

*Modeling approaches.* A primary assumption for all exploration activities is the availability of accurate-enough yet computationally efficient power/performance models.[128] Our approach mainly applies equation-based models that are first-order theoretical models (e.g., decoding gain) or, maybe later, calibrated models based on actual measurements (channel models). These models can easily be incorporated, reducing significantly the initial threshold in an exploration process. We distinguish two types of models: models that link performance and design parameters, and probability models (in the form of probability density functions, PDFs) that describe the likelihood $\Pr\{p_0\}$ of a particular value $p_0$.

The average transmitter power efficiency $E\{\eta\}$ is computed by weighting the instantaneous transmit power efficiency $\eta$(power $p_0$) for a discretized input variable in the limited definition range bounded by $\min\{p_0\}$ and $\max\{p_0\}$:

$$E\{\eta_{aPHY}\} =$$

$$\int_{\min\{p_0\}}^{\max\{p_0\}} \left( \int_0^{\max\{\Delta p(p_0)\}} \left( \eta_{\mathrm{aPHY}}(p_0 + \Delta p)\Pr\{\Delta p\}\Pr\{p_0\} \right) \mathrm{d}\Delta p\, \mathrm{d}p_0 \right). \tag{6.1}$$

Equation (6.1) represents a double discrete integration with the outer integration over the discrete transmit power steps $p_0$ and an inner integration over the differences between these steps. Since the distribution of transmit power steps is not necessarily equal, we have a nonconstant step width $\Delta p$. Knowledge of the actual PDF is of primordial importance.

*Link model.* Our intention is to explore the design space for a variety of user scenarios. An extensive investigation was carried out regarding different service mixes for multimedia home, small-office/home-office (SoHo), office, ad hoc conference, and lounge scenarios (Table 6.1).

---

[128] An example for such a model is given in **Fig. 6.14** for the case of a nonlinear amplifier. The model can initially be based on measurements (e.g., represented in table form), it can be approximated, or of analytical nature.

**Table 6.1.** Eight representative scenarios for 5-GHz WLAN

| Scenario | Goodput (Mbit s$^{-1}$) | Maximum distance (m) | Access | | | Path loss | | |
|---|---|---|---|---|---|---|---|---|
| | | | B | T | W | B | T | W |
| Multimedia home A | 18 | 15 | X | | | X | | |
| Multimedia home B | 18 | 15 | | X | | | | X |
| SoHo[129] A | 19 | 15 | X | | | | X | |
| SoHo B | 19 | 15 | X | | | | X | |
| Office A | 23 | 20 | X | | | | X | |
| Office B | 46 | 10 | | X | | | X | |
| Ad hoc conference | 19 | 20 | | X | | | X | |
| Lounge | 27 | 30 | X | | | | X | |

B, T, and W denote the classification into best, typical, and worst-case configurations

These scenarios take into account the correlation that exists between goodput request, distance, optimality of the AP placement (**Fig. 6.5**), and typical indoor multipath scenario.

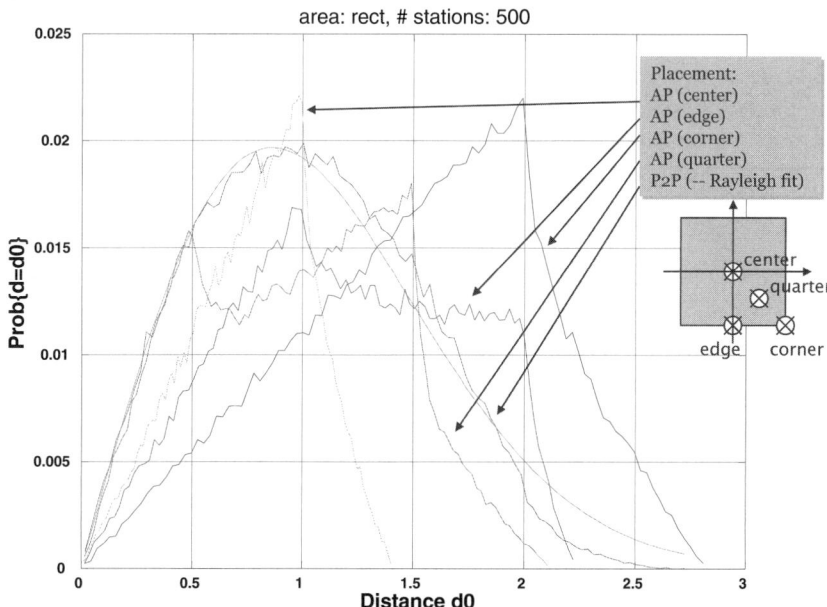

**Fig. 6.5.** Simulated power density function of the distance between the transmit and the receive antenna in a rectangular cell context, considering peer-to-peer (P2P) access and several access point (AP) locations

---

[129] Small-office/home-office.

**Fig. 6.5** shows the impact of receiver and transmitter placement for the example of a square room topology. The peer-to-peer (P2P) scenario results in a distance distribution with a fairly low average distance and can be approximated by a Rayleigh distribution. In an AP-based scenario, different AP placements result in large differences. The best placement is obviously in the center, exhibiting the smallest maximum distances and a fairly dense distribution. Quarter placement exhibits two peaks and edge placement a fairly flat plateau. The worst case is placement in a room corner with a high probability of large distances and the largest maximum distance.

Hence, room topology influences the outcome through AP placement parameters and path loss scenarios. Distance probabilities for several AP placements and uniform spatial terminal distribution were derived from 2D topology simulations and classified into three cases (B = best, T = typical, and W = worst). Similarly, three discrete path loss classes were derived based on reported broadband measurements in the 5-GHz covering outdoor, indoor, and large-room scenarios (e.g., [Medbo00]). Results were fitted to parameters $a$ and $b$ in the classical path loss model for distances $d$:

$$L(d) = L_0 + 10a \log_{10}(d/1m) + b, \tag{6.2}$$

where $L_0$ is the path loss for a distance of 1 m.

In this particular example, we assume a single link with an IEEE 802.11a peer-to-peer protocol employing a simple ARQ model with a retransmission bound. The resulting goodput $gp_{ARQ}$ depends on the packet error rate ($p_e$) involving retransmissions, the retransmission overhead $o_{pck}$, and the chosen net throughput $r$:

$$gp_{ARQ} = (1 - p_e)(1 - o_{pck})r + p_e(1 - p_e)(1 - o_{pck})\frac{1}{2}r. \tag{6.3}$$

Physical layer overhead and modulation/error correction coding have been included as defined in the standard.

Finally, the relation between the coded bit-error rate and signal-to-noise-and-distortion (SINAD) ratio for all relevant channel states is computed involving receiver thermal noise, noise figure (NF), and implementation loss (IL):

$$P_{Rx,dBm} = \left.\frac{S}{N}\right|_{Rx(dec),dB} + (N_0 B)_{Rx,dBm} + NF_{Rx,dB} + IL_{Rx,dB}. \tag{6.4}$$

The consideration of transmit power amplifier nonlinearity as a deteriorating effect on the SINAD is a particular extension to existing results [Zhao03]. The distance PDF($L(d)$) derived from access topologies combined with the required receive power determines the PDF of required average transmit power for a given distance $d$:

$$\mathrm{PDF}(\overline{P}_{\mathrm{Tx,dBm}}) = \overline{P}_{\mathrm{Rx,dBm}} + \mathrm{PDF}(L(d)_{\mathrm{dB}}). \tag{6.5}$$

*Signal model.* A significant extension toward physical layer signal models is required when it comes to power consumption models in the transmitter. Previous investigations mainly included performance aspects and neglected the instantaneous power distribution of an OFDM signal. In contrast to constant-envelope signaling such as GMSK, OFDM has a rather large peak-to-average power ratio (PAPR) [Tellado99] which requires a significant backoff margin at the transmit power amplifier; for practical class-A amplifiers, this leads to very low-power efficiencies (below 5%). Power efficiency also depends strongly on the maximum available transmit power; practical technology leads to different upper power limits (see **Fig. 6.3**). In our investigation, we include these effects and combine them with particular device models.

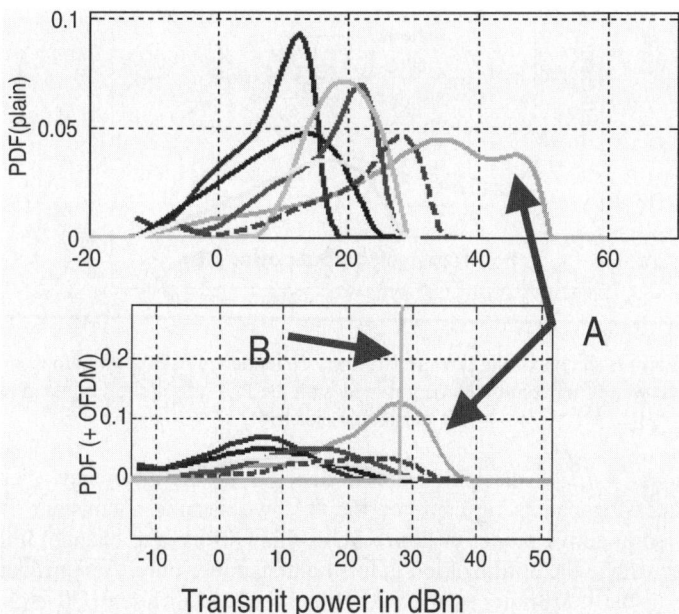

**Fig. 6.6.** Power limits and the signal shape have a large impact on the transmit power PDF (*upper*: unconstrained constant envelope, *lower*: 29-dBm power-limited OFDM) for different scenarios

Fig. 6.6 illustrates the variation between the PDFs of the transmit power in dBm for different scenarios; the upper plot assumes a constant-envelope signal and the lower plot takes into account the actual amplitude distribution of an OFDM signal and a saturating amplifier model with a limit of 29 dBm. We can clearly see that the OFDM signal nature results in a spreading of the PDF over a wider transmit power range (introducing a higher probability for peaks at higher output powers) unless these peaks get close to the 29-dBm limit and actually become saturated (the PDF shape appears truncated). Accounting for the OFDM signal shape together with a 29-dBm power limit leads to a roughly doubled probability at the power limit (*A*). (*B*) indicates an infeasible goodput of 46 Mbit s$^{-1}$; the bottleneck is not transmit power but the MAC system overhead was too high. Besides the effect of digital clipping to reduce the PAPR, future extensions will study the impact of filtering and predistortion techniques. The computation of this graph is presented in more detail in the context of the computational model below.

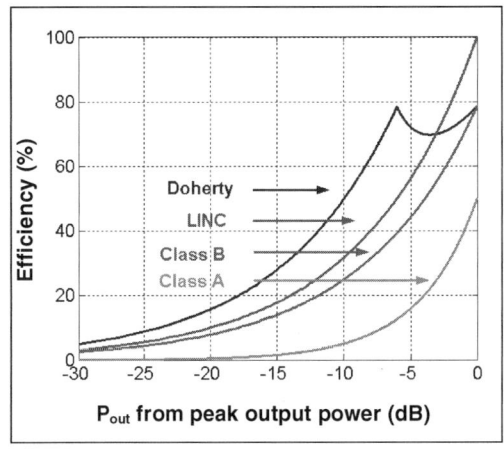

**Fig. 6.7.** Various shapes of the power efficiency dependency of PAs on output power lead to a different weighting result when combined with the PDF of required transmit power in a particular scenario

*Device model.* Finally, dissipated power depends on the efficiency of the entire signal processing chains. Except for the PA, we assume a constant Tx power consumption in active mode; in future SDR, this will have to change. Idle time is not foreseen since the optimization is for the transmitter only. A significant power consumer in the transmitter is the PA. We embedded different PA architectures [Raab02] ranging from conventional class-A and class-B to more evolved Doherty topologies. Their efficiency dependency on transmit power largely differs (**Fig. 6.7**); when weighted with the transmit power PDF derived from previous models, they offer a large tradeoff space. This degree of freedom is exploitable but only full design space exploration shows if a particular choice results in an actual performance/power gain.

### Evaluation

The availability of the individual performance/power models is not sufficient to yield conclusions. These models must be integrated into an evaluation framework.

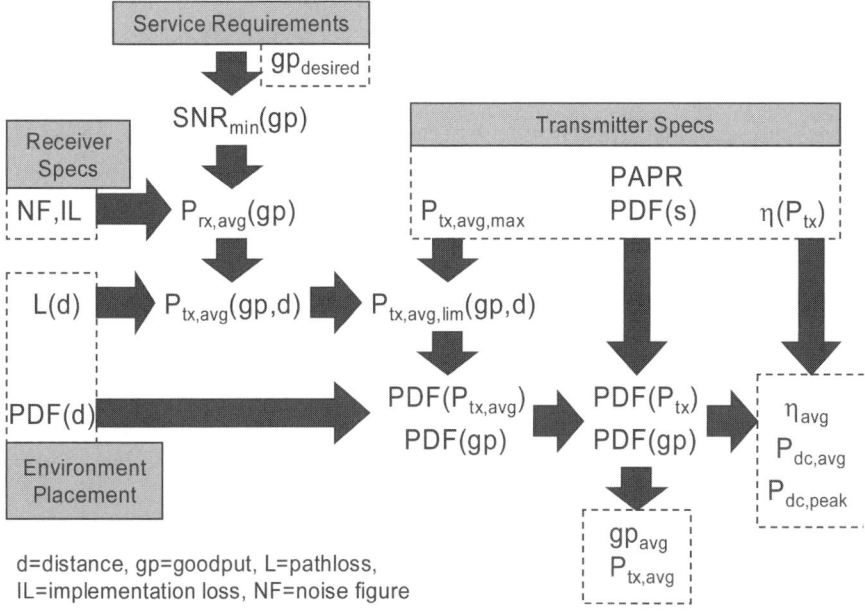

**Fig. 6.8.** The computational graph links specifications to relevant quality/performance and cost parameters

The result is a computational flow (**Fig. 6.8**) with both scalar and probabilistic input variables. The derivation of this flow starts from component specifications (receiver and transmitter), service requirements, and information about environment and placement. Its basic idea follows the classical link analysis which is shown in the upper-left part with minimum SNR, receive and transmit powers, and path loss $L(d)$; then follows the important extension with the actual signal properties (PAPR, PDF($s$)) in the upper-right corner and the statistical environment/placement information. The lower-right part performs a statistical treatment of this information. The final outputs are average values (averaged in the context of a scenario) but intermediate probability distributions remain accessible. Evaluation occurs in a parameter sweep across the required parameter ranges with an evaluation of the computational graph for each parameter set. Computational efficiency is hence important and influences, next to model availability, the complexity of applicable models. In most cases, we computed approximations to complicated performance/power dependencies a priori resulting in computationally simple look-up tables or equations. The entire flow is tool independent but for the practical tests it has been implemented in a MATLAB framework which allows convenient pre- and postprocessing for the modeling and exploration results.

*Results*

The previously described models for WLAN have been incorporated into our framework. The eight scenarios in Table 6.1 were combined with three transmit power limits of 16, 23, and/or 29 dBm corresponding to system-on-chip (SoC), system-in-a-package (SiP), and discrete power amplifiers. This results in $8 \times 3 = 24$ combinations. The entire design space has been evaluated and 16 out of 24 scenarios[130] are shown (**Fig. 6.9**). Total MATLAB simulation time amounted to less than 5 min (ca. 18 s per scenario) on a standard PC.

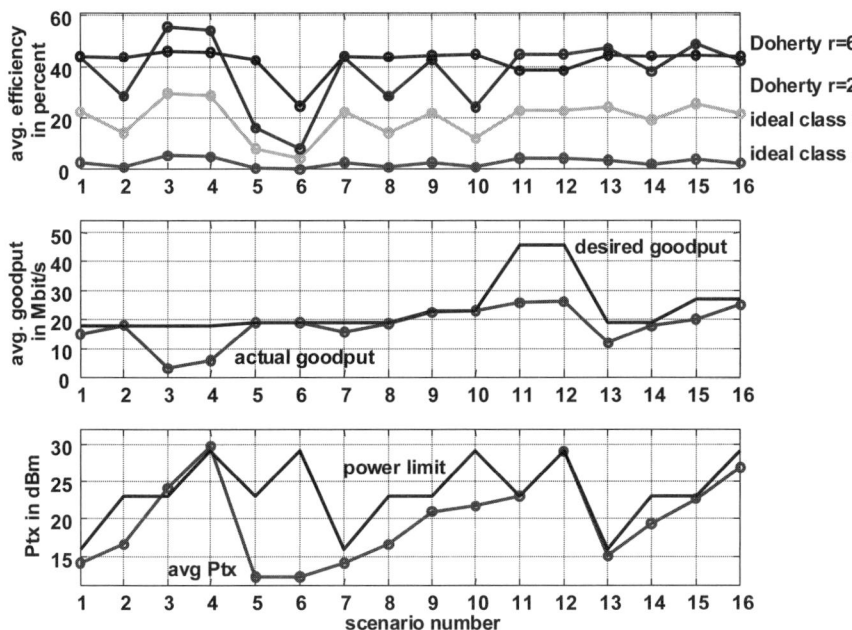

**Fig. 6.9.** Exploration reveals a huge variation across scenarios considering average power efficiency (*top*), actual vs. desired goodput (*middle*), and transmit power (*bottom*)

Results deserve a careful analysis since they reveal many dependencies. First, we can see that a more efficient Doherty PA does only pay off if we actually exploit its efficient region (a-6 is a counterexample). In most cases, however, we can win a factor 20 in efficiency compared to a standard class-A design (a-15). Second, we can identify scenarios where the power limit prohibits reaching the requested goodput (e.g., b-3), even despite allowing 29-dBm transmit power. Finally, we can also conclude (e.g., c-10) that the actual usage profile allows us to replace an expensive 29-dBm amplifier by a cheaper 23-dBm version. These were just a few

---

[130] The remaining eight scenarios were very similar to some of the other 16. The scenarios can be traced back to the original scenario in Table 6.1 by the desired goodput (graph in the middle) and the applied power limit (graph at the bottom).

examples. Simulation does not only report average numbers, but also indicates the probability for obtaining a particular goodput, e.g., as a function of the distance; goodput/distance (**Fig. 6.10**) or goodput/power cost tradeoffs can be made taking into account probability of occurrence for each case [VanDriessche03].

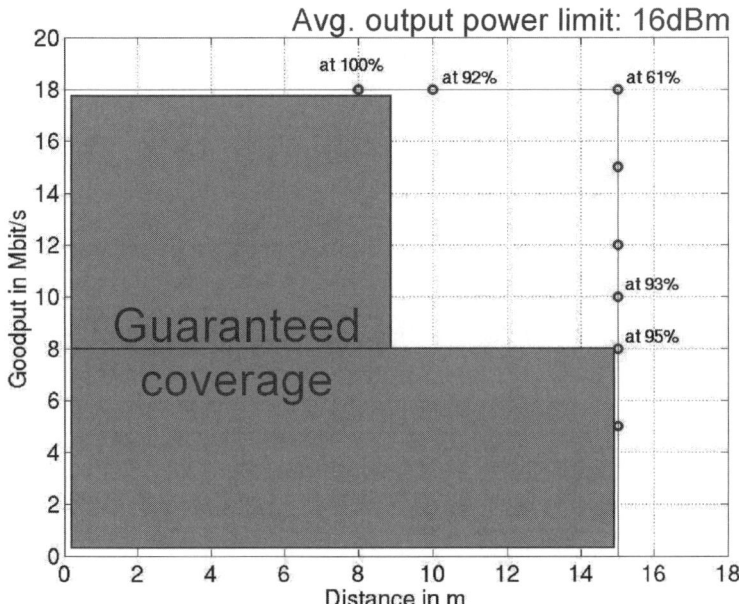

**Fig. 6.10.** A coverage analysis trades off goodput vs. distance.[131] This graph allows the validation of product features, e.g., coverage in case a 16-dBm PA is used ([VanDriessche03] © IEEE 2003)

In a second experiment, the tools capabilities were investigated to provide fast assessment of use cases. An inquiry with 25 participants from industry and IMEC was organized (**Fig. 6.11**). The form asks for expected goodput performance, cell radius, output power limits, and battery lifetime. Our software was used to traverse all requested scenario combinations in about 40 min. For the "today" scenario, the probability of success for reaching the desired goodput under the specified constraints was computed as well as the actual average goodput.

---

[131] A 15-m coverage may seem a low value for a 16-dBm PA; however, this graph does not show the feasibility of achieving a certain goodput *sometimes* but requests achieving this rate at a probability of 95% under statistical channel conditions. It can be seen that plain nonstatistical considerations overestimate actual achievable goodput.

| Requirements for the wireless link | | | | | | Number: ..... |
|---|---|---|---|---|---|---|
| **When on the market?** | **Application** | **Network; max. goodput in a single frequency band [in Mbit/s]** | **Cell radius [in m]** | **Terminal** | | |
| | | | | **Maximum output power [in dBm]** | **Power supply** | **Battery lifetime** |
| **Today (07/2002)** <br><br> 802.11a <br> 802.11b | ❑ Multimedia home <br> ❑ Small-office Home-office <br> ❑ Ad-hoc conference <br> ❑ Lounge | ❑ 5 <br> ❑ 10 <br> ❑ 20 <br> ❑ 30 <br> ❑ 40 <br> ❑ 50 | ❑ 5 <br> ❑ 10 <br> ❑ 15 <br> ❑ 20 <br> ❑ 30 | ❑ 10 <br> ❑ 16 <br> ❑ 23 <br> ❑ 30 | ❑ Mains supply <br> ❑ Car battery <br> ❑ Laptop battery <br> ❑ GSM battery | ❑ 2 hours <br> ❑ 5 hours <br> ❑ 1 day <br> ❑ 3 days <br> ❑ 1 week |
| **3 years from now (2005)** <br><br> 802.11a <br> 802.11b | ❑ Multimedia home <br> ❑ Small-office Home-office <br> ❑ Ad-hoc conference <br> ❑ Lounge | ❑ 5 <br> ❑ 10 <br> ❑ 20 <br> ❑ 30 <br> ❑ 40 <br> ❑ 50 | ❑ 5 <br> ❑ 10 <br> ❑ 15 <br> ❑ 20 <br> ❑ 30 | ❑ 10 <br> ❑ 16 <br> ❑ 23 <br> ❑ 30 | ❑ Mains supply <br> ❑ Car battery <br> ❑ Laptop battery <br> ❑ GSM battery | ❑ 2 hours <br> ❑ 5 hours <br> ❑ 1 day <br> ❑ 3 days <br> ❑ 1 week |
| **5 years from now (2007)** <br><br> 4G | ❑ Multimedia home <br> ❑ Small-office Home-office <br> ❑ Ad-hoc conference <br> ❑ Lounge | ❑ 5 <br> ❑ 10 <br> ❑ 20 <br> ❑ 30 <br> ❑ 40 <br> ❑ 50 | ❑ 5 <br> ❑ 10 <br> ❑ 15 <br> ❑ 20 <br> ❑ 30 | ❑ 10 <br> ❑ 16 <br> ❑ 23 <br> ❑ 30 | ❑ Mains supply <br> ❑ Car battery <br> ❑ Laptop battery <br> ❑ GSM battery | ❑ 2 hours <br> ❑ 5 hours <br> ❑ 1 day <br> ❑ 3 days <br> ❑ 1 week |

**Fig. 6.11.** Inquiry form[132]

### 6.1.2   Extension to Crosslayer Link-Level Optimization

System exploration so far covered mainly the physical analog and digital layer and only partially the MAC layer. An extension of the optimization across several layers is termed *crosslayer optimization* and is subject to intensive research [Schurgers02a, Schurgers02b, Ebert99, Kandukuri02, Givargis02, Shakkottai03]. Crosslayer optimization across the traditional OSI layers is subject to produce significant performance–energy savings since the existing standards at different OSI layers have often not been codesigned for optimum performance/power scalability.

In [Bougard03a], we have set a first step toward a novel framework that aims at enabling context-aware energy–performance tradeoff at run time while still minimizing complexity of the controller. This approach is based on ensuring the

---

[132] Assumptions for a car battery: 40 Ah; GSM battery: 1.05 Ah; and laptop battery: 3.0 Ah. For 1,200 mW, this is 400, 2.5, and 30 h, respectively.

constant consistency between the system configuration and the environmental and user-related conditions to achieve minimal energy consumption. This can be formulated as a combinatorial optimization problem. The variables of the problem are the low-level configuration knobs. The objective functions are the energy consumption and the performance. Constraints of the problem may be user requirements and on the other hand environment restrictions, e.g., radio channel conditions in the wireless context.

### Design-Time Run-Time Approach

To solve this problem, we have to follow a three-step procedure (**Fig. 6.12**) involving both design-time and run-time components. First, we develop an energy–performance model of the system. This is equivalent to the modeling approaches presented in Sect. 6.1. Second, based on simulation of the system under various configurations, we extract performance and energy as a function of the configuration.

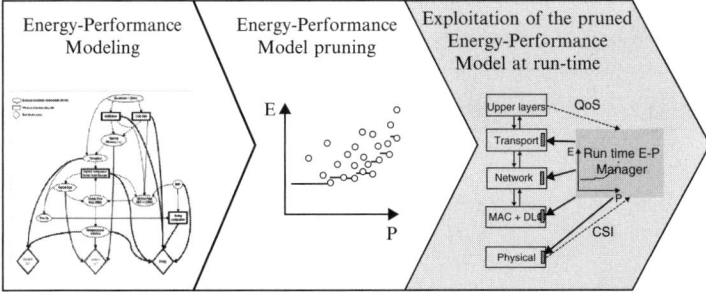

**Fig. 6.12.** System-wide design-time/run-time optimization ([Bougard03a] © IEEE 2003)

The resulting solution space typically contains a large number of suboptimal points with respect to one or the other axis. These points are not relevant and can be pruned from the set of points. Since we have at least two axes, optimization – except for degenerated cases – cannot result in a single-point solution. Instead, we face a so-called Pareto front or Pareto hyperplane which presents the range of solutions that are optimal in a multiobjective sense. An important goal is to reduce this number of points at design time to the minimum required ones since the third step involves the steering of all remaining configuration options at run time.

### System Under Test

The system under test (**Fig. 6.13**) consists of a physical and data link layer and exhibits configurable parameters for both transmitter and receiver (modulation scheme, code rate), in the transmitter (transmit power, backoff), in the receiver (number of Turbo decoder iterations), and in the data link layer (retransmission scheme).

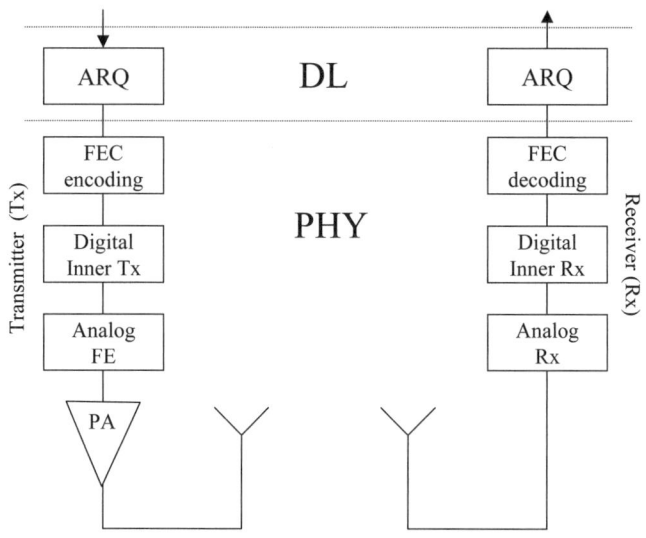

**Fig. 6.13.** System under scope

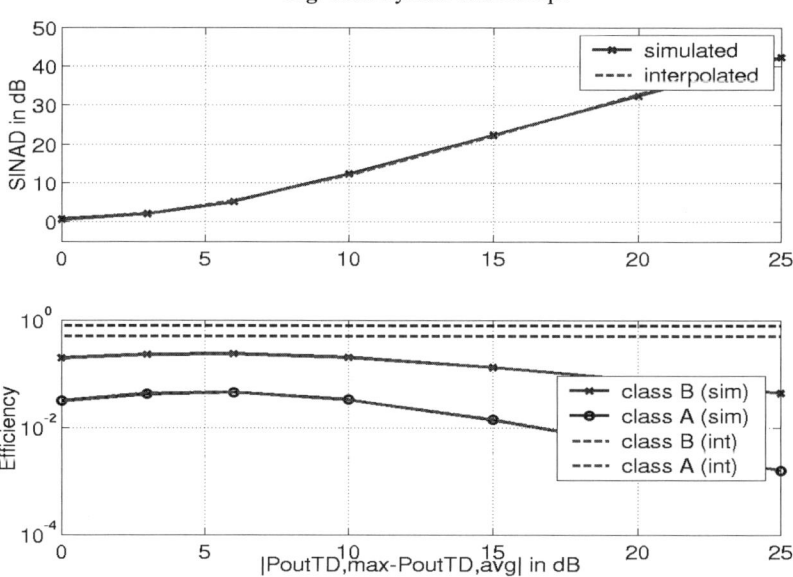

**Fig. 6.14.** Reducing the backoff at the PA results in higher efficiency at the cost of more distortion ([Bougard03a] © IEEE 2003)

As an example for the modeling, we illustrate the performance–power model for the power amplifier (**Fig. 6.14**). The PA was modeled using a simple third-order nonlinear AM/AM characteristic that flattens out after reaching its peak. Results shown were based on class-A operation; class-AB, class-B, and more advanced

PA configurations are under investigation. A crucial design-time parameter of the PA is the peak power and the class of operation. At run time, the backoff and hence the actual average transmit power can be varied and thus is considered as configuration knob. Due to the nonlinear AM/AM characteristic, a reduction of the backoff reduces the SINAD while increasing the power efficiency.

## Results[133]

For the system under test, we need to define the performance–cost metrics that we want to compare. Since we operate at the data link layer, goodput[134] is a good performance and energy consumption a good cost criterion. The resulting energy–goodput characteristics are given in **Fig. 6.15** for several combinations of

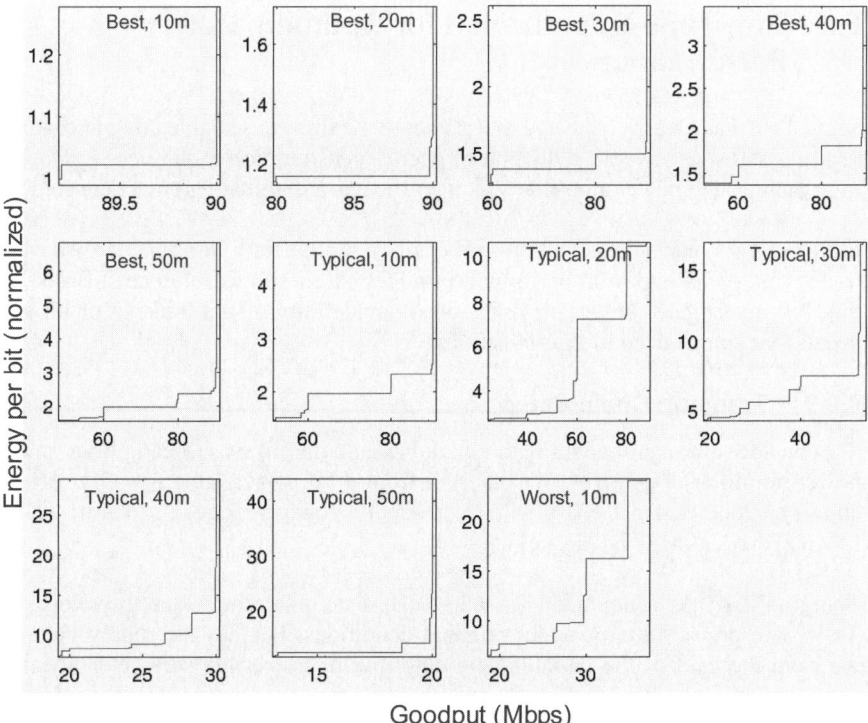

**Fig. 6.15.** Energy per bit vs. goodput characteristics for different environment scenarios (placement, distance). In some cases, significant differences in energy–performance scalability can be achieved (e.g., typical: 20/30/40 m or best: 50 m) ([Bougard04] © IEEE 2004)

---

[133] This work is based on a collaboration between B. Bougard and W. Eberle. Channel modeling and PA modeling originate from W. Eberle. MAC and baseband modeling originate mainly from B. Bougard.
[134] Error-free throughput.

placement and distance toward the access point. Traditionally, the system would have to be dimensioned by maximizing either the link capacity or the range, tolerating a constant residual packet error rate (e.g., 1%) and averaging on the propagation condition (e.g., 90% typical, 10% worst case).

Using our performance and energy model presented above fixing the packet size (512 bits), such an optimization would lead to energies from 77 nJ bit$^{-1}$ (maximum capacity, short range) to 300 nJ bit$^{-1}$ (maximum range). Comparing this value with the tradeoff curves, we can conclude that our approach would potentially enable an energy reduction by, at least a factor 2 and up to a factor 20, depending on the actual channel conditions.

## 6.2 Run-Time Optimization for Optimum Power–Performance

Section 6.1 has clearly indicated that power–performance scalable components are key to a statistical overall reduction of energy consumption in wireless terminals. In particular, the power amplifier was identified as a dominating power consumer.

In this section, we address the transmit chain in more detail. In particular, we want to derive a realistic power–performance model based on measurements. Based on this, we investigate the performance–power scalability of the transmit chain and discuss the run-time control implications.

### 6.2.1 Transmit Chain Setup

We consider a transmit chain with a variable-gain amplifier, a preamplifier, and a power amplifier (**Fig. 6.16**). We selected from a set of available 5–6 GHz power amplifiers one that offered a sufficient supply voltage range: the Hittite GaAs MMIC HMC280MS8G (**Fig. 6.17**).

The amplifier has a nominal gain of 18 dB, a nominal single-supply voltage of 3.6 V, and draws 480 mA under nominal conditions. Besides the supply voltage, we have the gain of the variable-gain amplifier as accessible parameter. Ideally, also the preamplifier would be adaptive[135] and we would have access to independent bias contacts at the PA [Saleh83, Asbeck01]. Since bias adaptation changes drastically and in a quite unpredictable way the internal operation of the power amplifier, we kept bias contacts to the recommended settings.

---

[135] A fully scalable design of a transmit chain except for the power amplifier is ongoing at IMEC.

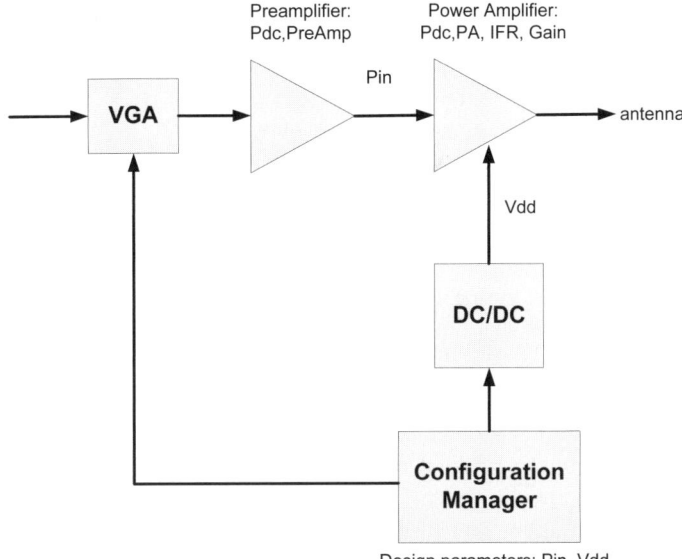

**Fig. 6.16.** The configuration manager controls input power and the PA supply voltage. Cost parameters are the power consumption of preamplifier and PA

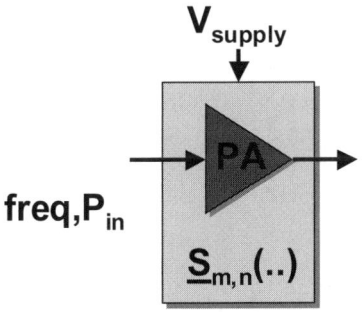

**Fig. 6.17.** A Hittite HMC280MS8G GaAs power amplifier was used as device under test

### 6.2.2    A Design-Time, Calibration-Time, and Run-Time Approach

Our goal is to derive an optimum configuration management with VGA gain and supply voltage as control parameters, power consumption of the chain as cost, and the SINAD ratio and output power $P_o$ as performance (quality) parameters. Since we assume a fixed gain in the preamplifier, we can also translate the VGA gain setting into an input power $P_{in}$ and use this as variable. Hence, we search for any achievable output power $P_o$ for configuration ($V_{dd}$, $P_{in}$) that results in the minimum DC power consumption $P_{dc}$:

For all
$$P_o \in [P_{o,\min}, P_{o,\max}]$$
with $|P_o - P_o(V_{dd}, P_{in})| < \delta_{tol}$

exists a tuple $:P_{dc,\min} = \min_{V_{dd}, P_{in}} \{P_{dc}(V_{dd}, P_{in})\}\Big|_{SINAD(V_{dd}, P_{in}) > SINAD_{\min}}$ . (6.6)
$(V_{dd,0}, P_{in,0})$

such that for all
$$P_{dc,\min} = P_{dc}(V_{dd,0}, P_{in,0})$$

In a second step, we determine each point from the Pareto front which satisfies the SINAD constraint. Note that this SINAD requirement relates to the BER and PER and is hence an input parameter. The total power consumption for the two-stage ($N = 2$) combination of power amplifier and preamplifier (pre) is defined as

$$P_{dc,total} = \sum_{k=0}^{N-1} P_{dc,k} \overset{N=2}{=} P_{dc,PA}(V_{dd}, P_{in}) + P_{dc,pre}(P_{in}).$$ (6.7)

The characteristics of the preamplifier are based on a class-A model which is a realistic assumption[136] for preamplifier stages. For the power amplifier, we performed extensive nonlinear measurements of the Hittite HMC280MS8G.

## 6.2.3   Measurements

Extensive measurements were carried out on the Hittite PA using a nonlinear vector network analyzer from HP.[137] We varied input power from −10 to 12 dBm in steps of 0.5 dBm and swept the supply voltage settings in nine steps (1.5, 1.8, 2.1, 2.5, 2.7, 3.0, 3.3, 3.6, and 4.0 V). The full nonlinear $S$ parameters and the DC power consumption were measured for all combinations. In a postprocessing step, we derived the intermodulation-free range based on the measurement result at the fundamental frequency (5.25 GHz) and the intermodulation products resulting from the third harmonic.

Besides these measurements in a controlled load situation, we also performed a number of measurements to study the effect of load changes on the PA output These load changes can be introduced by objects in the near field of the antenna which is not untypical for WLAN cards in, e.g., laptops or adapters in other mobile appliances. We found from dipole measurements[138] that, except for the situation where the antenna is located parallel to a metal plate, reflected power is less than 10 mW (or 5%) for an output power of 200 mW. These reflected powers do not harm or influence the power amplifier significantly and the effect can be

---

[136] To a lesser extent, also class-AB stages are used in practice.
[137] In collaboration with the ELEC department at the VUB, Brussels.
[138] Reported in IWT UniLAN, MEDEA+ A106, Deliverable D3.3. Author: W. Eberle.

ignored. For an exceptional case,[139] however, up to 125 mW were reflected which would require protection mechanisms.

### 6.2.4 Results

The optimization of the transmit chain model was performed for several minimum SINAD requirements. The power-added efficiency (PAE) of the amplifier for various levels of input power $P_{in}$ ranging from −10 to +15 dBm is given in **Fig. 6.18**; we also performed a sweep over the supply voltage from 1.5 to 4 V in nine steps. We can clearly see that the PAE has an optimum $f_o$ supply voltages around 2.7 and 3.0 V but also for higher supply voltages. However, for the higher supply voltages, the PAE is not an accurate measure of performance since it also includes unwanted saturation results.

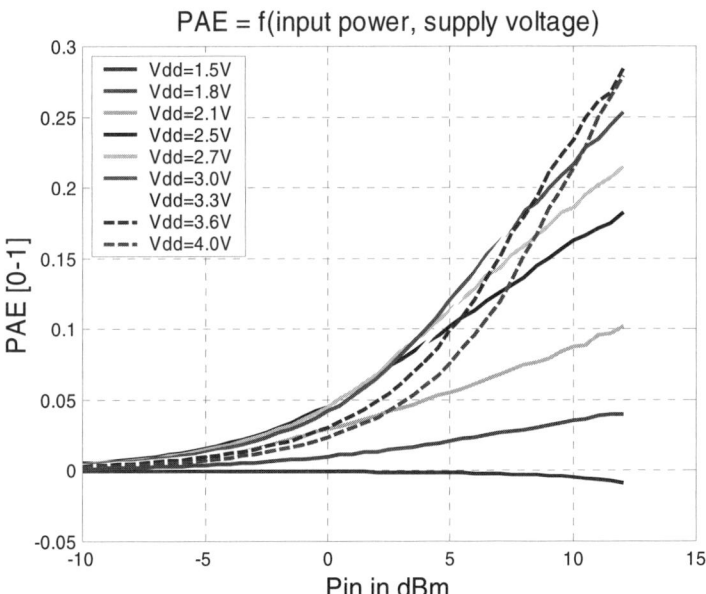

**Fig. 6.18.** Power-added efficiency (PAE) improves with increasing supply voltage ($V_{dd}$) and input power ($P_{in}$). The higher three gain settings operate the device beyond its guaranteed specifications and exhibit a different slope

---

[139] The exceptional case refers to placing the antenna only a few centimeters away from a large horizontal metal plate; this is a situation that can appear when, e.g., placing a phone, handheld, or laptop device on a metal table, on a metal file or shelf, or on a table under a metal shelf.

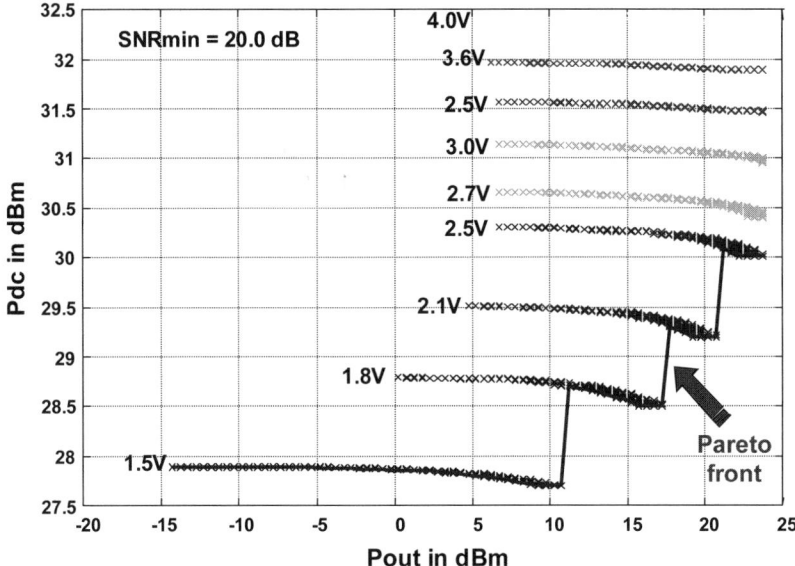

**Fig. 6.19.** Supply adaptation of the HMC280MS8G power amplifier exhibits four useful control settings (1.5, 1.8, 2.1, and 2.5 V) with a difference in power consumption of 70%

Hence, we actually have to measure the SINAD for each configuration. An example result for a minimum SINAD of 20 dB, sufficient for 16-QAM transmission, is shown in **Fig. 6.19**. We see that we only need four out of nine possible configurations of the power amplifier. For the configuration manager at run time, we need to invert the relationship in (6.8) since desired output power and minimum SINAD are the input parameters there:

$$(V_{dd}, P_{in}) = f(P_o, \text{SINAD}_{min}). \tag{6.8}$$

Power consumption does not appear directly anymore since the Pareto criterion in (6.6) guarantees that, for each combination of input parameters, always the configuration with the minimum DC power consumption is provided.

## 6.3   Summary and Discussion

This chapter raised the design space scope from digital component design (Chap. 4) and mixed-signal component design (Chap. 5) to system-level and communications link-level design. Moreover, we particularly addressed the tradeoff between performance and corresponding power consumption. In this context, we proposed:

- An approach to modeling abstraction covering jointly the analog/digital and physical/MAC layer scope
- A methodology and computational flow for fast exploration of the performance–power behavior of a wireless transceiver system for a set of service, environment, and device parameters; this technique has been used for classical design analysis but also proved its usability for early investigations in the product definition phase
- An extension of this methodology to a true, generic crosslayer optimization technique applicable in a wider context
- Design-time/run-time adaptation techniques to arrive at a power–performance optimal configuration of the transmit chain with a particular focus on the dominant power consumer: the power amplifier

This chapter has introduced an exploration methodology and a resource control approach for wireless systems with a focus on transmit-side examples. A very similar methodology and run-time approach have been used already in Chap. 5 in the context of mixed-signal receiver optimization. The generalization of these concepts is described in Chap. 7.

# 7 Methodologies for Transceiver Design

*If the only tool you have is a hammer, you tend to see every problem as a nail.*
Abraham Maslow, 1908–1970.[140]

Actual application design is an ideal source of inspiration for essentially *required* design methodologies. Practical experience allows the definition of clear requirements that stimulate both the development of new techniques and the adaptation of existing ones (Chap. 2).

This approach establishes the underlying rationale for this chapter. During the application-oriented research in this chapter and Chaps. 4–6, we faced particular inefficiencies in existing design flows, such that we investigated improvements to available methods and the introduction of new techniques. In general, we can observe that in each chapter we developed or applied *additional* methods *and* reused the ones developed earlier.

We believe that this *incremental* approach to codesign of design methodology and application design automatically leads to a focus on the *essential* design problems. A risk of this approach is that only ad hoc solutions are produced that may not be generic enough to be reusable. However, this can be avoided by considering the methodology and tool development as a concurrent process with its own output.

---

[140] Among many interesting things, Maslow noted while he worked with monkeys, was that some needs take precedence over others. On the other hand, availability of particular tools tends to influence the approach humans take to solve a problem. He took this idea and created his now famous hierarchy of needs starting from basic deficit needs to being needs. Being needs, also called self-actualization and growth motivation, forms the top. Maslow was one of the pioneers to bring the human being back into psychology, and the person back into personality during the 1960s phase of behaviorism and physiological psychology. It has to be noticed that, in terms of a scientific basis, the cognitive development theory [Piaget72] has replaced much of Maslow's work but Maslow's message that psychology should be, first and foremost, about people retains its relevance.

Guaranteeing the actual concurrent execution of both the design and the design support processes is mainly a management challenge.[141] Moreover, our *incremental* does not exclude that the problem description for an essential design problem can also be taken out of the design flow and then treated in a classical tool-oriented EDA context.

A consequence is that diverging sources of problems also require investigation of rather diverse techniques. To put them into perspective, we classify them according to the time of use, design space, and abstraction level. **Fig. 7.1** shows the classification and the transitions across these dimensions.

- *Time of use*
  We distinguish between design time and run time. Design time describes the classical range of techniques applied before or during implementation. Run-time techniques are techniques that involve control, adaptation, or reconfiguration of the *implemented* device. Further, we distinguish between the first-time use and reuse of components.
- *Design space*
  From functional and architectural point of view, we address the classical digital hardware and the analog/RF domains. However, we extend this with the application design space.
- *Abstraction level*
  We roughly differentiate between the behavioral level, an IP module, and the implementation level. This abstraction is not based on *formal* criteria but about their usability to capture design information for system-level design, design reuse, and implementation.

We distinguish between design-time, run-time, and reuse methodologies. As design-time techniques, we developed mixed-signal design space exploration, mixed-signal codesign, digital design refinement, and analog behavioral modeling. Reuse is described for the digital design space and run-time techniques were developed for the mixed-signal design space. Analog design reuse has not been considered here since it requires a look at circuit design and technology-related aspects which go beyond our scope here.

This chapter is organized as follows. Section 7.1 establishes a practical digital design flow. We introduce the underlying OCAPI technology, discuss the necessary modifications and extensions, and conclude with a (re)use analysis based on two ASICs and an FPGA design. Section 7.2 extends the scope to mixed-signal system simulation and behavioral modeling. We introduce the underlying FAST technology and discuss our cosimulation approach and particular modeling techniques. Section 7.3 constitutes a design-time run-time partitioning approach for mixed-signal systems. We motivate behavioral modeling, design space

---

[141] "Do not change priorities and recruit the tool designer as circuit designer, if deadlines approach."

exploration, pruning of complexity, and the synthesis of a run-time controller structure. The chapter ends with a summary and discussion of results in Sect. 7.4.

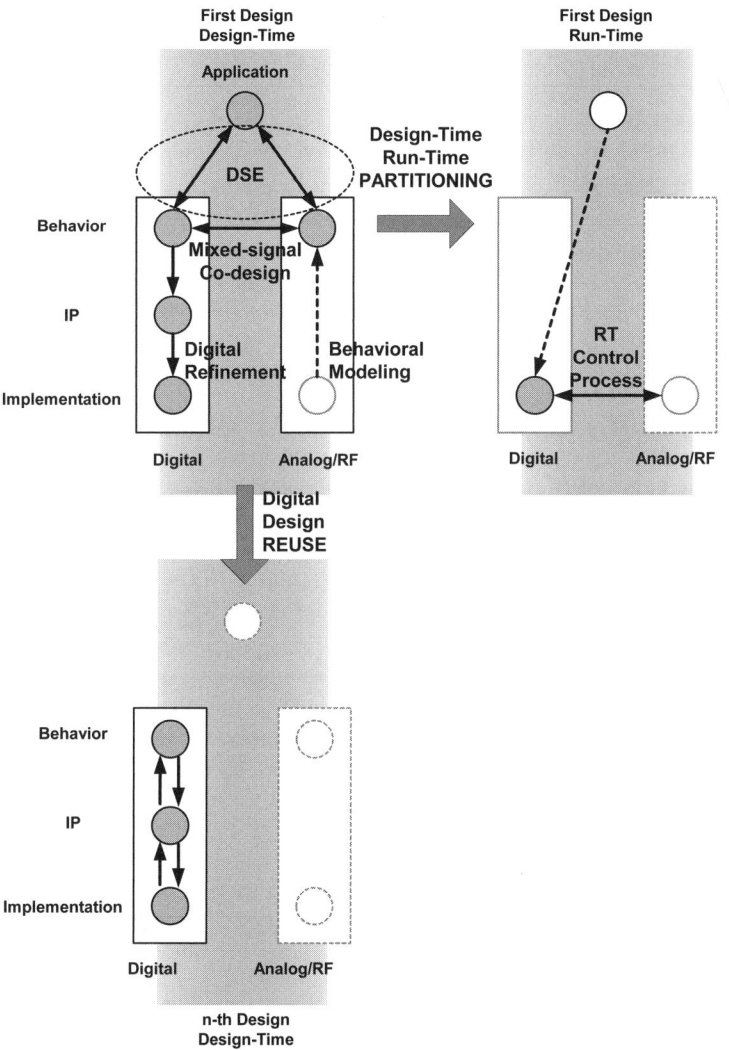

**Fig. 7.1.** We distinguish between design-time, run-time, and reuse methodologies. As design-time techniques, we developed mixed-signal design space exploration, mixed-signal codesign, digital design refinement, and analog behavioral modeling. Reuse is described for the digital design space and run-time techniques were developed for the mixed-signal design space

# 7.1  A Practical Digital Design Flow

For the design of a flexible baseband transceiver as described in Chap. 4, we need a design environment that offers:

- Means for describing the design at several abstraction layers to allow efficient algorithm–architecture codesign [Zhang01]; as a minimum, we see dataflow semantics for the functional exploration and register transfer semantics for the architecture exploration
- A refinement strategy for quantization allowing a mix of, e.g., fixed-point and floating-point style at different abstraction levels
- A refinement strategy for communication between computational blocks (interface refinement)
- An output format fitting standard hardware description languages such as VHDL [VHDL] or Verilog [Verilog] synthesis and layout flows and hence toward a pathway to silicon implementation [Bryant01]

The traditional design process of such a telecom system starts at the level of MATLAB models. Extensive simulations allow us to decide on the algorithms and system parameters that meet the system requirements. From this high-level algorithmic specification, HDL coding is started without any intermediate design steps. All decisions related to architecture and implementation are coded immediately in the register transfer (RT)-level HDL code suited for synthesis. Because of the low simulation speed of the RT-level HDL code, exploration of architectural tradeoffs is limited and algorithmic changes imply a long and cumbersome iteration over the MATLAB specification. For simulation speedup, often an intermediate implementation in C is used. However, translations between languages require manual code rewriting, an error-prone and lengthy process.

We investigated and classified a number of available tool solutions such as MATLAB [Aue01], DSP Canvas [Murthy01], CoWare,[142] OCAPI [Schaumont98], and Synopsys Design Compiler with respect to the design abstraction levels they cover (**Fig. 7.2**). Ptolemy [Buck94, Lee96] offers a large variety of models but also fails on delivering a path to implementation. The desired comodeling for algorithm architecture optimization and a smooth path toward implementation was only offered by OCAPI developed at IMEC [Schaumont98].

Today, several flavors of C- and C++-based solutions and, in particular, SystemC [SystemC, Prophet99, Verkest00, Panda01, Siegmund01, Pasko02] are quite popular for architecture-level modeling but at the time of our baseband designs, little was available. UML [UML] is often considered as a high-level specification language but – from our experience – too abstract to capture most of the signal processing interactions appearing in communications designs. Moreover, the specification is not

---

[142] CoWare was evaluated in 1997/1998 time frame. Hence, we do not refer to the recent ConvergenSC framework.

the point where most design iterations take place; instead, we need an efficient solution for the level where co-optimization of algorithms and architecture is possible. Hence, UML would only be used as a representation language with adequate representation means for, e.g., control processes but without explicit benefits for the representation and transformation of signal processing aspects.

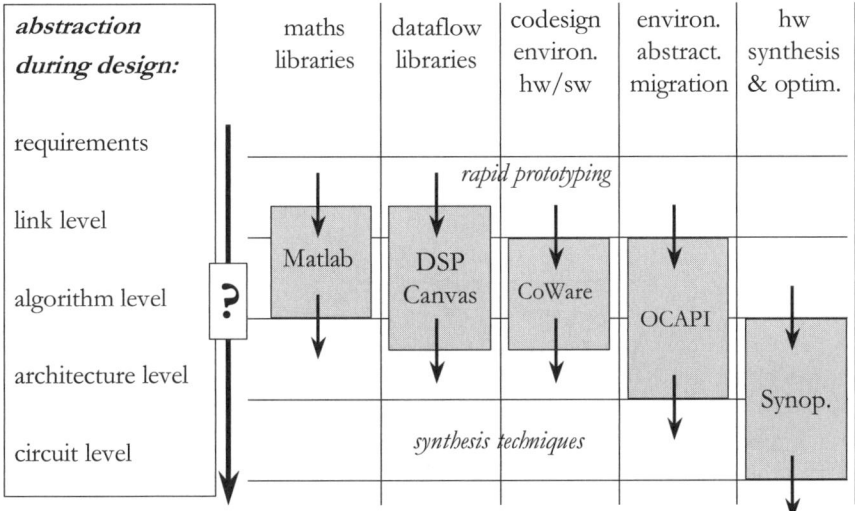

**Fig. 7.2.** A complete design chain requires the combination of several tools. Important for the selection of OCAPI was the overlap with both algorithm and architecture level where performance and cost are traded against each other

This section continues with a short introduction to OCAPI. Then, we describe the significant extensions we added to OCAPI for an improved design efficiency. We conclude with our experiences on use and reuse of our extended OCAPI-based design flow across two completed designs (Festival and Carnival).

### 7.1.1 A Digital Design Flow Based on OCAPI

OCAPI is a C++-based dataflow and RT simulator for digital circuits featuring also a direct path to HDL-based implementation [Schaumont98, Vernalde99]. OCAPI's dataflow simulation is based on FIFO queues that connect C++ objects. Each C++ object has a set of firing rules related to the inputs and state information. Dataflow simulation is mainly used to explore algorithmic and architectural options during digital system design. A round robin scheduler is used, exploiting the firing rule informations provided by each block. OCAPI also includes a register transfer description style. Both styles can be mixed. The smartness of OCAPI is that it only defines a few objects that are related to hardware. To get the expressive power, it relies entirely on the C++ programming language which gives the designer convenient extension capabilities (Sect. 7.1.2). OCAPI has also been used for other telecommunication ASIC designs such as an upstream cable modem [Schaumont99b] or a multirate digital-IF processor [Pasko00].

In OCAPI, an object-oriented C++ model is gradually refined starting from a high-level behavioral description using dataflow semantics (**Fig. 7.3**). Further steps include data-type refinement (from floating point to fixed point) and the introduction of timing by means of an RT description. The translation between these two abstraction levels is essentially manual. RT-level simulation is linked to the clock domains and uses a cycle-based event scheduler. At the RT level, this results in a combination of finite state machines and datapaths (FSMDs). Semantically, FSMDs can be directly translated into HDL syntax, allowing automatic generation of HDL code.

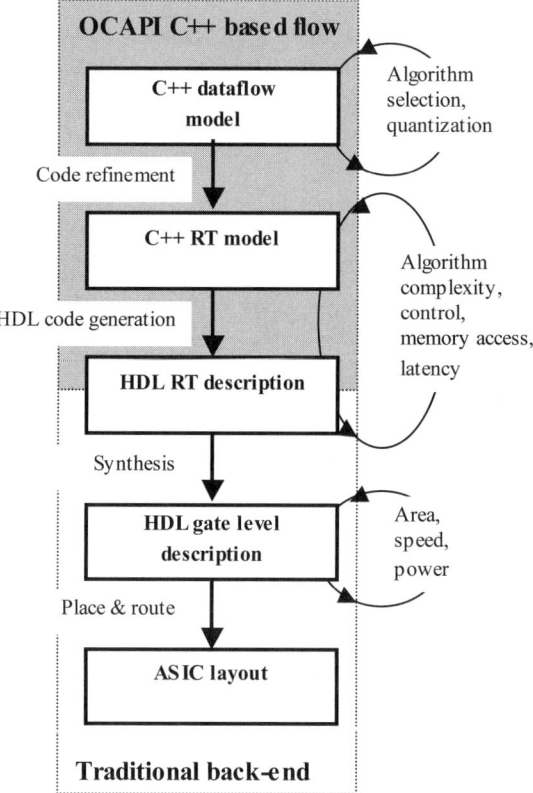

**Fig. 7.3.** C++-based design flow on top of a traditional back-end flow. The OCAPI flow covers algorithm exploration down to RTL implementation in an open C++ environment ([Verkest01a] © IEEE 2001)

Algorithm selection and studies on implementation loss due to quantization were mainly performed in dataflow. Control, complexity, latency, and detailed memory access were treated at the C++ RT and HDL RT levels. Overall data transfer and storage exploration (DTSE) was, however, analyzed already at a partially refined architectural model that still used dataflow notations for interunit communications.

Most DTSE optimization could be manually performed and did not require tool assistance (Sects. 4.3.1 and 4.3.2).

Nowadays, SystemC [SystemC] has become established and can be considered as a replacement for OCAPI at various abstraction levels. However, SystemC itself does not improve the translation between different abstraction levels. This requires still the creativity of the designer or the designer to rely on library support from the vendor.[143]

## 7.1.2   Extensions to OCAPI During the Design Phase

When using OCAPI for the design of the Festival and Carnival ASICs, we encountered a number of shortcomings with respect to functionality but also with respect to efficiency.

### Extensions to the Design Flow

Because of the object-oriented nature of the OCAPI environment, all meaningful design concepts (FSM, state, transition, clock, register, etc.) are modeled as objects and hence can be readily manipulated by the designer. To that end, the designer adds extra methods to the already defined classes (much in the same way that one would traditionally write complex scripts to browse through design databases to extract relevant information). These methods have direct access to the objects of interest. This principle was at the basis of our extensions. All extensions were expressed in the form of a set of new class libraries under the term *UNIFY*.

Extensions to the design flow and to OCAPI as a tool were required to further improve the design efficiency, especially with respect to the path to silicon implementation [Niemann98]. OCAPI foresees the generation of HDL code per design entity but did not come with sufficient support on simulation, synthesis, and test script generation. The conventional approach is to reidentify specific signals such as clocks, reset, or test signals from the HDL code. This is an error-prone way that can be avoided by top-down propagation of design knowledge or designer's intentions. Next, we added support for hierarchical code and script generation, such that a complete design unit (e.g., the 130-kGate equalizer) could be automatically simulated and synthesized at HDL level. Importantly, full support for multiple, inverted, and gated clocks[144] was added including synthesis script support for constraining these blocks.

A simple call to a makefile would start the compilation of the OCAPI/UNIFY code for a particular design unit and result in:

---

[143] Modern ESL tools based on SystemC such as CoWare's ConvergenSC come with libraries with several abstraction levels for particular third-party building blocks such as processors, busses, etc.
[144] As an extension to the clocking capabilities described in [Rijnders00].

- Simulation and output vector generation in C++
- Instantiation of the entire HDL design hierarchy
- Code generation for all structural and nonstructural HDL entities
- Generation of simulation scripts for HDL simulation
- Generation of synthesis scripts for HDL synthesis including constraint passing for I/O ports and clocks as well as full scan-chain insertion
- Bottom-up first and top-down incremental synthesis for a hierarchical design unit

Design templates are provided to the user both for the C++ description of the design units and the makefile. The designer adds nonfunctional parameters such as clock frequency, test options, etc., into the C++ description. The designer does not have to create or modify scripts or resulting HDL code.

We introduced a series of code quality improvements in the semantical translation from C++ to HDL code to speedup the synthesis process such as merging of multiple signal-flow graphs for a reduction of HDL processes, and improved conditional statements.

For the convenient automatic instantiation of different on-chip SRAM memories, we implemented an extension with a behavioral C++ model combined with a parametrizable HDL wrapper that included test and BIST support. Similarly, for operators such as a real division, a behavioral model and a matching HDL implementation was designed.

All extensions were reusable in both designs and led to a significant design speedup since they could be shared between several designers and amortized over two consecutive designs. The effort for the development of the generic tool library extensions accumulated to about 20% of the time of the chip architect during the first chip design (about eight person-weeks full-time) and about 5% of the time of the same person during the second chip design for extensions (about two person-weeks full-time).

### 7.1.3   Experience of (Re)Use

Success and applicability of design methodologies, particularly questions concerning reuse, cannot be adequately answered without involving actual design. Since design reuse is all about reducing design effort, increasing reliability and yield, and hence reducing cost and risk, already small but recurrent and annoying practical issues can prevent a methodology from being applicable.[145]

---

[145] Applicability of a methodology is, in general, hardly measurable in an objective way since it largely depends on psychological concerns of the user. A tool, even showing some improvement in a key point capability, e.g., synthesis quality, will not be accepted by users if it shows annoying deficiencies on other points, e.g., coding style or practical tool handling. This especially applies if the user faces deficiencies at the beginning or during a

In this section, we comment on reuse methodologies and user's experience with a series of four digital WLAN designs (Festival ASIC, Carnival ASIC, technology transfer code, and FlexCop FPGA). All of them reached final hardware implementation, three as an ASIC, one mapped to an FPGA. Improved versions of the Carnival OCAPI dataflow model using extensive object-oriented C++ methods were used for the cosimulation case with FAST in Sect. 7.2.3 and for an upgraded IEEE 802.11a standard-compliant baseband transceiver model in technology transfer and consultancy for a startup company.[146]

### Architectural Reuse

In [Vermeulen00], reuse possibilities in the context of hardware design are subdivided into three categories: the highest layer denotes process control, the middle layer contains loop definition and indexing, and the lowest layer describes reuse of scalar objects such as optimized IP. We actually found that large reuse was made at all levels. At the process control layer, we could reuse the generic control communication and data communication; and clock gating concepts in the Festival and Carnival ASICs allowed reuse across the entire design. For example, the burst controller and all design units operate in exactly the same way in both designs. Inside individual design units, we recognized large reuse at the loop/index layer since most units were designed with parametrizable subcomponents such as address generators or VLIW controllers (mapper, SSR, FFT). Reuse percentages between Festival and Carnival ASICs are given per design unit in **Fig. 7.4.**

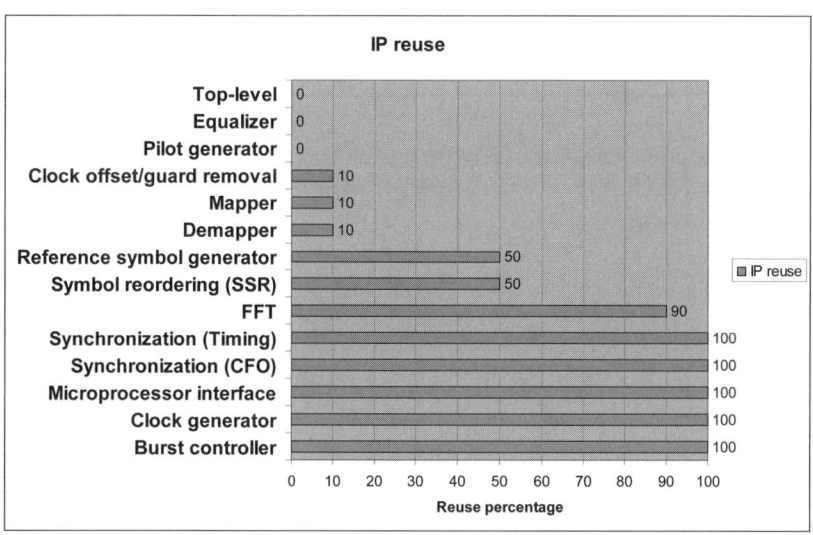

**Fig. 7.4.** Reuse percentage per Festival design unit in the Carnival design

---

long phase of the design: feeling of "the tool could/should be improved this way," "need for using additional tools, e.g., scripting."

[146] Resonext Communications, Inc.; later acquired by RFMD, Inc.

*Level of Reuse*

Project constraints did not allow the systematic creation of all modeling instances for all design units. Some design units that existed already in VHDL were modeled as behavioral models at the C++ dataflow model only. In some cases, simplified models were used. Table 7.1 provides an overview of the available abstraction levels per design unit. Both dataflow and architectural C++ models were largely reused in the second design and most VHDL units were automatically generated. Even VHDL blocks such as the FFT, microprocessor interface, and clock generator could be largely reused.

*Design Technology Reuse*

During the Festival design, UNIFY was developed as an extension library to OCAPI to improve design efficiency. Through its generation capabilities, UNIFY guaranteed reuse of the synthesis scripting and simulation flow as well as support for memory and IP wrappers. Only simple adaptations were required to reflect the change of processing technology and memory IP. The same instantiation semantics could be reused in C++. The object-oriented concept of OCAPI allowed also the creation and multiple reuse of a library of specific signal processing functions that used object manipulation techniques [Schaumont99a].

**Table 7.1.** Available abstraction levels for each design unit in the Carnival chip

|                               | C++ dataflow         | C++ architectural | VHDL RTL       |
|-------------------------------|----------------------|-------------------|----------------|
| Top level                     | Yes, but simplified  | No                | Yes            |
| Microprocessor interface      | Yes, but simplified  | No                | Yes            |
| Clock generator               | No                   | No                | Yes            |
| Burst controller              | Yes                  | Yes               | Yes, generated |
| FFT                           | Yes                  | No                | Yes            |
| Symbol reordering (SSR)       | Yes                  | Yes               | Yes, generated |
| Equalizer                     | Yes                  | Yes               | Yes, generated |
| Mapper                        | Yes                  | Yes               | Yes, generated |
| Demapper                      | Yes                  | Yes               | Yes, generated |
| Synchronization (timing)      | Simplified           | Yes               | Yes, generated |
| Synchronization (CFO)         | Simplified           | Yes               | Yes, generated |
| Synchronization (guard/clock) | Yes                  | Yes               | Yes, generated |
| Reference symbol generator    | Yes                  | Yes               | Yes, generated |
| Pilot generator               | Yes                  | Yes               | Yes, generated |

## Comments on Design Entry

Several designers were involved in the design of the Festival and Carnival ASICs. Background ranged from software engineering, systems, to algorithm, and HDL designers. The adoption of C++ required a careful review and adaptation of the OCAPI coding styles to familiarize both algorithm designers and HDL designers with the tool. Designers were hesitant about the usage of C++ [Bryant01, Gupta01]. The original coding style was largely based on SW engineering practices. As part of the extensions described in Sect. 7.1.2, we added a consistent and largely self-checking C preprocessor macrolayer on top of the original description style. The macrolayer together with the functional extensions resulted in a well-documented design methodology with clear coding style examples that actually resulted in the adoption of C++ during the first design and a smooth reuse in the second design. Besides the tool and documentation, clear methodology was crucial for dissemination and for convincing designers, which is also confirmed in [Sakiyama03].

## Comments on Code Size

We have also found the commonly faced code explosion during model refinement, but we have not suffered from it during the refinement process. The scalability and the code generation capabilities of our C++ design approach have saved us from costly iterations involving code rewriting. Dataflow and architectural description in C++ resulted in a significant reduction of the code size compared to the corresponding, automatically generated VHDL and Verilog counterparts (**Fig. 7.5**). The difference in line counts between the semantically equivalent RT-level C++ code and RT-level VHDL code is due to the abstraction mechanism offered by OCAPI. The class libraries used to represent a design at the RT-level C++ code encapsulate the concept of FSMDs (FSM with datapath) in a very concise manner.

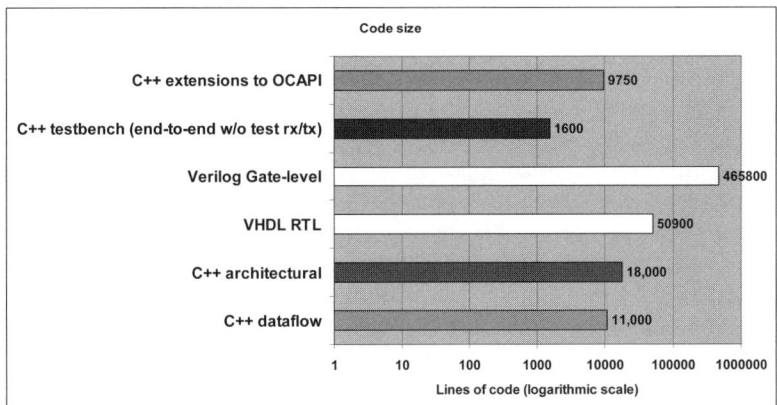

**Fig. 7.5.** The C++ dataflow and architectural description are significantly more compact than the generated RTL code. Our generic C++ extensions to OCAPI (UNIFY) represent a one-time effort that was shared across multiple designs

Because the system model is used for an end-to-end simulation, the OFDM transceiver is instantiated twice in the model. One of the two instances, however, is not refined toward an implementation and should be counted as part of the testbench. Hence, the extra 11,000 lines in the row correspond to the testbench.

## 7.2   Mixed-Signal System Simulation

Mixed-signal simulation, from a CAD point of view, requires the adoption of different simulation concepts, tailored to the specific needs of either the analog or the digital designer. This is certainly true, when focusing at the lower abstraction levels of design and bottom-up design, where circuit simulation is used [Zuberek92]. However, modern communication systems have become increasingly more complex, such that a top-down approach is required to evaluate system performance appropriately and early enough in a project time frame. In this case, behavioral simulation is required where the abstraction levels of both analog and digital circuitry are raised just enough for acceptable *behavioral* accuracy, but at much more favorable simulation complexity and simulation time.

First, we briefly review the design challenges in a mixed-signal system context (Sect. 7.2.1). Then, we introduce FAST, a high-level equivalent baseband simulator for analog systems (Sect. 7.2.2). In Sect. 7.2.3, we describe the coupling of FAST with OCAPI to enable a mixed-signal system-level simulation. We conclude with a specific modeling approach that exploits efficiently the system-level problem structure to speedup mixed-signal system simulations (Sect. 7.2.4).

### 7.2.1   Design Challenges and State of the Art

Front-ends of telecom transceivers perform the combination of downconversion, removal of interference by filtering, channel selection and amplification in the receive path, and upconversion and amplification in the transmit path [Wambacq02a]. These functions are distributed over the analog and digital domain. The amount of digital signal processing is steadily growing and its benefits in digital compensation were clearly illustrated in Chaps. 5 and 6. To predict the effectiveness of complicated digital compensation schemes, the analog and digital subsystems need to be simulated together [Wittmann03]. Transistor-level simulations are not feasible for this purpose. Even a cosimulation of digital blocks at a higher abstraction level with analog blocks at the transistor level, as, e.g., possible in Saber [Getreu90] or similar tools, is not feasible since telecom systems typically require Monte Carlo simulations that cover one or many packets consisting of thousands of bits [Jeruchim92].

Since time to market is crucial, early cosimulation of analog and digital high-level models would be beneficial to reduce the risk of design mistakes [Baltus03]. The main EDA players offer a bunch of high-level simulation tools for mixed-signal communication systems. Examples are SPW, COSSAP, ADS, and ORCA

[Crols95]. Still, research on high-level simulation is ongoing [Vandersteen00, Vassiliou99, Gielen02] to increase the simulation efficiency even more. In addition, mixed-level cosimulation becomes more solid, such as a cosimulation of circuit-level models in SpectreRF and system-level models in SPW [Moult98, Chen99].

Still, high-level simulation is not yet widely accepted for analog and mixed-signal systems. Analog designers have a blind confidence in circuit-level SPICE-type modeling. This reluctance has several reasons. First, high-level models of analog blocks often depend on low-level details. This complicates the construction of high-level models. In spite of their limited acceptance, high-level simulations increase the design productivity [Wambacq02b]. In [Halim94], Chadwick also questions the fact whether simulation speedup should be the only metric. Time consumed for modeling and describing a design are equally important.

A fast analog simulation environment has been provided already through FAST (Sect. 7.2.2). In addition, we develop here an efficient cosimulation approach with the digital dataflow simulator OCAPI (Sect. 7.1.1) and an efficient way of modeling typical mixed-signal interactions encountered in system-level simulations.

## 7.2.2    Fast System-Level Front-End Simulation (FAST)

Analog systems cover a large range of frequencies, in the WLAN case starting from the RF at 5–6 GHz over possibly intermediate frequencies down to baseband signals around DC. In Chaps. 5 and 6, we mentioned that many front-end components exhibit nonlinear behavior which results in harmonics and out-of-band intermodulation products [Wambacq98]. To cover all frequencies and imperfections in an adequate, accurate, and still efficient way, a joint optimization of the signal representation, signal processing, and software implementation is required. FAST offers such an approach [Vandersteen00]. This approach is a considerable improvement on top of classical approaches as described in [Mayaram00, Allen90]. A similar approach based on damped exponential base functions has been developed by [Vanassche01].

FAST uses a specific multirate multicarrier (MRMC) representation of signals. Each modulated carrier is represented with a complex equivalent baseband low-pass model. Harmonics and out-of-band intermodulation products are explicitly addressed as individual, separable signals.

FAST has been implemented in C++ and offers its functionality in the form of a library with execution kernel (EK) functions, dataflow semantics to connect these EK blocks, and a dataflow scheduler with efficient vector processing capabilities. A basic API toward MATLAB allows setup, configuration of blocks, and simulation.

The designer describes his/her front-end architecture through mathematical models based on the signal processing functions in the EK libraries. Functions range from simple mathematical operations to FIR and IIR filters, FFTs, and polynomial or Volterra series computations. The complexity of the designer entry depends mainly on the availability of models in libraries. Intrinsically, only mathematical base functions are provided. In the context of our development, we developed a number of simple models representing the behavior of typical analog/RF components such as nonlinear amplifiers, mixers, etc. The software automatically translates this description into an MRMC graph, optimizes it through graph rewriting techniques [Goffioul02], and executes it using static dataflow semantics. Vector processing is used when possible but sample-by-sample processing is automatically used when required, e.g., in feedback loops. This combined approach results in a speedup[147] against ADS or MATLAB-based simulation up to a factor of 700 for a WLAN front-end.

For the analog designer, the description of the behavior of analog blocks using basic mathematical functions may appear unusual. However, this can be solved by the introduction of macros that represent the classical linear and nonlinear behavioral models, e.g., a saturating third-order model of an amplifier or the conversion matrix for a mixer. Linear blocks can be characterized by their transfer function or system matrices (impedance matrix, $S$-parameter matrix). Hence, mismatch between blocks is taken into account. Nonlinear blocks are characterized by their polynomial nonlinear I/O relationship. In most cases, an initial translation of parameters into the mathematical form is required which can be performed as a preprocessing step in MATLAB.

The existing FAST implementation was taken as a starting point for the development of a mixed-signal cosimulation with the OCAPI dataflow computational model described in Sect. 7.1.1, as will be described next.

### 7.2.3  Extension to Mixed-Signal Cosimulation (FAST–OCAPI)

For an efficient mixed-signal system simulation, we need to build a joint view at the analog and digital parts that translates into a modeling and simulation approach adapted to the analog, the digital, and the system designer's needs. In existing tools (e.g., Ptolemy alone, ADS or SPW), we find three shortcomings that discourage full system-level simulations: poor modeling support, low simulation efficiency for mixed-signal, or a missing direct link toward digital implementation.

Based on the excellent domain-specific capabilities of FAST for analog/RF and OCAPI for digital system description, we decided to link the two simulators. For mixed-signal simulations involving two simulators, we can either perform cosimulation with a common cosimulation backplane or perform a true mixed-

---

[147] A superheterodyne receiver front-end with 256-carrier OFDM was simulated.

signal simulation using a single kernel. We decided for keeping separate FAST and OCAPI domains since they differ strongly in functional granularity, and the interaction between the two domains is only a fraction of the overall dataflow communication. Note that cosimulation at the system level can often resort to vector processing which reduces the switching overhead between both programs considerably.

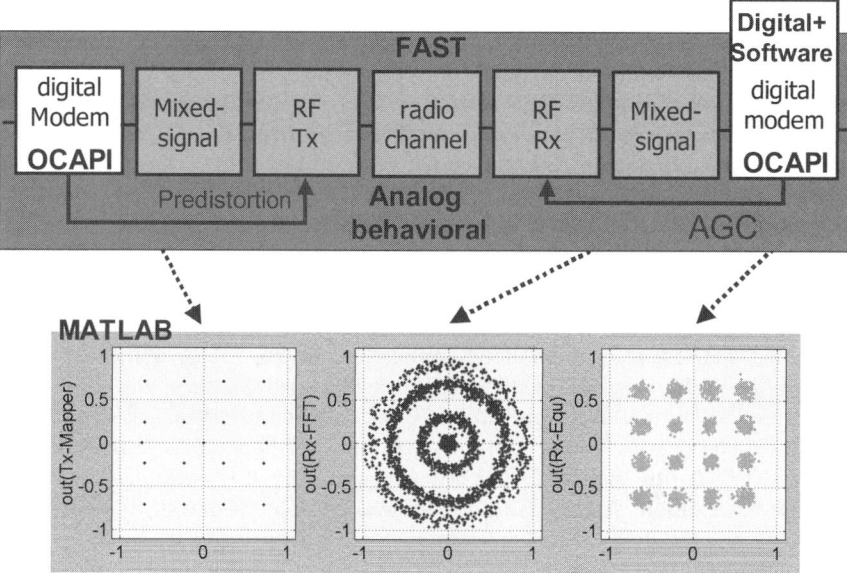

**Fig. 7.6.** Typical case of system simulation over a mixed analog/digital end-to-end link. Besides the data traveling in one direction, we may also find feedforward or feedback loops for digital compensation purposes ([Eberle03] © IEEE/ACM 2003)

**Fig. 7.6** shows a typical case of an end-to-end mixed analog/digital system simulation with a partitioning of functionality between OCAPI and FAST. The simulation chain consists of two digital portions described in OCAPI, and five analog portions described in FAST. A frequency-selective channel and AWGN are applied in the radio channel model. Predistortion establishes an additional feedforward link between digital and analog and automatic gain control (AGC) establishes a feedback link between digital and analog. All these interactions are well supported. Simulation results show constellation diagrams for the original transmit constellation (16-QAM and zero carriers), the constellation at the receiver before equalization, and the constellation at the receiver after equalization.

## Data Passing Between FAST and OCAPI

Since both OCAPI and FAST are based on the same dataflow scheduling along with generalized firing rules, the coupling interface has to implement queue management and data-type adaptation only. To maximize simulation speed and minimize computing platform dependencies, an interface class EKOCAPI with direct access to both its OCAPI and FAST I/O queues was preferred over a memory pipe, system pipe, or file-based solution.

The two simulators OCAPI and FAST are coupled through their FIFO queue interfaces. This requires a translation of the interface semantics and syntax. The syntax basically refers to the data types. OCAPI uses a real-valued class dfix type while FAST allows us to use either double or complex double data types. dfix is internally based on a double representation, such that a mapping of the real-valued type can be accomplished without any conversion loss. Fixed-point attributes assigned to the dfix type are, however, lost. A complex interface is currently realized by splitting a complex FAST pipe into two real-valued FAST pipes and connecting them to two OCAPI queues.

## Scheduling

Maximum independence between OCAPI and FAST partitions is achieved by slaving all OCAPI schedulers as subprocesses under a FAST master scheduler in a hierarchical way (**Fig. 7.7**). Both OCAPI and FAST partitions can be even developed and tested as stand-alone applications before the system integration. This preserves the localities of the dataflow scheduling in every partition leading to a simpler system-level schedule. At instantiation time, an object of the EKOCAPI class defines the connections between OCAPI and FAST partitions. At run time, i.e., during the simulation, it handles I/O queue management including multirate adaptation at the partition boundaries. The result is a distributed, hierarchical scheduling with lean communication at partition boundaries only, which translates into a low coupling overhead. The OCAPI scheduler is slave to the FAST scheduler to maintain the efficiency of the FAST block scheduling. Typically, vector processing can be much more efficiently applied in the domain of analog behavioral modeling than in the digital domain where block granularity and heterogeneity are usually larger (control-dominated multirate issues).

Between the OCAPI and FAST scheduler, a semantics translation is required. While FAST uses a block scheduling with priorities depending on the queue lengths at inputs and outputs, OCAPI relies on firing rules implemented by the user mainly applied to the inputs. The translator is called once in every scheduling cycle (referring to the master scheduler in FAST) and maps OCAPI queues to FAST pipes and vice versa depending on the pipe sizes defined in FAST.

The scheduler is prepared for vector-based and sample-by-sample scheduling. With the definition of appropriate initial delays in the digital OCAPI-based subsystem, mixed-signal connections are not limited to forward connections. The coupled FAST–OCAPI simulator can be used to implement a feedback topology as well, for example modeling AGC with digital steering of an analog variable-gain amplifier. The insertion of an appropriate initial number of samples (corresponding to the loop delay) in the feedback dataflow queues (or pipes) makes sure that the dataflow scheduler does not end in a deadlock situation at the start of the simulation. Of course, block-based dataflow handling becomes limited locally to the number of samples injected like this.[148]

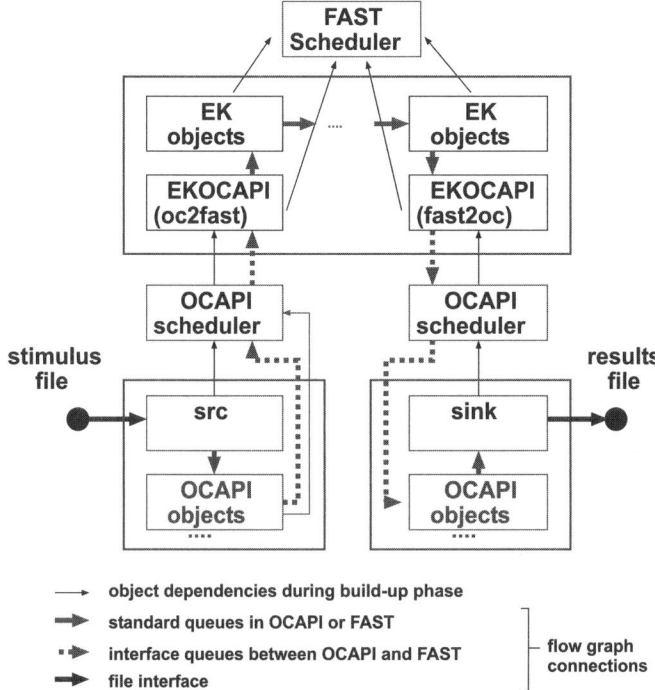

**Fig. 7.7.** A two-level scheduling approach is used. The FAST scheduler acts as a master and directly coordinates all FAST EK objects. Digital subsystems are hooked up as EKOCAPI objects with a locally embedded scheduler. The example shows a typical connectivity case for end-to-end simulation with digital stimuli read, processed in a digital–analog–digital chain and received bits stored in a results file ([Eberle03] © IEEE/ACM 2003)

### Software Architecture and User Interaction

During the development of the system model, its simulation and evaluation, the designer distinguishes the following phases:

- System description: describing the structure of the system as a network of building blocks including hierarchy

---

[148] Important limitations apply to block-based operations like FFT primitives as part of FAST models. FFT block size and delays in feedback loops must be properly dimensioned to avoid deadlock situations. The correct dimensioning is currently up to the designer; the tool only informs about the status of all pipes in the case of a deadlock. Note also the choice for a too large FFT block size means also that the transient behavior with respect to the loop feedback would not be modeled correctly; hence, a smaller block size should be chosen.

- System configuration: configuring parametrizable building blocks for a given evaluation task
- System interfaces: interfacing different building blocks with each other and with the testbench
- System scheduling and simulation: determining the order of execution, initial conditions and stop conditions with a dynamical schedule
- Result extraction/visualization: extracting simulation results such as queue statistics, output values, state information from building blocks or queues

The designer specifies the front-end architecture in C++ based on FAST primitives or derived classes that represent analog/RF macros based on FAST primitives. The C++ description is compiled and linked into a dynamically linked library (DLL) that can be executed from within MATLAB with or without OCAPI, depending on whether there is a digital subsystem involved in the system description. Configuration and simulation can be controlled from the MATLAB interface.

The cosimulation enhancements of FAST and OCAPI have been integrated, together with the tools themselves, into a framework with MATLAB and Java (**Fig. 7.8**). MATLAB is used as user interface for specification, model configuration, and result visualization. In a newer version, code generation was implemented in Java.

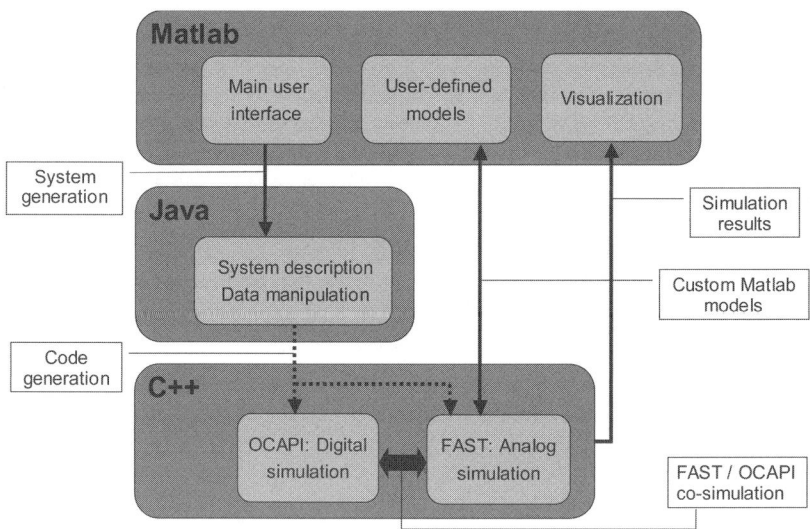

**Fig. 7.8.** Integration of the C++-based OCAPI and FAST tools with cosimulation enhancements

*Application Examples*

Several system cases were designed in the cosimulation tool framework and successfully demonstrated:

- A mixed-signal receiver demonstration around a zero-IF front-end and a digital OFDM receiver modeled at a high abstraction level (see also Chap. 5)
- A complete end-to-end simulation chain for a wireless LAN link with a receive front-end exhibiting all relevant analog nonidealities; real circuit parameters were used. The complete end-to-end simulation is able to simulate three OFDM symbols per second ($3 \times 80 = 240$ samples) (see also Chap. 5)
- A complete end-to-end link with a detailed receive front-end model and mixed-signal AGC in a feedback configuration **(Fig. 7.9)** (Sect. 7.2.4); the mixed-signal AGC part has been described in Chap. 5

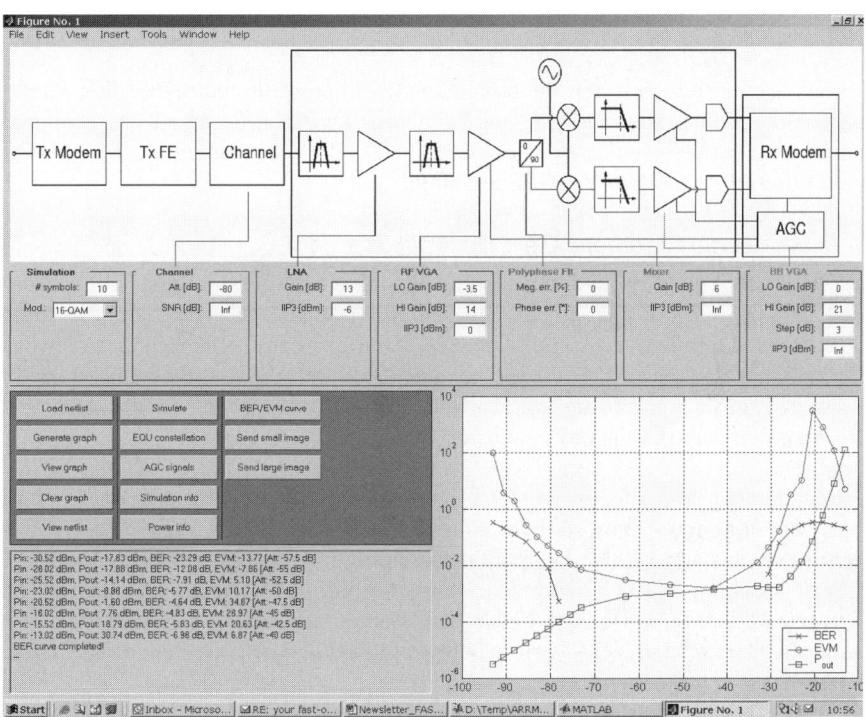

**Fig. 7.9.** Cosimulation of system-level properties such as EVM and BER across an end-to-end link with a mixed-signal feedback automatic gain control loop and a detailed receiver front-end model

**Fig. 7.9** shows a customized user front-end and end-to-end simulation chain, shown in the upper part. Several blocks can be parameterized through the user interface with parameters shown below the chain diagram. The lower-left part shows a set of action buttons and a status window. The lower-right part gives a snapshot of obtainable results, in particular bit-error rate (BER), error vector magnitude (EVM), and output power plots as a function of the receive power. The typical front-end behavior with acceptable BER and EVM for medium receive powers is observed; for high receive powers, saturation effects are obtained; for low receive powers, noise reduces receive quality.

### 7.2.4  Efficient Mixed-Signal Modeling Techniques

Efficient modeling of mixed-signal feedback topologies, as they occur in the case of mixed-signal AGC loops[149] with digital estimation and distributed analog steering, is particularly tricky and often slow in available simulation tools [Busson01, Busson02, Gardner96]. In particular, the modeling of this topology must exhibit adequate accuracy when describing higher-order effects of the front-end while still exhibiting sufficient simulation speed to perform simulations over entire packets. Guaranteeing adequate accuracy is a well-known challenge [Leenaerts01].

In this section, we describe the simulation of a complete end-to-end link with a particular focus on the modeling and simulation approach for an RF variable-gain amplifier in the signal path of a receiver that is controlled in a feedback configuration by a digital AGC [Eberle03].

*A Mixed-Signal Automatic Gain Control Loop*

We will use a wireless LAN end-to-end link as a system under test. The end-to-end link consists of the mixed-signal transmitter and receiver, and the radio channel model in between. The boundary of our example is the interface between multiple-access control (MAC) and physical layer (PHY), where payload data are presented in the form of data packets. This allows us to model complete transmission bursts at the physical layer.

The radio channel model provides its signal to the receiver, consisting of a zero-IF receiver front-end followed by a digital OFDM baseband demodulator. The receive chain starts with an RF section containing LNA, controllable RF VGA and filters, providing an RF section to a pair of direct downconversion mixers. The downconverted in-phase and quadrature signals are filtered and amplified in controllable baseband VGAs before being digitized in A/D converters (**Fig. 7.10**).

First, we address the modeling of the RF VGA itself. Second, we integrate it into a model of the entire RF section. Next, we describe the architecture of the digitally controlled AGC loop. Finally, the implementation of the models (e.g., the

---

[149] Feedback loops from the digital to the analog part can also appear in the context of other digital compensation techniques. However, in contrast to the AGC case, they can often be translated into purely digital techniques without a feedback to the analog front-end.

translation of the continuous-time blocks such as linear transfer functions and $S$ parameters into digital filters) is addressed.

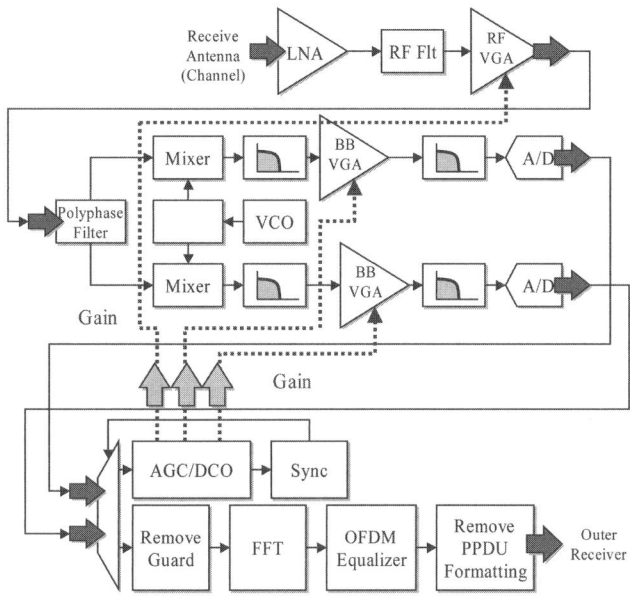

**Fig. 7.10.** Receiver topology with distributed gain control for baseband VGA and RF VGAs ([Eberle03] © IEEE/ACM 2003)

## Modeling Abstraction of the VGA

The VGA model was intended to reflect the outcome of measurements of an existing 5-GHz BiCMOS RF VGA design. This circuit can switch between two gain values (low and high), by selecting one out of two differential pairs that each have a different bias current. High- and low-gain setting are controlled digitally. The dependency of the input matching $S_{11}$ on the gain setting was specifically targeted due to its large impact on the gain. $S$-parameter characterization was performed at $f_0 = 5.25$, 10.5, 15.75 GHz for low and high gains ($S_{21} = -244$ dB, $S_{21} = 6.76$ dB, $S_{11} = -10.9$ dB, $S_{11} = -5$ dB).

The VGA is treated as a three port with external ports $(x_1, x_2)$, $(y_1, y_2)$, and $(c_1, c_2)$. These are the RF input, RF output, and control input, respectively. Indices 1 indicate incident waves and 2 indicate reflected waves.

The VGA model starts from an ideal three-port model (ports $u$, $v$, $w$) encapsulated by a frequency-dependent set of $S$ parameters and a cubic nonlinearity at the output. In steady state, the time-domain multiplication can be treated as a constant multiplication instead of a convolution in the frequency domain. When considering the digital receiver later, we will see that we are not interested in modeling transient effects accurately. Consequently, we only have to model these two steady states accurately, each corresponding to one gain setting. For simulation efficiency, we can now translate the time-variant three port into two time-invariant two ports, one for each gain setting. In each model, the gain control node translates into an internal constant (**Fig. 7.11**).

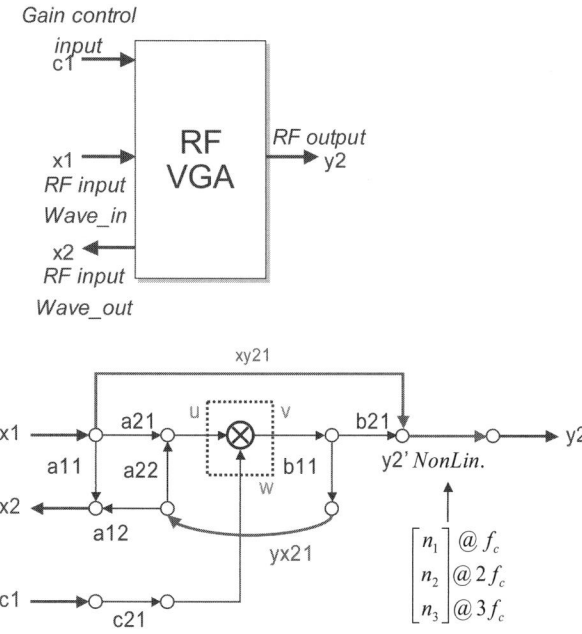

**Fig. 7.11.** RF VGA black-box model with relevant ports (*above*) and the white-box computational model (*below*) ([Eberle03] © IEEE/ACM 2003)

We can neglect intermodulation products between RF signal and gain control since transients at the gain control input are neither too short nor too long. Moreover, gain changes are issued from the digital receiver which inserts a tolerance time period for any sort of transient effect following a gain change. Hence, we do not need to model the transient effects accurately. We can consider the VGA output $y_2'$ explicitly as linear in its two steady states and model the output compression behavior of $y_2$ as a polynomial nonlinearity following $y_2'$.

Between internal nodes ($u$, $v$, $w$) and external nodes, $S$-parameter networks are placed to represent the effect of input and output impedances at RF. The input dependency on the actual gain and RF feedthrough effects are modeled using explicit branches $yx_{21}$ and $xy_{21}$, respectively. Matching the measurements, reflected waves at the baseband gain control input and the output were neglected, resulting in a set of two transfer functions:

$$y_2' = \left( \frac{xy_{21}}{c_1} + \frac{a_{21}b_{21}c_{21}}{1 - a_{22}yx_{21}b_{11}c_{21}c_1} \right)(c_1x_1),$$

$$x_2 = \left( a_{11} + \frac{a_{21}a_{12}yx_{21}b_{11}c_{21}c_1}{1 - a_{22}yx_{21}b_{11}c_{21}c_1} \right)x_1,$$

(7.1)

where $(a_{11}, a_{12}, a_{21}, a_{22})$ represent the signal path input $S$ parameters, $(b_{11}, b_{21})$ represent the signal path output $S$ parameters, and $(c_{21})$ represent the control path input $S$ parameters.[150]

Finally, these equations can be resolved such that the RF VGA input–output behavior is represented by two transfer functions per gain setting $c_1$:

$$y_2' = \beta_{xy}(c_1)x_1,$$

$$x_2 = \beta_{xx}(c_1)x_1.$$

(7.2)

### Embedding of the VGA Model in the RF Chain

The RF receiver section starts with the antenna followed by the LNA, a bandpass filter and the previously described RF VGA. The RF VGA finally connects to the downconversion mixer. This section can be treated as a three port with the antenna and the gain control signal as inputs and the mixer input as output.

For behavioral simulation, frequency-dependent $S$-parameter models can be used to describe the linear time-invariant behavior only. Our RF VGA model, however, is time-variant and nonlinear. Hence, we cannot apply linear system theory to translate the cascade into an I/O transfer function (**Fig. 7.12**).

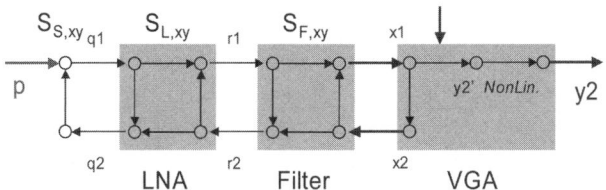

**Fig. 7.12.** A cascade of LNA, filter, and the already discussed RF VGA represent the RF section between antenna and downconversion mixer
([Eberle03] © IEEE/ACM 2003)

Still, we can represent the steady-state responses for the VGA in high- and low-gain mode and hence transform the VGA into two two-port models. The model allows us to decouple the cascade into a linear section up to node $y_2'$ followed by

---

[150] Not all matrix fields need to be modeled for sufficient accuracy.

a nonlinear transfer function to produce final output $y_2$. The linear part can be solved numerically when $S$ parameters are given. Our specific case is also analytically tractable. Skipping the lengthy result for the general case, we only report the result (7.3) for the special case when conjugate matching is applied to the LNA input and the filter output and no return loss in the VGA ($\beta_{xx} = 0$) is present:

$$\Rightarrow y_2' = \frac{S_{F_{21}} S_{L_{21}} \beta_{xy}}{\left(1 - S_{L_{11}}^2\right)\left(1 - S_{S_{22}}^2\right)} p. \tag{7.3}$$

Obviously, the forward gain consists of the cascaded forward gains of its components and the input and output matching losses only.

### Mixed-Signal Integration with the Digital Gain Control Loop

The AGC loop contains two modeling problems: crossing the analog/digital boundary in both directions and correct implementation of the loop delay.

We propose a digital architecture containing a run-time controller that handles all saturation scenarios and a configuration mapper that uses both run-time information obtained through digital estimators and also design-time information. Digital estimators provide signal power and DC offset estimates during acquisition phase. An extended cascade analysis at design time provides the optimum front-end configuration for each RF input power level to the configuration mapper. The configuration mapper controls the VGAs in the front-end. This control signal closes a mixed-signal feedback loop. AGC and DCO become thus subject of both the overall impulse response of the involved analog forward and feedback paths, and the digital implementation and algorithm delays.

Our approach is seen as an instantiation of a generic method for the coevaluation of a digital portion with an analog/RF portion in which particular digital knowledge is added to simplify the analog/RF modeling complexity. In general, designer knowledge is required to reinstantiate this technique in other cases. Hence, while the overall cosimulation and execution of the models is supported by the FAST–OCAPI simulation engine, the modeling itself is not supported by automation steps.

Our solution is noncontinuous in time and thus insensitive to the shape of the impulse response, but it requires the analysis of the overall impulse response. The first 2 µs of the acquisition preamble are subdivided into three phases, each consisting of an estimation and compensation phase (**Fig. 7.13**). The configuration mapper will adapt the VGA gain at the start of each compensation phase. For the quality of the digitally estimated gain (in the context of the digital AGC), it is better to reduce the number of estimation samples to those that are stable than to take into account samples that are affected by the gain change transients.

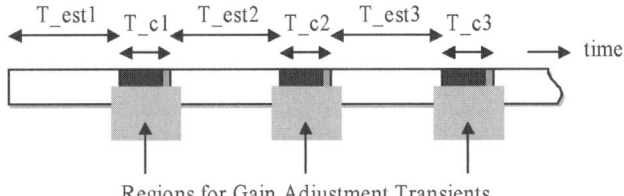

Regions for Gain Adjustment Transients

**Fig. 7.13.** The compensation phase length depends on the gain adjustment, transients
of the VGA, the filters, and the digital implementation delays
([Eberle03] © IEEE/ACM 2003)

Using our more accurate joint analog/digital receiver model, we can now refine
the results of the statistical cascade analysis. It is mainly due to the correct
modeling of nonlinearities and of the cascaded nonlinearities that the accuracy of
this joint analog/digital simulation is improved compared to a statistical cascade
analysis. Moreover, from the mixed-signal model, we can obtain timing parameters such as loop delay and settling time, which is not possible with the cascade
analysis.

### Implementation Aspects in the Cosimulation in FAST and OCAPI

The computational graph of the RF section shows a typical model consisting of
library kernel blocks with basic signal processing functions and queues linking
these blocks to each other. In this example, two filters (EKFIR) implement the
transfer functions for high and low gain of the RF section between the antenna
input $x_1$ and the VGA output $x_2$. Their respective results are then interpolated
depending on the transient weights of the step response on the gain control input
$c_1$. The nonlinear part is not shown (**Fig. 7.14**).

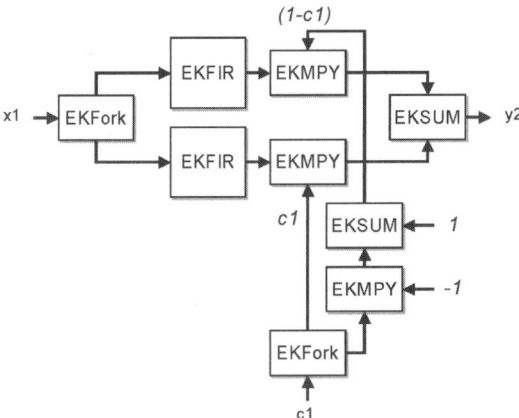

**Fig. 7.14.** The computational graph of the RF section links execution kernel library objects
with signal queues ([Eberle03] © IEEE ACM 2003)

*Simulation Performance and Results*

A full transmit/channel/receive end-to-end simulation for 80,000 payload bits takes about 25 s on a Pentium III with 512 MB RAM including all pre- and postprocessing. Results can be immediately translated into figures showing relevant design information. For example, **Fig. 7.15** shows the front-end behavior for a sweep of the relative RF input power from −100 to +50 dB.[151] We can observe high BERs for low and high input power levels corresponding to noise and saturation problems, respectively. The estimated signal power illustrates this by its saturation behavior at both edges. We can observe when the RF VGA switches from high to low gain and the actual baseband VGA setting over the complete RF input range.

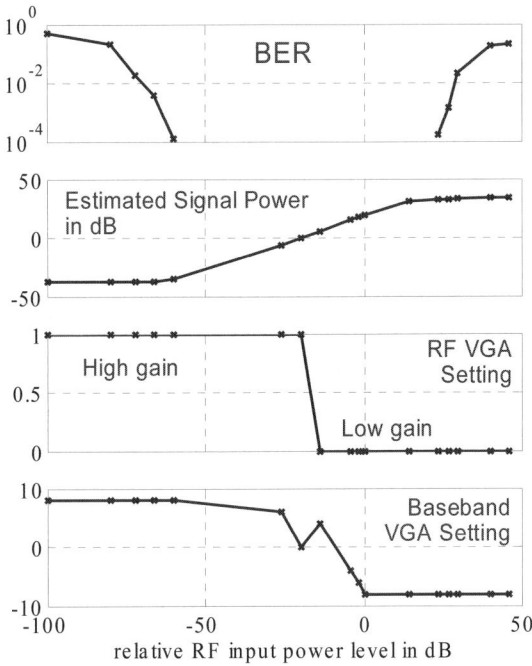

**Fig. 7.15.** The full model allows the verification of the actual receive range of the receiver in detail. Saturation and noise effects are revealed ([Eberle03] © IEEE/ACM 2003)

Besides the optimum timing, mixed-signal AGC (Chap. 5) allows also tradeoffs regarding the required gain range. Gain range, step size, and tolerance of the VGA have an impact on the average signal level at the A/D converter. The dependency of the BER on the RF input signal level has been analyzed with 10-bit A/D quantization, −8 to +8 dB gain range in 2-dB steps for the digital baseband VGA.

---

[151] These numbers do not refer to actual design values but originate from an initial feasibility study. Relevant numbers are, e.g., shown in **Fig. 7.9**.

The run-time estimator determined the optimum setting for each power level. For low input power levels, quantization and noise effects become visible while, for high input power levels, saturation effects come into play. The linear signal power estimator becomes biased both for weak and strong input signals, urging the need for the other nonlinear branch of our AGC algorithm (Chap. 5).

To prove efficiency and usefulness of our approach, we selected two representative cases of design exploration tasks that a system designer faces in wireless communications systems. End-to-end link bit-error analysis involving the analog nonidealities and a mixed-signal optimization problem around the VGA accuracy was successfully demonstrated. Results of these experiments were presented above. In all simulations, the mixed analog/digital AGC loop is involved at the beginning of each receive burst to find the optimum gain within the constraints specified by the designer.

## 7.3  Design-Time Run-Time Techniques

Design-time run-time (DT/RT) techniques were introduced in Chap. 5 for the AGC/DCO design and in Chap. 6 for the transmit chain control. We consider DT/RT techniques as a metaconcept in design such as divide and conquer, hierarchical decomposition, division of communication and computation, etc. This concept can be applied to a large variety of problems from different domains.

### 7.3.1  Multiobjective Design-Time Optimization

Optimization is an activity that aims at finding the best (i.e., optimal) solution to a problem. For optimization to be meaningful there must be an objective function to be optimized and there must exist more than one feasible solution, i.e., a solution which does not violate the constraints. In our case, we face multiobjective optimization problems [Papalambros00, Allais43, Pareto06, Lampinen00]. Following the Pareto optimality principle, the result of such an optimization is not a single solution if more than two orthogonal objectives exist but a set of optimal points situated on a Pareto hyperplane (also Pareto front).

In classical system optimization, a single solution is targeted at design time which often describes the worst-case behavior. However, in the case of conflicting constraints such as performance (quality) and power (cost), such a decision cannot be made at design time since it depends on input variables that are only determined at run time. Still, at design time, pruning of all possible solutions in the Pareto design space to those on the Pareto front is performed, which results in a significant reduction of the number of configurations to store.

In Chaps. 5 and 6 based on the joint AGC/DCO example and the transmitter example, respectively, we have shifted the complexity of this optimization step to design time (**Fig. 7.17**).

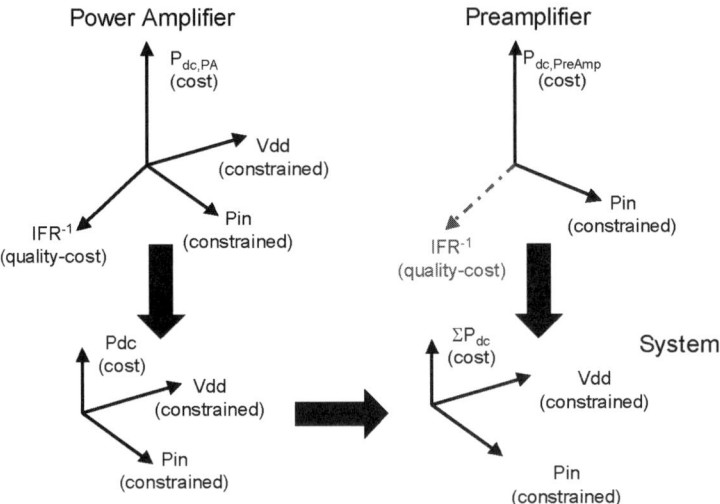

**Fig. 7.16.** Merging of two separate quality–cost characteristics into a single Pareto front eliminates an axis and can reduce the overall complexity. In the given example, $P_{in}$ is a joint parameter for the preamplifier and the power amplifier which is eliminated through the merger

Figure 7.17 illustrates the distribution of tasks between design time and run time for the AGC/DCO example. The linking element between both phases is the look-up tables linking configurations and input variables.

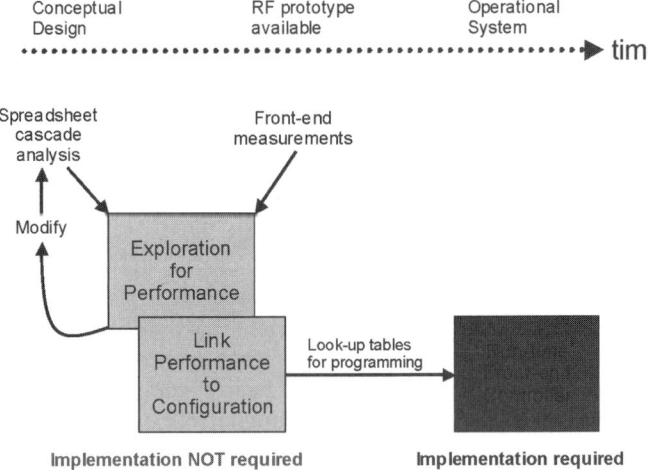

**Fig. 7.17.** Flow across design time and run time for the AGC/DCO example

## 7.3.2 An Architecture for Run-Time Control Assisted by Design-Time Knowledge

The database information extracted at design time can be used at run time to reach optimal configuration decisions. Based on the two examples at the receive (Chap. 5) and transmit side (Chap. 6), we derived a generic architectural template for a configuration manager (**Fig. 7.18**), for which also a patent has been filed [Eberle02d].

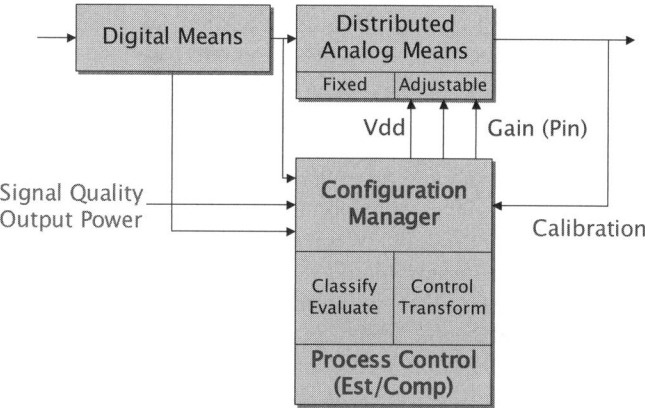

**Fig. 7.18.** Instantiation for the transmit case of a generic architecture template of a run-time configuration manager assisted by design-time information

At design time, a design space exploration is carried out as a function of all run-time accessible parameters. The optimum configuration for each desired input signal occurrence is computed. An important aspect is the inclusion of tolerance margins in the decision process; this allows an a priori analysis of the effect of imperfections on the input signal estimates or on the output control means (see AGC in Chap. 5). Tolerance margins typically lead to a reduction of the useful granularity of configurations since neighboring configurations may not lead to distinguishable solutions. Additional quality (e.g., minimum transmit SINAD) or cost constraints (e.g., power consumption) can be added in the optimization process.

At run time, a loop process is executed. Input signals are digitally estimated and classified using – in general – nonlinear functions. A particular algorithm may be selected based on this classification. The algorithm computes an index to the optimum configuration and retrieves the corresponding steering parameters from the database. The optimum configuration is applied. Finally, the process state is updated compared to, e.g., an optimization goal.

Our architecture embeds estimation, compensation, and calibration into a single process and allows the adaptation of the process timing or process steps. Our approach addresses in particular the control of distributed analog means, for example distributed resources in a front-end architecture.

## 7.4  Conclusions

In this chapter, we have addressed design methodologies and design technology developed for interdisciplinary design in the field of digital, mixed-signal, and system design. In particular, we contributed:

- A practical and effectively used C++-based design flow based on OCAPI and a proprietary library extension for a smooth transition to HDL design and high-level design entry; based on two actual ASIC designs and other design experiments, we also gained experience in reuse patterns
- A practical and fast cosimulation solution for the fast system-level codesign of analog/digital subsystems; the tool was demonstrated for end-to-end link-level simulations illustrating the impact of front-end nonidealities on the digital receiver
- Effective modeling techniques in the case of mixed-signal functionality that exploit system-level knowledge to reduce simulation time; this was exemplified on the system-level simulation for a mixed-signal AGC loop
- A design-time/run-time methodology and architecture for digital control of analog and mixed-signal blocks; two cases were illustrated: AGC in a receiver and optimum power–performance configuration of a transmit chain

All design methodology and design technology described here have been developed based on a concrete request that occurred during the design of the WLAN components and system that is at the basis of our work (**Fig. 7.19**). All techniques and tools were demonstrated using actual design or prototype demonstrators. The digital design solutions were codesigned and applied during the Festival and Carnival ASIC designs. The mixed-signal solutions were demonstrated with several software simulation demonstrators at various occasions inside and outside IMEC. The design-time/run-time techniques instantiated in the AGC context were used in system demonstrators; their instantiation for the PA is currently introduced in demonstrations.

In the meanwhile, several of the practical ideas especially in the digital design flow have been taken up by tool vendors and integrated, e.g., into SystemC-based digital and system modeling tools or SystemC-based synthesis. Still, a hesitance can be noted with respect to explicitly expressing signal properties such as clock and test mechanisms at an earlier stage in the flow or using a common design base. Similarly, little or inconsistent support for linking the models with scripting solutions is found. Both lead to an additional step in communication which can result in errors and additional design time.

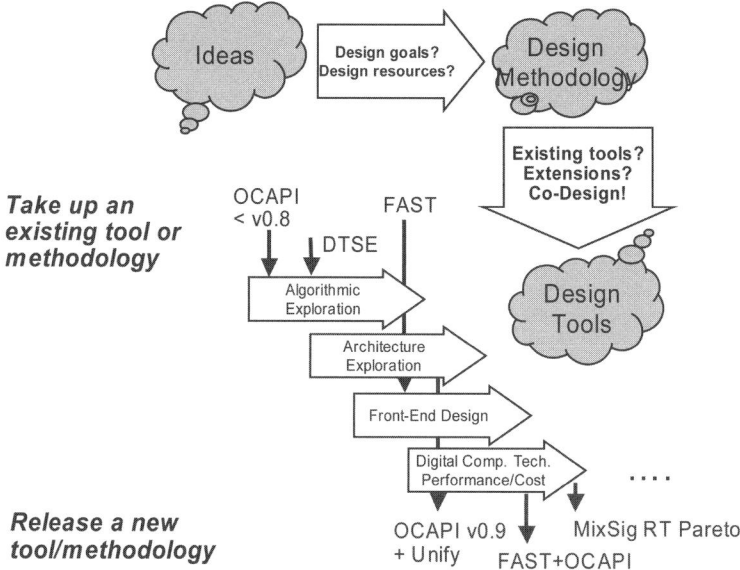

**Fig. 7.19.** The development of design methodology and design technology followed the actual design steps and was triggered by design problems encountered in practice. Our solutions focused on interdisciplinary extensions of existing point tools in the cases of OCAPI/UNIFY and FAST/OCAPI

Based on our mixed-signal system-level simulation and modeling work, we described a concept for a mixed-signal design flow [Wambacq00]. In this approach, the system-level simulation allows early top-down exploration but also bottom-up integration of measured or modeled circuits and devices. Our system-level simulation can easily be coupled to circuit-level-based model extraction, e.g., DISHARMONY [Dobrovolný01], models based on generalized Volterra series [Goffioul02], or models derived using a best-linear approximation approach [DeLocht05]. However, further integration is beyond the research scope and more a question of the EDA industry to offer a solution based on integration of these modular tool components since it mainly requires standardization of tool interfaces with respect to these models of computation. On the other hand, the adoption of novel EDA modeling and description techniques by designers has been and is still known as a fairly slow process [Diesing94, Halim94].

The application of multiobjective (Pareto) optimization in the context of wireless systems is rather new and mainly driven by the increasing drive for lower power consumption at equal or increasing performance. Substantial extensions to the initial work on multiobjective optimization on wireless systems were carried out in the work of Bougard.[152] A large number of mathematical tools have become

---

[152] See [Bougard04]. A separate Ph.D. dissertation on Bougard's work is subject to appear in 2006.

available through advances in evolutionary computing and multiobjective optimization. Still, quite a number of fundamental and generic questions can be asked when it comes to multiobjective optimization as a practical design approach or in an application-specific context [Lampinen00]: How should the Pareto points be preferably distributed over the Pareto front? Can we synthesize them? How do we select them? How to select an initial set of points and extend it? Should extreme values for individual objective functions be included in the Pareto set as absolute comparisons? How do we compare two Pareto sets if the size of the sets is different? This is a question for specific metrics in the case of multiobjective optimization. How to express the quality of a Pareto solution as a function of such metrics?

# 8 Conclusions and Further Research

*The carpenter is not the best*
*who makes more chips than all the rest.*
Arthur Guiterman, 1871–1943.[153]

*As for the future, your task is not to foresee it, but to enable it.*
Antoine de Saint-Exupéry, 1900–1944.[154]

Time has come to close the circle between the obtained results and the proposed objectives. Citing Sect. 1.2, we identified two main objectives. In the context of OFDM-based 2G wireless LAN and similar future flexible broadband transmission schemes, we strived to:

- Develop functionality and scalable architectures enabling low-cost low-power implementation including the mitigation of all relevant practical nonidealities
- Establish a mixed-signal exploration and design flow that enables an efficient process for this design task, ultimately speeding up time to market

We now summarize the major techniques that we have developed to meet these objectives. Section 8.1 reviews our contributions in the application domain and draws conclusions. Section 8.2 constitutes the same in the design methodology and technology context. An integral part of research is that along with the above answers a whole set of new questions has escaped the box of Pandora. Ideas for continuing and new promising future research are presented in Sect. 8.3.

---

[153] Well, at first sight, this may sound strange in the microelectronics business, where more chips might mean more money. At second sight, however, the challenge lies in the tremendous increase in complexity *per* chip. In that sense, this quote from the *wooden times* still holds perfectly in our *silicon era*.

[154] Not astonishingly for a man of deed, Saint-Exupéry considers getting his hands dirty to pave the path for future generations a necessary and honorable step.

## 8.1  Contributions to Application Design

Design for portable consumer-oriented applications is subject to cost and power consumption constraints. The demand for improved performance and features while satisfying severe constraints is high. The multitude of emerging applications creates a significant pressure to design scalable and (re)configurable solutions. Hence, we have aimed at architectural and functional concepts that prove to be scalable in performance, power consumption, and cost.

We have advocated for a crossdisciplinary approach of design and hence evaluate our contributions based on their principal property: scalability and flexibility for the sake of performance/power improvement.

We have shown that design for scalability and flexibility across several levels is key for a performance/power-efficient system. Moreover, our results prove that flexibility and scalability do come both at an affordable design and manufacturing cost. We have addressed fundamental issues of scalability and flexibility at nearly[155] all relevant levels of design at the physical layer:

- *Scalability at the link level with respect to application and environment*
  In Chap. 6, we have introduced a mixed-signal run-time/design-time concept for a performance/power-driven link-level optimization of the transmitter. This approach used flexibility at several communication layers and across digital and analog.
- *Scalability and flexibility at the mixed-signal terminal level*
  In Chap. 5, we have illustrated a scalable digital-driven design-time/run-time concept for AGC and DC offset compensation in the receive process of the terminal. We have also shown that this technique is applicable for several front-end architectures. Similarly, in Chap. 6, we have shown digitally controlled solutions for trading off power vs. performance for the transmitter front-end.
- *Scalability at the digital chip level*
  In Chap. 4, we have presented a baseband transceiver architecture concept based on a distributed multiprocessor architecture that has proven its scalability and reusability in two ASICs and one FPGA design with WLAN as driver application.
- *Scalability and flexibility at the digital block level*
  In Chap. 4, we have introduced novel solutions for the receiver acquisition, for reuse in the transmit/receive data reordering processes, an efficient scalable FFT, and algorithm/architecture tradeoffs for the equalization. We have illustrated that a significant amount of flexibility

---

[155] We did not investigate the *actual design* of analog scalable blocks, though we analyzed practical cases of existing analog/RF components. A paradigm shift to scalable analog block design is motivated in the section on future work.

at several conceptual levels can be incorporated in a *parametrizable ASIC* at acceptable design cost.[156]

Particularly for OFDM-based WLAN, proof was needed that an efficient digital solution would be feasible as well as that front-end specifications could be relaxed to an affordable level. Our digital and mixed-signal solutions significantly have contributed to reaching this goal. The resulting ASIC designs – Festival and Carnival – were both the first integrated WLAN baseband processor designs at the time. Particularly, the first design (Festival) was recognized as one of the ISSCC milestones in the field of Signal Processing in the ISSCC 50th Anniversary Virtual Museum[157] in 2002.

## 8.2 Contributions to Design Methodology and Technology

We conceive design methodology and technology as a *practical* means to achieve an improvement in design efficiency in a particular application domain. As a consequence, developed methodologies have first been applied in an ad hoc way or to a particular problem only. Later, methodologies were proven with a design iteration or a second design. In a few cases, methodologies were effectively instantiated in prototyping tools or as enhancements to existing prototyping tools.

In all cases and hence in alignment with the design rationale devised in Chap. 2, we have investigated the role, applicability, usefulness, and reusability of a particular technique. Moreover, we have assessed the process of development, transfer, and use of a particular methodology.

This section summarizes three aspects. First, we have presented a practical system-oriented mixed-signal design flow. This includes our contributions to digital design refinement, to mixed-signal cosimulation, and to behavioral models for analog components. Second, we have extended the problem of design refinement to design space exploration for (re)configurable mixed-signal systems. The multiobjective optimization challenge has been approached with a design-time/run-time partitioning methodology. Finally, we comment on methods to introduce and exploit interdisciplinary knowledge and skills in the design process.

---

[156] Acceptable design cost mainly refers to design complexity. This implies additional control complexity. Additional datapath complexity is very limited. Functional scalability is part of the specification: this would imply parallel selectable implementations when we compare with dedicated designs. Given the number of desired configurations, this becomes simply impossible.

[157] See http://www.sscs.org/History/isscc50/signal/index.html.

### 8.2.1    A Practical System-Oriented Mixed-Signal Design Flow

In Chap. 7, we have introduced a system-oriented mixed-signal design flow (**Fig. 8.1**). This flow aims at a codesign approach for analog and digital design from the transaction level on to the register transfer level and its sample-based[158] equivalent on the analog side. We have not been able to instantiate all ingredients and transitions of this flow. Instead, we focused on three transitions:

1. Digital design refinement from the system level to RTL
2. Mixed analog/digital cosimulation at the system level
3. Embedding of behavioral gray-box models for analog components in the system-level simulation

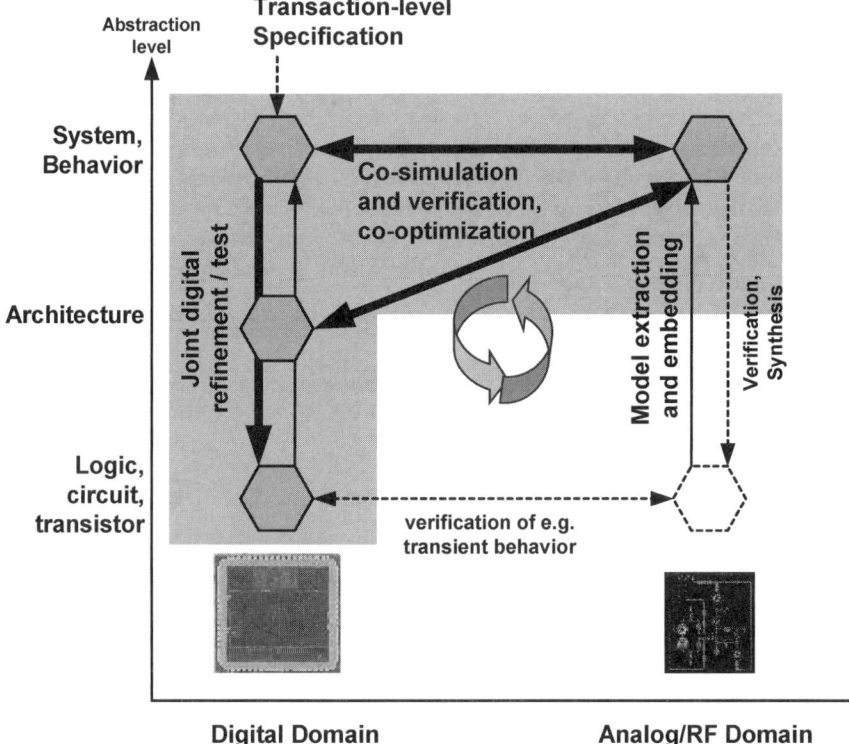

**Fig. 8.1.** We have proposed instantiations for part of the complete mixed-signal design flow with a major focus on the mixed-signal system level and digital refinement challenges. Since research is still needed for other partial flows, we were forced to use traditional or ad hoc techniques there

---

[158] We refer here to a model that is accurate with respect to the digital sampling process; i.e., we do not need to describe the behavior more accurate than seen by the analog-to-digital converter.

The center of activities was chosen to be on the digital and system-level side, since we identified the first major risk factors as the VLSI design for OFDM and later as a mixed-signal codesign for digital compensation of front-end nonidealities. Digital refinement and the mixed-signal cosimulation methodology were embedded as major extensions to the existing purely digital and purely analog in-house research tools OCAPI and FAST, respectively. Both extensions required modifications in the tool kernels invisible to the user and user-accessible libraries or language features. Note that we mainly addressed the hardware design part. For HW/SW codesign, we can refer to existing, but here not exploited capabilities in OCAPI/OCAPI-XL [Vanmeerbeeck01].

Obviously, the lower-right analog counterpart in **Fig. 8.1** is a desirable enhancement but was not the scope of our research. Research, in particular missing aspects of this flow, is suggested in Sect. 8.3. Note, however, that, since not yet supported parts of the flow such as transient measurements, component characterization, or circuit-level model extraction were needed in actual designs and tests, these were performed in an ad hoc way.

### 8.2.2 Methodologies for (Re)Configurable Mixed-Signal Design

Multiobjective design space exploration results in a Pareto-optimal front of configurations. This design-time approach has been widely used in engineering. However, we have extended this idea with two important practical items for the field of integrated communication systems:

1. A process for a clear separation of the design-time exploration and pruning, a run-time calibration procedure, and a run-time controller architecture; this reduces design effort and limits architectural complexity.
2. An extension of this approach beyond the digital domain into digitally controlled mixed analog/RF-digital systems. First, with the embedding of analog/RF in the solution, we introduced the mandatory step of run-time calibration. Second, we considered quality side constraints at design time *and* run time, which raises the impact of this control loop, linking it to quality-of-experience (QoE) aspects that are steered by the user.

**Fig. 8.2** illustrates the steps of this process from design time to run time and across the different design domains.

We have illustrated these two extensions with two applications. First, the entire process has been demonstrated in an integrated way for a generic automatic gain control and DC offset compensation for the receiver front-end in Chap. 5. Second, a process for application scenario-aware transmitter requirements specification and architecture selection has been developed in Chap. 6. A generalization was presented in Chap. 7.

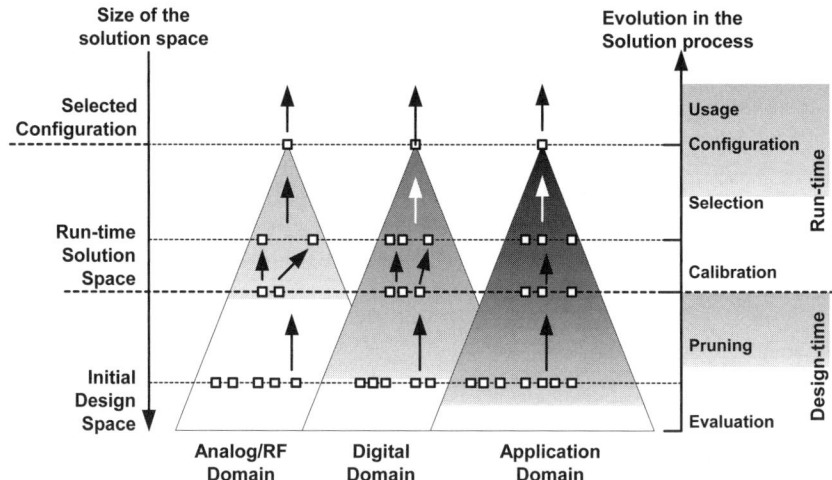

**Fig. 8.2.** Illustration of the process steps from design space exploration at design time to the usage of the optimum configuration given run-time constraints at run time

### 8.2.3   Crossdisciplinary Approach in System Design

Both application- and methodology-oriented research were subject to the design rationale expressed in Chap. 2. This means that we have considered and reviewed all contributions with respect to their role and impact in a *single* design process. Notable consequences of this approach were:

- The crossdisciplinary chain of obtained solutions, motivated by changing research domain and direction toward the *next closest major risk factor* whenever the previously identified major problem was solved. This motivated our research to cross the algorithm–architecture boundary in Chap. 4, the digital–analog/RF boundary in Chap. 5, and finally the application design space boundary in Chap. 6.
- The consequent codevelopment of design methodology with the application design. As a result, the developed methodology was carefully balanced between immediate applicability for the designer and suitability for more than one application.

Our approach honors the principles of goal-oriented design at two levels. First, we did not consider a static object as driver application (object oriented), but took the WLAN specification as a starting point for obtaining *design solutions with desired properties*, particularly scalability and flexibility. Second, we conceived the design process as a whole in which each design step and decision was clearly motivated by its impact in reaching the final application goal and the methodology used. This allows a backtrace to evaluate the actual usefulness of particular design techniques.

It should be noted that the codesign of application design and methodology also carries a risk on planning. Intrinsically, design methodology development is an initial investment the benefits of which should be reaped later. This has an impact on planning and may accumulate milestones at the end of the design, since productivity increases are assumed due to methodology and tools. For smaller design teams and novel innovative products, this approach seems definitively advantageous. But also, larger design teams benefit from local methodology development, notably a *customer-owned system design process* as a domain- or product-specific differentiator, which can hardly be provided by external CAD vendors alone. For a CAD vendor, establishing and supporting customer-specific system design processes are, in general, not a profitable business.

## 8.3  Further Research

Research is often perceived as a never-ending quest for a better and *deeper* understanding of a particular problem. Its Latin origin[159] may bring up the view of a never-ending spiral around *one* focal point. But, despite approaching the ultimate goal ever closer, surprises keep on appearing and new research needs to be initiated to explore the problem further.

At first sight, this may remind us of the picture of a Russian Matryoshka doll. Having peeled off yet another shell, what may come next? But today's challenges are not only in the depth: there is not only yet another shell inside, but we may actually question, is not there another shell outside of the problem – a *widening* of context – that we should consider. Interdisciplinary research obviously aims at looking beyond the current focal point. Flexibility and (re)configurability in the context of multiobjective optimization goals do not really aim at a *single* point; instead, their goal is the derivation of a minimal set of optimum points suitable to cover a bounded but varying context.

This work was inspired by such an interdisciplinary mindset. As the previous section may have illustrated, this opens a great new world of opportunities for research. Our research has hopefully contributed some pieces of the puzzle, but we found even more opportunities worthwhile to explore. These will be addressed here. On the one hand, this may stimulate new research in the form of future projects or Ph.D. topics. On the other hand, it may add thoughts and ideas to the discussion around *how* engineering as a discipline will position itself in a more and more interdisciplinary world.

This section is structured by analogy with the entire work. First, we discuss suggestions for application design. Next, we address design methodology and technology. An outlook on interdisciplinary challenges beyond the *pure* scope of engineering concludes this section.

---

[159] *re + circare* = to go around again and again.

## 8.3.1 Suggestions for Application Design

During the joint exploration of application, algorithm, and architecture design space, many interesting combinations of requirements or candidate solutions that deserve further investigation were revealed. From those, some represent obvious extensions to the work we have already done. Others are more visionary and require a long-term research approach to evaluate their feasibility and to develop a clear roadmap. First, we address some explicit extensions before we motivate research in two longer-term topics.

### Apparent Extensions to Multistandard and Multiantenna Systems

Future communication systems beyond 3G (4G) will be characterized by a horizontal communication model, which combines different access technologies [Mohr00]. Already today, we see examples of this evolution such as discrete MIMO-based WLAN [Sampath02], MIMO chipsets from several vendors,[160] or combinations of WLAN and cellular [Luo03]. The need to integrate compliant solutions for multiple standards, bands, modes, and services into low-power terminals will lead to a significant complexity increase in application, function, and architecture design space exploration, shifting essentially the current focus of design complexity from lower to higher abstraction layers [Fodor03, Kalliokulju01].

Despite a challenge, access to multiple transmission modes or techniques also increases the solution space and offers interesting tradeoffs. **Fig. 8.3** illustrates results of a preliminary study for MIMO-WLAN. We assume availability of both MIMO transmit and receive processing (e.g., [Khaled05]) and a performance/power-scalable forward error correction (e.g., the Turbo decoder from [Bougard03b]). The optimum configuration for the mobile terminal actually depends on the network topology and the radio channel conditions; in a star-topology, processing complexity in both transmit and receive mode is best shifted to the access point. In a peer-to-peer scenario, transmit and receive power consumption is best balanced between all terminals. Clearly, AP and MT configurations exploit the entire flexibility range offered.

This preliminary example requires all of the techniques proposed before:

- Performance/power scalability through algorithm–architecture codesign
- Digital compensation techniques to improve performance/cost despite increasing front-end requirements

---

[160] By the time of writing, announcements of MIMO chipset implementations, e.g., from AirGo Networks, swept the market. Yet, with the increased complexity of a MIMO solution, comparability of existing SISO and MIMO chipsets with respect to capacity and energy efficiency remained under heavy discussion [Mannion03].

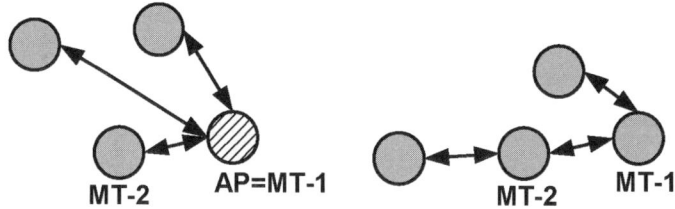

**a) Star topology with dedicated AP**
**(AP operates from mains supply)**

**b) Peer-to-peer topology**
**(MTs operating on battery supply)**

| Access Scenario | Channel state | MT-1 or AP | | MT-2 | |
|---|---|---|---|---|---|
| | | MIMO proc. configuration | relative power cons. | FEC configuration | relative power cons. |
| Star | "good" | Rx | 1.51 | Lite | 1.00 |
| Star | "bad" | Tx | 1.81 | Full | 1.26 |
| Peer-to-peer | "good" | Tx | 1.25 | Lite | 1.26 |
| Peer-to-peer | "bad" | Rx | 1.51 | Full | 1.56 |

**c) Relative power consumption in the optimum configuration**

**Fig. 8.3.** (a) and (b) The optimum system configuration depends on the network topology. (c) The power balance changes between 81% more power at MT-1 and a nearly balanced power consumption, while the optimum configurations exploit all configuration combinations

- Design-time/run-time techniques to guarantee sufficient adaptivity and robustness at run time while shifting a maximum of design complexity to design time
- Coexploration of application and design space to derive appropriate product specifications

*Design for Scalability*

The previous item illustrated the benefit of scalability with an ad hoc example. In practice, we need a more systematic approach to include scalability at the right place and of the right amount, such that the system will not be overdesigned.

It is important to mention that, in this work, we deliberately did not investigate scalability at the circuit level for its additional complexity. However, our functional and architectural solutions have been designed for compatibility with scalable circuit design techniques. It can exploit available analog or digital circuit-level scalability in a seamless way.

In the digital domain, our interface-centric distributed multiprocessor design can easily be extended to the globally asynchronous locally synchronous (GALS) design style suitable for large SoC designs. Design changes would only occur at the physical level. First, clock and reset tree insertion would be replaced by distributed clock generation. Second, besides clock gating, supply voltage control can be applied selectively to synchronous islands [Benini99]. Moreover, within synchronous islands, standard-cell-based operators can easily be replaced by custom-designed ones, particularly for dedicated word-level operations. Further specialization of these islands opens up the space of reconfigurable architectures [Srikanteswara03].

In the analog domain, scalability comes mostly at the price of performance, power consumption, or cost eventhough, to a certain extent, component deficiencies can be reduced by compensation techniques. This may allow the extension of analog/RF component design from classical single-operating-point (SOP) to multiple-operating-point (MOP) design. Run-time techniques also allow in situ calibration and adaptation. These techniques may be used to increase yield, extending the range of design-for-manufacturing (DFM) techniques from design time to run time. Both DSP-based calibration and analog self-calibration have been identified as significant research issues by the SRC [SRC00]. Why not combining them? Importantly, the move to MOP design raises the question of what metrics are important for *scalable* analog components? Obviously, the supply voltage reduction accompanying the technology scaling trend toward smaller dimensions poses a severe threat to *circuit scalability*. Specific figure-of-merit (FOM) metrics for analog/RF circuits have recently been added to the ITRS roadmapping process [Brederlow01], but these metrics are based on the SOP paradigm and represent worst-case assumptions. The question is how to quantify scalability?

## Cooperative Devices

Scalability at the network level may lead either to more parallelism such as in a multilayer networking approach [Mohr00] or more sequential paradigm such as in multihop networking. The latter requires much less infrastructural support and is motivated by the *ambient intelligence*[161] idea [Weiser91].

Both schemes require that terminals become more aware of their environment [Sperling03]. From a control theory point of view, this means that the obser vability of environmental parameters is increased. In cellular operator-owned networks, significant state of the art is present such as power control or handover management, which mitigates interference and increases interoperability. However, while controllability of networking resources may be enabled by operator-owned infrastructure, this is not necessarily guaranteed for ad hoc multihop networking. Limited controllability may lead to inefficient usage of capacity or even instability in multihop networks despite adequate observability.

---

[161] From [Weiser91]: "The most profound technologies are those that disappear. They weave themselves into the fabric of everyday life until they are undistinguishable from it."

Hence, performance and thus QoE in ad hoc or multihop networks depend largely on *cooperation* between devices. Essentially, this cooperation is in the hands of the users. For example, users may allow the use of their terminal as a relay station or deny it. Reasons for a denial may vary from saving battery power for later personal usage but also simply for security aspects. This means that both technical and nontechnical aspects can largely influence the operation and efficiency of the network. How can we guarantee QoE in such a scenario? Are there incentives for the user to offer his/her terminal for relay purposes? What would be an intelligent mechanism that automates these decisions based on simple overall user-defined or learning-based rules? Finally, do we want this process to be automated at all?[162]

### 8.3.2 Suggestions for Design Methodologies and Technology

*Scalable Behavioral Models in the Analog/RF Domain*

Advances in scalable analog/RF components are necessary to support run-time flexibility. Traditional design techniques focused mainly on SOP optimization. Therefore, modeling techniques and design technology are tailored for SOP design. However, scalable components require modeling and design techniques that efficiently cover MOP components. Essentially, the complete process from measurement to gray-box behavioral model extraction requires an adaptation toward the MOP paradigm.

For the behavioral modeling, we can identify three challenges. First, local approximation or linearization techniques will likely not work anymore in the MOP case. Second, the division in a design-time and run-time control component and the sensitivity of a component to external influence factors may require run-time estimation and adaptation of MOP models or model parameters (Chaps. 5 and 6, [Asbeck01]). Third, models with a wide coverage will essentially need to include information on the higher-order statistical distribution of parameters. Otherwise, extensive design space exploration and optimization may come up with an optimum configuration with very low design margins, which unavoidably leads to a low yield after manufacturing. Hence, process variations must be adequately modeled and/or included in the calibration process. Note that MOP behavioral modeling requires a mature SOP characterization and modeling approach in place, on which it can build upon further, e.g., [Harame03].

Essentially, the task of behavioral modeling is to aid the designer in extracting adequate models. This demands a designer who is aware of his/her *actual* model accuracy needs in a given situation and, based on this, trusts gray-box behavioral modeling. Yet, the behavioral level of modeling is still frequently perceived as *less accurate* than circuit-level modeling [Getreu90].

---

[162] From [Sperling03]: "Social interactions are the focus of our existence. We are social animals, and for any technology to be useful, it must eventually support socialization; otherwise it will not survive."

So, what are practical MOP models for which an efficient run-time calibration can be performed? Can we extract parametric information at design time and thus reduce the run-time complexity? How do MOP models scale with the number and location of operating points?

### Quick Incremental Design vs. Run-Time Flexibility

Nowadays, replacement of portable devices such as mobile phones often takes place on a yearly basis. Hence, the short lifetime of such devices may not require a significant amount of run-time flexibility and reconfigurability. Updates or upgrades up to 1 year may be foreseeable. For the designer, this results in a tradeoff between once in a while designing a very flexible device or quickly producing incremental derivatives on a yearly basis or shorter.

Essentially, this equals a tradeoff between design-time and run-time effort and cost. Incremental design requires a design flow that shortens time and effort required to iterate on the exploration and optimization process and optimally reuses the already existing predecessor of a device. Incremental design is strongly connected to systematic reuse at different levels. Interesting techniques for digital reuse with ASICs can be found for functional extensions in [Vermeulen02] or component configuration in [Schaumont99a]. However, methodology is missing that adds the analog/RF dimension and embeds these techniques in a hetero-geneous architectural context.

Incremental design appears orthogonal to the global multiobjective optimization approach proposed for design space exploration. Furthermore, the quality of an incremental design in practice reduces with the number of iterations compared to a redesign from scratch. Note that this requires a design process management that specifically covers multiple designs and design evolution.[163] Hence, two research issues are identified. First, research is needed into methodologies that can efficiently explore a remaining or modified design space given an initial, but partial set of constraints and configurations. Second, based on the status and the incremental evolution of a design, what are appropriate criteria to go for a redesign from scratch instead of yet another incremental design step?

### A Synthesis and Configuration Process for Run-Time Controller Design

When the design-time/run-time approach is combined with incremental design, it is essential to support this concept with an at least partially automated top-down synthesis and reconfiguration process. Incremental design may force adaptations of the system to smaller or larger changes in particular resources. At design time, for instance, an upgrade of the application may require the reoptimization of the system, taking into account:

---

[163] Citing [Colwell04]: "Insidiously, only the core architects and the project leaders could see the real culprit: the cost of complexity."

- Reuse of already available (= fixed) resources
- Added or removed resources
- Reuse of connectivity and interfacing

At configuration and run time, we need a flexible architecture and processing that:

- Adapts to a change in the environment, e.g., to a different radio channel
- Adapts to changes in the surrounding architecture, e.g., to a different front-end partitioning
- Adapts to application constraints, e.g., to a different type of traffic and expected QoS

The designer needs essentially synthesis support to traverse the different steps of the optimization process without error-prone manual transitions:

- At design time for the design space exploration and pruning
- At design time to split the process into a design-time, configuration-time, and run-time component including their interaction, based on the pruned results
- At design time to generate the fixed and reconfigurable parts of the run-time controller
- At configuration and run time to reconfigure or reprogram the controller

For the lower-level steps, existing FSM controller synthesis techniques can be reused [Benini96]. A step toward higher abstraction is protocol controller synthesis [Siegmund02]. However, more research is needed to analyze, optimize, and synthesize the higher-level hierarchy and architecture of the control scheme. Note that a hierarchical controller concept is mandatory to address ratios of 10,000–100,000 in *reaction time*[164] between clock cycle-based front-end control and frame-based MAC control.

### 8.3.3    Impact Beyond Engineering

More and more, designers of complex communications solutions get in contact with service requirements and application scenarios. Nontechnical requirements from marketing and sales start interfering with technical limitations. The aspect of functional flexibility is a tradeoff between design margins and upgrading or installation cost. We briefly address two interdisciplinary challenges, which interact between business and engineering, for the design of future multistandard devices: *flexibility in standardization* and *product differentiation*.

---

[164] For HiperLAN/2, 2-ms MAC-frame length vs. 50-ns ADC sample duration results in a ratio of 40,000.

*Going for a Flood of Standards or a Few Optimized and Flexible Standards*

Standardization in wireless applications has been a major success factor to bring technical solutions into the market. On the one hand, the consumer market is particularly sensitive to compliant and backward-compatible devices and standards. On the other hand, proprietary formats and solutions along with the right marketing push have often had more impact on the market than the technologically better solution. The wireless world today faces a flooding of standardization activities. Many standards *in spe* show significant overlap in application target or specifications. Moreover, device compatibility between different standards or interoperability is not always properly taken into account. Examples are the overutilization of the 2.4-GHz band through various home networking standards [Sherif02] or the discussion about regulation for ultra-wideband (UWB) standards usage in already used frequency bands [Nakagawa03]. For manufacturers, the main question will be to select and support an appropriate set of standards and to provide upgrading possibilities.

Interaction between different players such as manufacturers, operators, application developers, and the customer may significantly increase once we target global optimization across multiple services [Pereira01]. Add to this a global picture of resource management within and across multiple networks [Zander00, Mohr00] and the idea of flexible, market-driven spectrum auctioning [Ikeda02].

Our examples of mixed-signal codesign and application/design space exploration in Chaps. 5 and 6 have shown that already the codesign in a limited problem space provides substantial performance/power improvements. Given the *huge* problem space described before, these results represent only first steps into the right direction. However, they show that, for efficient operation, future systems may require availability and access to particular *knobs*, extensive and efficient communication between different layers, and even smart cooperation between different devices in a network. Most of these capabilities are not standard compliant, especially when they go across standardization layers or working groups. Many of these aspects also require an implementation-aware standardization. At the cost of a delayed standardization process and a more complicated interoperability test definition, the resulting standard may, however, offer improved operation and more possibilities for proprietary product differentiation.

Citing Goodman in [Wickelgren96], "flexible standards is an oxymoron. If it's too flexible, it's not a standard." Hence, a fundamental question is how can we introduce "just-enough flexibility" in the standardization process of today? Is it enough to optimize just the transition or handover between different standards? How do we build in sufficient interoperability capabilities between standards? Can standardization help in standardizing this transition process at least to increase interoperability?

## *Product Differentiation and Metrics in a Multistandard World*

Product differentiation is a key aspect in the consumer world. Certainly, price is a major selling factor while advanced technical features often cannot be marketed effectively in understandable terms for the end customer. Indeed, we have to "sell the application, not the network" [Rose01]. So, what is more appealing to the customer, given two flexible, upgradable communication devices to choose from? For mobile phones, talk and standby time have become quite well-understood metrics for the customer. By what will we replace this for a user-dependent mix of services? How do we measure flexibility or upgradability? How can we illustrate and guarantee improved QoE to the customer?

Besides finding good metrics, the challenge even increases when we move to more distributed solutions. In a cooperative environment such as a multihop network, performance *and* cost and hence also QoE depend more and more on the environment. Even worse, this environment will not be under control of a single operator such as in a cellular system. "Connectivity may really become a commodity" [Saracco03]. Clearly, the shift toward service-centric business is mandatory, since the quality of the single device is of reduced importance: the properties of the environment have a much larger impact on QoE. So, what are the final product differentiators?

Finally, how many *different* devices do you have to build? Is "one device that does it all" [Savage03] sufficient. This is a particularly interesting tradeoff since it covers the entire problem space from the most flexible to the least flexible solution and marketing may deliberately delimit the design space for the technical solution while design or manufacturing cost may actually result in a different space. Do they overlap? In how many different subspaces, and accordingly derivative products, should we subdivide for a product range with optimum success?

# Glossary

## Mathematical Notation

| | |
|---|---|
| $a$ | scalar |
| $\hat{a}$ | estimated value of a parameter, here for a scalar |
| $\mathbf{x}_n$ | vector of dimension $n$; $n$ is optional |
| $\mathbf{X}_{m \times n}$ | matrix of dimension $m \times n$; $m$ and $n$ are optional |
| $\mathbf{X}^*$ | complex conjugate of matrix $\mathbf{X}$ |
| $\mathbf{X}^{-1}$ | inverse of matrix $\mathbf{X}$ |
| $\mathbf{X}^T$ | transposed matrix $\mathbf{X}$ |
| $|x|$ | absolute value of $x$ |
| $E\{x\}$ | expectation of a random variable $x$ |
| $E\{x[k], k\}$ | expectation of a random variable $x$ with respect to $k$ |
| $x(t)$ | continuous-time baseband equivalent signal |
| $x[t]$ | discrete-time baseband equivalent signal |
| $X(\omega)$ or $X(f)$ | continuous frequency-domain signal |
| $X[\omega]$ or $X[f]$ | discrete frequency-domain signal |
| $R\{x\}$ | real part of $x$ |
| $I\{x\}$ | imaginary part of $x$ |

## Acronyms and Abbreviations

| | |
|---|---|
| 2G | second generation, *mobile voice and messaging communications* |
| 2.5G | between 2G and 3G |
| 3G | third generation, *mobile multimedia communications* |
| 4G | fourth generation, *worldwide roaming between data, voice, and multimedia communication standards* |
| AAC | alternating adjacent channel |
| ABM | abstract behavioral model |
| ABR | available bit rate |
| ac | analog current |

| | |
|---|---|
| AC | (a) adjacent channel |
| | (b) autocorrelation |
| ACF | autocorrelation function |
| ACK | acknowledge |
| ACPR | adjacent channel power regrowth |
| ACU | address calculation unit |
| ADC | analog-to-digital converter |
| ADSL | asymmetric digital subscriber line |
| AFC | (a) analog frequency control |
| | (b) automatic frequency control |
| AFE | analog front-end |
| AGC | automatic gain control |
| ALU | arithmetic and logic unit |
| AP | access point |
| API | application programming interface |
| ARQ | automatic retransmission query |
| ASIC | application-specific integrated circuit |
| ASIP | application-specific instruction processor |
| ASSP | application-specific standard product |
| ATM | asynchronous transfer mode |
| ATPG | automatic test pattern generation |
| ATS | absolute timing sequence |
| AWGN | additive white Gaussian noise |
| BAN | body area network |
| BB | baseband |
| BER | bit-error rate |
| BiCMOS | bipolar CMOS |
| BIST | built-in self-test |
| BPF | bandpass filter |
| BS | basestation |
| BSI | burst state information |
| CA | collision avoidance |
| CAD | computer-aided design |
| CAGR | compound annual growth rate |
| Carnival | *name of IMEC's second baseband OFDM ASIC* |
| CBR | constant bit rate |
| CC | crosscorrelation |
| CCDF | complementary cumulative density function |
| CCF | crosscorrelation function |
| CDF | cumulative distribution function |
| CDMA | code division multiple access |
| CFB | Cartesian feedback |
| CFO | carrier frequency offset |
| CMOS | complementary metal oxide semiconductor |
| COFDM | coded OFDM |
| CORDIC | coordinate rotation digital computer |

CP .......................................... cyclic prefix
CPE........................................ common phase error
CPN ...................................... common phase noise
CQFP ................................... ceramic quad flat pack package
CRC ...................................... cyclic redundancy check
CSD ...................................... canonical sign digit
CSE ....................................... common subexpression elimination
CSI ........................................ channel state information
CSMA .................................. carrier-sense multiple access
CSMA/CA ........................... CSMA with collision avoidance
CW ........................................ continuous wave
DA ......................................... data aided
DAB ...................................... digital audio broadcasting
DAC ...................................... digital-to-analog converter
DC ......................................... digital current
DCO ...................................... DC offset
DCR........................................ direct-conversion receiver
DD ......................................... decision directed
DDI......................................... dynamic datapath information
DECT .................................... digital-enhanced cordless telecommunications
DFC ...................................... data format converter
DFE ....................................... decision feedback equalizer
DFM ...................................... design for manufacturing
DFT ....................................... (a) discrete Fourier transformation
                                            (b) design for test
DICORE ............................... digital compensation for radio enhancement,
                                            *name of an IMEC internal project*
DIFS ..................................... distributed interframe space
DLC ...................................... data link control
DLL ....................................... data link layer
DM ........................................ design methodology
DMA ..................................... direct memory access
DMT....................................... discrete multitone
DR .......................................... dynamic range
DRAM ................................... dynamic random-access memory
DRC ...................................... dynamic range control
DSE........................................ design space exploration
DSL......................................... digital subscriber line
DSP ....................................... (a) digital signal processor
                                            (b) digital signal processing
DSSS ..................................... direct sequence spread spectrum
DT........................................... design technology
DTSE ..................................... data transfer and storage exploration
DUT ...................................... device under test
DVB ...................................... digital video broadcasting
DVB-H .................................. DVB handheld

DVB-T ............................... DVB terrestrial
ECP .................................... estimation and compensation phase
EDA ................................... electronic design automation
EDGE ............................... enhanced data rates for global evolution
EER ................................... envelope elimination and restoration
EIRP ................................. effective isotropic radiated power
EK ..................................... execution kernel
ESL ................................... electronic system level
ETSI .................................. European telecommunications standards institute
EVM ................................. error vector magnitude
FB ..................................... feedback
FBAR ................................ film bulk acoustic resonator
FCC ................................... federal communications commission
FD ..................................... frequency domain
FDM .................................. frequency division multiplex
FDMA ............................... frequency division multiple access
FEC ................................... forward error control
Festival .............................. *name of IMEC's first OFDM ASIC and name of the corresponding project*
FF ..................................... flip-flop
FFT ................................... fast Fourier transformation
FIFO ................................. first-in first-out
FIR .................................... finite impulse response
FOM .................................. figure of merit
FPGA ................................ field-programmable gate array
FSK ................................... frequency shift keying
FSM ................................... finite state machine
FSMD ................................ finite state machine and datapath
GaAs ................................. Gallium arsenide
GaN ................................... Gallium nitride
GOPS ................................ giga-operations per second
GPRS ................................ general packet radio service
GSM .................................. global system for mobile communications
HDL ................................... hardware description language
HDTV ............................... high-definition television
HF ..................................... high frequency
HiperLAN/2 ...................... high-performance radio local area network, type 2
HLS ................................... high-level synthesis
HPF ................................... high-pass filter
HSDPA .............................. high-speed downlink packet access
HW .................................... hardware
I/O ..................................... input–output
I/Q ..................................... in/quadrature phase
IBI ..................................... interblock interference
IBO ................................... input backoff
IC ...................................... integrated circuit

| | |
|---|---|
| ICE | interchannel error |
| ICI | intercarrier interference |
| IDFT | inverse discrete Fourier transformation |
| IEEE | Institute of Electrical and Electronics Engineers |
| IF | intermediate frequency |
| IFFT | inverse fast Fourier transformation |
| IFS | interframe spacing |
| IIR | infinite impulse response |
| IL | implementation loss |
| IM | intermodulation |
| IP | (a) Internet protocol |
| | (b) intellectual property |
| IR | infrared |
| ISDN | integrated services digital network |
| ISI | intersymbol interference |
| ISM | industrial–scientific–medical |
| ISO | international standards organization |
| ITRS | international technology roadmap for semiconductors |
| ITU | international telecommunications union |
| JEDEC | joint electron device engineering council |
| L1 | OSI layer 1; physical layer |
| L2 | OSI layer 2; data link control layer |
| L3 | OSI layer 3; network layer |
| LAN | local area network |
| LINC | linear amplification with nonlinear components |
| LNA | low-noise amplifier |
| LO | local oscillator |
| LOS | line of sight |
| LP | low power |
| LPF | low-pass filter |
| LS | least squares |
| LSB | least significant bit |
| LSE | least square error |
| LTS | long training sequence |
| LUT | look-up table |
| MAC | (a) medium access control |
| | (b) multiply accumulate |
| MC | (a) multiple carrier |
| | (b) Monte Carlo |
| MC-CDMA | multiple-carrier code division multiplex |
| MCM | (a) multichip module |
| | (b) multicarrier modulation |
| MEMS | microelectromechanical system |
| MIMO | multiple-input multiple-output |
| MLSE | maximum-likelihood sequence estimator |
| MMIC | monolithic microwave integrated circuit |

MMSE .............................. minimum mean-square error
MoC ................................ model of computation
MOP ............................... multiple-operating point
MPEG ............................. motion pictures experts group
MPI................................. microprocessor interface
MRMC............................. multirate multicarrier
MSB ................................ most significant bit
MSE ................................ mean square error
MT .................................. mobile terminal
MUX ............................... multiplexer
NDA ................................ non-data aided
NF ................................... noise figure
NIC .................................. network interface card
NLOS .............................. nonline of sight
OAM ............................... operation, administration, maintenance
OBO ................................ output backoff
OEM ................................ original equipment manufacturer
OFDM .............................. orthogonal frequency division multiplexing
OSI .................................. open systems interconnection
PA .................................... power amplifier
PAN ................................. personal area network
PAPR ............................... peak-to-average power ratio
PC .................................... personal computer
PCB .................................. printed circuit board
PCF .................................. point coordination function
PDA ................................. personal digital assistant
PDF .................................. probability density function[165]
PDU ................................. protocol data unit
PER .................................. packet error rate
PETRARCH ...................... power-efficient transmitter architectures, *name of*
                                                 *an IMEC internal project*
PGA ................................. programmable gain amplifier
PHY ................................. physical layer
PICARD ........................... *name of an IMEC internal prototyping platform*
PIFS ................................. point coordination function interframe spacing
PL .................................... payload
PLCP ................................ physical layer convergence protocol
PLL .................................. phase-locked loop
PLME ............................... physical layer management entity
PMD ................................. physical medium dependent
PN .................................... pseudorandom noise
PP .................................... peak to peak
PQFP ............................... plastic quad flat pack package
PSD .................................. power spectral density

---

[165] Also probability distribution function.

PSK ..................................... phase-shift keying
PWM ................................... pulse-width modulation
QAM ................................... quadrature amplitude modulation
QoE .................................... quality of experience
QoS.................................... quality of service
QPSK .................................. quaternary phase-shift keying
RAM ................................... random-access memory
RF....................................... radio frequency
RFID................................... radio frequency identification
RLC .................................... radio link control
RMS ................................... root mean square
ROM ................................... read-only memory
RRC .................................... radio resource control
RRM ................................... radio resource management
RSSI ................................... receive signal strength indicator
RTL .................................... register transfer level
RTS..................................... relative timing sequence
Rx ...................................... receiver
SAP ..................................... service access point
SAW ................................... surface acoustic wave
SC ...................................... single carrier
SCR .................................... system clock reference
SDF .................................... synchronous dataflow
SDMA ................................ space division multiple access
SDR .................................... software-defined radio
SDRAM ............................. synchronous DRAM
SFG.................................... signal-flow graph
Si........................................ silicon
SIFS ................................... short interframe spacing
SiGe ................................... Silicon–Germanium
SINAD ............................... signal-to-noise-and-distortion ratio
SiP ..................................... system in a package
SIR...................................... signal-to-interference ratio
SISO ................................... single-input single-output
SLIF ................................... system-level interface
SNR .................................... signal-to-noise ratio
SoC .................................... system on chip
SoHo .................................. small-office/home-office
SOP .................................... single-operating point
SQNR .................................. signal-to-quantization noise ratio
SR....................................... software radio
SRAM ................................. static random-access memory
SRR .................................... software-reconfigurable radio
SSR..................................... symbol-based sample reordering
STS ..................................... short training sequence
SVD .................................... singular value decomposition

SVP ..................................... silicon virtual prototype
SW ...................................... software
SyRS ................................... system requirements specification
TCM ................................... task concurrency management
TCP ..................................... transmission control protocol
TDD ................................... time-division duplex
TDM ................................... time-division multiplex
TDMA ............................... time-division multiple access
TLP ..................................... task level parallelism
TPS ..................................... transmission parameter signaling
transceiver ........................... *a device that acts as OSI layer L1 covering both transmission and reception*[166]
TV ....................................... television
Tx ........................................ transmitter
UML ................................... unified modeling language
UMTS ................................ universal mobile telecommunication system, *defined in the RACE Mobile Definition project in 1986*
UTRA .................................. UMTS terrestrial radio air interface
UWB ................................... ultra-wideband
VBR .................................... variable bit rate
VC ....................................... virtual component
VCO .................................... voltage-controlled oscillator
VGA .................................... variable-gain amplifier
VHDL .................................. VLSI hardware description language
VLIW ................................... very long instruction word
VLSI .................................... very large-scale integration
VoD ..................................... video on demand
VoIP .................................... voice over IP
VPN ..................................... virtual private network
VSI ...................................... virtual socket interface
VSIA .................................... VSI alliance
VSWR .................................. voltage standing wave ratio
W-CDMA ............................ wideband code division multiple access
Wi-Fi ................................... wireless fidelity (see http://www.wi-fi.org)
WiMAX ................................ worldwide interoperability for microwave access, *see also IEEE standard 802.16*
WLAN ................................. wireless local area network
WLL .................................... wireless local loop
WMAN ................................. wireless metropolitan area network
WPAN .................................. wireless personal area network
WWRF .................................. wireless world research forum
xDSL ................................... *collective name for digital subscriber line techniques*

---

[166] Note that, in the analog/RF community, transceiver often is used for the nondigital front-end part only.

ZF ...................................... zero-forcing

## Parameter Naming Conventions

*Note.* For the orientation of the reader, we provide the typical unit scale in brackets.

| | |
|---|---|
| $\Delta f_{sc}$ | carrier frequency offset (kHz) |
| $\Delta f_{clk}$ | clock frequency offset (kHz) |
| ** $\sigma_\tau$ | rms delay spread (ns) |
| ** $\overline{\tau}$ | mean excess delay (ns) |
| $\omega$ | angular frequency |
| $B_{coh}$ | coherence bandwidth of the radio channel (MHz) |
| $B_{sig}$ | nominal signal bandwidth (MHz) |
| $E_{b,total}$ | total dissipated energy per bit (nJ bit$^{-1}$) |
| $E_{b,Tx}$ | transmitted energy per bit (nJ bit$^{-1}$) |
| $f_c$ | carrier frequency (GHz) |
| $f_s$ | sampling frequency (MHz) |
| $h(t)$ | channel impulse response |
| $H(t)$ | power delay profile |
| $M$ | constellation size |
| $N_{sc}$ | number of subcarriers in an OFDM symbol |
| $N_d$ | number of data subcarriers in an OFDM symbol |
| $N$ | noise power (dBm) |
| $N_0$ | spectral noise power density |
| $p_b$ | bit-error probability, *often referred to as bit-error rate* |
| $P_{total}$ | total instantaneous power dissipation |
| $P_{Tx}$ | total instantaneous transmit power |
| $S_{mn}$ | $S$ parameter from port $m$ to port $n$ |
| $T_{coh}$ | coherence time of the radio channel (ms) |
| $T_s$ | duration of an OFDM symbol (μs) |
| $V_{DD}$ | positive digital power supply voltage |

## List of Software Programs

| | |
|---|---|
| ADS | RF and microwave simulator, with DSP options (Agilent Technologies) |
| CAPSIM | C-based physical and link layer simulator (XCAD Corp.) |
| CoCentric System Studio | algorithmic and DSP design environment (Synopsys) |
| ConvergenSC | mixed abstraction-level digital (HW/SW) simulation and analysis environment (CoWare) |
| COSSAP | DSP development (Synopsys) |
| Design Compiler | digital circuit synthesis software (Synopsys) |

DSP and System Canvas .... block diagram-based design environment for DSP (Angeles Design Systems)

FAST ................................ behavioral-level modeling and simulation tool for analog and RF (IMEC)

MATLAB .......................... integrated technical computing environment (MathWorks)

OCAPI ............................. behavioral and RT-level modeling and simulation tool for hardware–software codesign (IMEC)

Saber, Saber HDL .............. mixed-signal multitechnology simulator (Synopsys, formerly Analogy)

Simulink ........................... interactive multidomain block diagram-based simulator (MathWorks)

SPICE .............................. circuit simulation software (University of California at Berkeley)

SPW .................................. algorithm and DSP development (CoWare, formerly Cadence); the tool has been renamed into CoWare Signal Processing Designer

UNIFY .............................. library extension to OCAPI version 0.9 (IMEC)

# Bibliography

[Abidi97]      A. A. Abidi, Direct-conversion radio transceivers for digital communications, *IEEE J. Solid-State Circuits*, 30(12):1399–1410, 1997

[Abidi00]      A. A. Abidi, G. J. Pottie, and W. J. Kaiser, Power-conscious design of wireless circuits and systems, *Proc. IEEE*, 88(10):1528–1545, 2000

[ADSL]         ITU-T G.992, Asymmetrical Digital Subscriber Line (ADSL) transceivers, July 1999

[AHD00]        *The American Heritage Dictionary of the English Language*, 4th edition. Boston: Houghton Mifflin, 2000 (http://www.bartleby.com)

[Aiello03]     G. R. Aiello and G. D. Rogerson, Ultra-wideband wireless systems, *IEEE Microwave Mag.*, 4(2):36–47, 2003

[Akansu98]     A. N. Akansu, P. Duhamel, and X. Lin, Orthogonal transmultiplexers in communication: A review, *IEEE Trans. Signal Process.*, 46(4):979–995, 1998

[Alard87]      M. Alard and R. Lassalle, Principles of modulation and channel coding for digital broadcasting for mobile receivers, *EBU Technical Review*, no. 224, pp. 168–190, August 1987

[Allais43]     M. F. C. Allais, *A la recherché d'une discipline économique: L'économie pure*. Paris: Ateliers Industria, 1943

[Allen90]      P. E. Allen, B. Chan, and W. M. Zuberek, Comparison of mixed analog–digital simulators, in *Proceedings of IEEE International Symposium on Circuits and Systems*, New Orleans, USA, May 1990, pp. 101–104

[Armour00]     S. Armour, A. Nix, and D. Bull, Complexity evaluation for the implementation of a pre-FFT equalizer in an OFDM receiver, *IEEE Trans. Consum. Electron.*, 46(3):428–437, 2000

[Asbeck01]     P. M. Asbeck, L. E. Larson, and I. G. Galton, Synergistic design of DSP and power amplifiers for wireless communications, *IEEE Trans. Microwave Theory Techn.*, 49(11):2163–2169, 2001

[Aue01]        V. Aue, J. Kneip, M. Weiss, M. Bolle, and G. Fettweis, MATLAB based codesign framework for wireless broadband communication DSPs, in *Proceedings of IEEE International Conference on Acoustics, Speech, and Signal Processing*, Salt Lake City, USA, May 2001, pp. 1253–1256

[Baas99a]          B. Baas, A low-power, high-performance, 1024-point FFT processor, *IEEE J. Solid-State Circuits*, 34(3):380–387, 1999

[Baas99b]          B. Baas, An approach to low-power, high-performance, Fast Fourier Transform processor design, Ph.D. Dissertation, Stanford University, Palo Alto, USA, February 1999

[Bae94]            J. Bae, V. K. Prasanna, and H. Park, Synthesis of a class of data format converters with specified delays, in *Proceedings of International Conference on Application Specific Array Processors*, San Francisco, USA, August 1994, pp. 283–294

[Bae98]            J. Bae and V. K. Prasanna, Synthesis of area efficient and high-throughput rate data format converters, *IEEE Trans. VLSI Syst.*, 6(4):697–706, 1998

[Bahai99]          A. R. S. Bahai and B. R. Saltzberg, *Multi-Carrier Digital Communications – Theory and Applications of OFDM*. New York: Kluwer, 1999

[Baltus03]         P. G. M. Baltus, Efficient RF Design Methods, presented at IMEC Design Technology Seminar, 14 March 2003

[Bannon94]         L. Bannon, R. Keil-Slawik, and I. Wagner, eds., *A Multidisciplinary Foundation for System Design and Evaluation*. Germany: Schloss Dagstuhl, 1994

[Bell76]           T. E. Bell and T. A. Thayer, Software requirements: Are they really a problem? in *Proceedings of International Conference on Software Engineering*, San Francisco, USA, 1976, pp. 61–68

[Bello65]          P. A. Bello, Selective fading limitations of the Kathryn modem and some system design considerations, *IEEE Trans. Commun. Technol.*, 13(3):320–333, 1965

[Benini96]         L. Benini and G. De Micheli, Automatic synthesis of low-power gated-clock finite-state machines, *IEEE Trans. CAD*, 15(6):630–643, 1996

[Benini99]         L. Benini and G. De Micheli, System-level power optimization: Techniques and tools, in *Proceedings of International Symposium on Low Power Electronics and Design*, San Diego, USA, August 1999, pp. 288–293

[Bergland69]       G. Bergland, Fast Fourier transform hardware implementations – An overview, *IEEE Trans. Audio Electroacoust.*, 17(2):104–108, 1969

[Bertran91]        E. Bertran and J. M. Palacin, Control theory applied to the design of AGC circuits, in *Proceedings of Mediterranean Electrotechnical Conference*, Ljubljana, Slovenia, May 1991, pp. 60–70

[Bickerstaff98]    M. Bickerstaff, T. Arivoli, P. J. Ryan, N. Weste, and D. Skellern, A low-power 50-MHz FFT processor with cyclic extension and shaping filter, in *Proceedings of Asia and South Pacific Design Automation Conference*, Yokohama, Japan, February 1998, pp. 335–336

[Bidet95]          E. Bidet, D. Castelain, C. Joanblancq, and P. Senn, A fast single-chip implementation of 8192 complex-point FFT, *IEEE J. Solid-State Circuits*, 30(3):300–305, 1995

[Bingham90]        J. Bingham, Multicarrier modulation for data transmission: An idea whose time has come, *IEEE Commun. Mag.*, 28(5):5–14, 1990

[Bisdikian01]      C. Bisdikian, An overview of the Bluetooth wireless technology, *IEEE Commun. Mag.*, 39(12):86–94, 2001

[Blackard93]       K. L. Blackard, T. S. Rappaport, and C. W. Bostian, Measurements and models of radio frequency impulsive noise for indoor wireless communications, *IEEE J. Sel. Areas Commun.*, 11(7):991–1001, 1993

[Bolle98]        M. Bolle, D. Clawin, K. Gieske, F. Hofmann, T. Mlasko, M. Ruf, and
                 G. Spreitz, The receiver engine chip-set for digital audio broadcasting,
                 in *Proceedings of URSI International Symposium on Signals, Systems,
                 and Electronics*, Pisa, Italy, September 1998, pp. 338–342

[Bolsens97]      I. Bolsens, H. De Man, B. Lin, K. Van Rompaey, S. Vercauteren, and
                 D. Verkest, Hardware/software co-design of digital telecommunication
                 systems, *Proc. IEEE*, 85(3):391–418, 1997

[Borel99]        J. Borel, Design Automation in MEDEA: Present and Future, *IEEE
                 Micro*, 19(5):71–79, 1999

[Bougard03a]     B. Bougard, G. Lenoir, W. Eberle, F. Catthoor, and W. Dehaene,
                 A new approach to dynamically trade off performance and energy
                 consumption in wireless communication systems, in *Proceedings of
                 IEEE Symposium on Signal Processing System*, Seoul, South Korea,
                 August 2003, pp. 298–303

[Bougard03b]     B. Bougard et al., A scalable 8.7-nJ/bit 75.6-Mbit/s parallel
                 concatenated convolutional (turbo-) CODEC, in *IEEE International
                 Solid-State Circuits Conference Digest*, San Francisco, USA, February
                 2003, pp. 152–153

[Bougard04]      B. Bougard, S. Pollin, G. Lenoir, W. Eberle, L. Van der Perre,
                 F. Catthoor, and W. Dehaene, Modeling of energy-scalable wireless
                 local area network transceivers, in *Proceedings of IEEE Workshop on
                 Signal Processing Advances in Wireless Communication*, July 2004

[Bouras03]       I. Bouras et al., A digitally calibrated 5.15–5.825 GHz transceiver for
                 802.11a wireless LANs in 0.18 μm CMOS, in *IEEE International
                 Solid-State Circuits Conference Digest*, San Francisco, USA, February
                 2003, pp. 352–353

[Brakensiek02]   J. Brakensiek, B. Oelkrug, M. Bücker, D. Uffmann, A. Dröge, M.
                 Darianian, and M. Otte, Software radio approach for re-configurable
                 multi-standard radios, in *Proceedings of IEEE International Symposium
                 on Personal, Indoor, and Mobile Radio Communications*, 2002, pp.
                 110–114

[Brederlow01]    R. Brederlow, W. Weber, J. Sauerer, S. Donnay, P. Wambacq, and
                 M. Vertregt, A mixed-signal design roadmap, *IEEE Des. Test Comput.*,
                 18:34–46, 2001

[Brockmeyer99]   E. Brockmeyer, C. Ghez, J. D'Eer, F. Catthoor, and H. De Man,
                 Parametrizable behavioral IP module for a data-localized low-power
                 FFT, in *Proceedings of IEEE Workshop on Signal Processing Systems*,
                 October 1999, pp. 635–644

[Bryant01]       R. E. Bryant et al., Limitations and challenges of computer-aided
                 design technology for CMOS VLSI, *Proc. IEEE*, 89(3):341–365, 2001

[Büchi97]        M. Büchi and W. Weck, A plea for grey-box components, TUCS
                 Technical Report No. 122, Turku Centre for Computer Science, August
                 1997

[Buck94]         J. T. Buck, S. Ha, E. A. Lee, and D. G. Messerschmitt, Ptolemy: A
                 framework for simulating and prototyping heterogeneous systems, *Int.
                 J. Comput. Simul.*, 4:155–182, 1994

[Busson01]       P. Busson, P.-O. Jouffre, P. Dautriche, F. Paillardet, and I. Telliez, A
                 complete single-chip front-end for digital satellite broadcasting, in
                 *Proceedings of International Conference on Consumer Electronics*, Los
                 Angeles, USA, June 2001, pp. 112–113

**[Busson02]**  P. Busson et al., Satellite tuner single chip simulation with ADS (http://eesof.tm.agilent.com/pdf/st.pdf)

---

**[Callaway02]**  E. Callaway et al., Home networking with IEEE 802.15.4: A developing standard for low-rate wireless personal area networks, *IEEE Commun. Mag.*, 40(8):70–76, 2002

**[Catthoor88]**  F. Catthoor et al., Architectural strategies for an application-specific synchronous multiprocessor environment, *IEEE Trans. Acoust. Speech Signal Process.*, 36(2):265–284, 1988

**[Catthoor90]**  F. Catthoor, D. Lanneer, and H. De Man, Efficient microcoded processor design for fixed rate DFT and FFT, *J. VLSI Signal Process.*, 1:287–306, 1990

**[Catthoor98a]**  F. Catthoor, S. Wuytack, E. De Greef, F. Franssen, L. Nachtergaele, and H. De Man, System-level transformations for low data transfer and storage, in *Low Power CMOS Design*, A. Chandrakasan and R. Brodersen, Eds. New York: IEEE, 1998, pp. 609–618

**[Catthoor98b]**  F. Catthoor, D. Verkest, and E. Brockmeyer, Proposal for unified system design meta flow in task-level and instruction-level design technology research for multi-media applications, in *Proceedings of International Symposium on System Synthesis*, Hsinchu, Taiwan, December 1998, pp. 89–95

**[Cetin97]**  E. Çetin, R. C. S. Morling, and I. Kale, An integrated 256-point complex FFT processor for real-time spectrum analysis and measurement, in *Proceedings of Instrumentation and Measurement Technology Conference*, Ottawa, Canada, May 1997, pp. 96–101

**[Chang69]**  R. W. Chang, Synthesis of band limited orthogonal signals for multichannel data transmission, *Bell Syst. Techn. J.*, 45:1775–1796, 1969

**[Chang70]**  R. W. Chang, Orthogonal frequency-division multiplexing, US Patent 3,448,455, filed November 1966, issued January 1970

**[Chang96b]**  J.-M. Chang and M. Pedram, Energy minimization using multiple supply voltages, in *Proceedings of International Symposium on Low Power Electronics and Design*, Monterey, USA, August 1996, pp. 157–162

**[Chang99]**  H. Chang, L. Cooke, M. Hunt, G. Martin, A. McNelly, and L. Todd, *Surviving the SOC Revolution: A Guide to Platform-Based Design.* Boston: Kluwer, 1999

**[Chen94]**  K.-C. Chen, Medium access control of wireless LANs for mobile computing, *IEEE Network*, 8(5):50–63, 1994

**[Chen99]**  J. Chen, D. Feng, J. Philips, and K. Kundert, Simulation and modeling of intermodulation distortion in communication circuits, in *Proceedings of IEEE Custom Integrated Circuits Conference*, San Diego, USA, May, 1999, pp. 5–8

**[Chen01]**  J. C. Chen and J. M. Gilbert, Measured performance of 5-GHz 802.11a Wireless LAN systems, Technical Report, Atheros Communications, 27 August 2001

**[Chia91]**  S. T. S. Chia, Network architectures for supporting mobility in a Third Generation mobile system, in *Proceedings of IEEE International Symposium on Personal, Indoor, and Mobile Radio Communications*, September 1991, pp. 236–240

[Chow91a]     J. S. Chow, J. C. Tu, and J. M. Cioffi, Performance evaluation of a multichannel transceiver system for ADSL and VHDSL services, *IEEE J. Sel. Areas Commun.*, 9(6):909–919, 1991

[Chow91b]     J. S. Chow, J. C. Tu, and J. M. Cioffi, A discrete multitone transceiver system for HDSL applications, *IEEE J. Sel. Areas Commun.*, 9(6):895–908, 1991

[Chung01]     S. T. Chung and A. J. Goldsmith, Degrees of freedom in adaptive modulation: A unified view, *IEEE Trans. Commun.*, 49(9):1561–1571, 2001

[Ciborra94]   C. Ciborra, *From Thinking to Tinkering*. New York: Wiley, 2004

[Cimini85]    L. J. Cimini, Analysis and simulation of a digital mobile channel using orthogonal frequency division multiplexing, *IEEE Trans. Commun.*, 33:665–675, 1985

[Claasen99]   T. A. C. M. Claasen, High speed: Not the only way to exploit the intrinsic computational power of silicon, in *IEEE International Solid-State Circuits Conference Digest*, San Francisco, USA, February 1999, pp. 22–25

[Claessen94]  A. Claessen, L. Monteban, and H. Moelard, The AT&T GIS WaveLAN air interface and protocol stack, in *Proceedings of IEEE International Symposium on Personal, Indoor, and Mobile Radio Communications*, September 1994, pp. 1442–1446

[Clawin98]    D. Clawin et al., Architecture and performance of an alternative DAB receiver chip set, in *Proceedings of European Microwave Conference*, October 1998, pp. 645–650

[Colwell04]   B. Colwell, Design fragility, *IEEE Comput. Mag.*, 37(1):13–16, 2004

[Côme00]      B. Côme, R. Ness, S. Donnay, L. Van der Perre, W. Eberle, P. Wambacq, M. Engels, and I. Bolsens, Impact of front-end non-idealities on bit error rate performances of WLAN–OFDM transceivers, in *Proceedings of IEEE Radio and Wireless Conference*, Denver, USA, September 2000, pp. 91–94

[Côme04]      B. Côme, D. Hauspie, G. Albasini, S. Brebels, W. De Raedt, W. Diels, W. Eberle, H. Minami, J. Ryckaert, J. Tubbax, and S. Donnay, Single-package direct-conversion receiver for 802.11a wireless LAN enhanced with fast converging digital compensation techniques, in *IEEE MTT-S International Microwave Symposium Digest*, Fort Worth, USA, 6–11 June 2004

[Cooley65]    J. W. Cooley and J. W. Tukey, An algorithm for machine calculation of complex Fourier series, *Math. Comput.*, 19:297–301, 1965

[Coombs99]    R. Coombs and R. Steele, Introducing microcells into macrocellular networks: A case study, *IEEE Trans. Commun.*, 47(4):568–576, 1999

[CRC59]       Collins Radio Co., Collins kineplex systems, in *Collins Radio Company 1959 General Catalog*, 1959, pp. 67–70

[Crols95]     J. Crols, S. Donnay, M. Steyaert, and G. Gielen, A high-level design and optimization tool for analog RF receiver front-ends, in *Proceedings of IEEE/ACM International Conference on Computer-Aided Design (ICCAD)*, November 1995, pp. 550–553

[Crols98]     J. Crols and M. Steyaert, Low-IF topologies for high-performance analog front-ends of fully integrated receivers, *IEEE Trans. Circuits Syst. II*, 45(3):269–282, 1998

[Crow97]        B. P. Crow, I. Widjaja, L. G. Kim, and P. T. Sakai, IEEE 802.11
                wireless local area networks, *IEEE Commun. Mag.*, 35(9):116–126,
                1997

[Czylwik97]     A. Czylwik, Comparison between adaptive OFDM and single carrier
                modulation with frequency domain equalization, in *Proceedings of
                IEEE Vehicular Technology Conference*, Phoenix, USA, 1997, pp.
                865–869

---

[d'Hainaut86]   L. d'Hainaut, Interdisciplinarity in general education, in *International
                Symposium on Interdisciplinarity in General Education*, UNESCO,
                May 1986 (http://www.unesco.org/education/pdf/31_14.pdf)

[DAB]           ETSI ETS 300 401, Digital Audio Broadcasting (DAB); DAB to
                Mobile, Portable and Fixed Receivers, February 1995 (created: May
                1992)

[Daffara96]     F. Daffara and O. Adami, A novel carrier recovery technique for
                orthogonal multicarrier systems, *Eur. Trans. Telecommun.*, 7:323–334,
                1996

[Dai03]         W.-J. Dai, D. Huang, C.-C. Chang, and M. Courtoy, Silicon virtual
                prototyping: The new cockpit for nanometer chip design, in
                *Proceedings of Asia and South Pacific Design Automation Conference*,
                January 2003, pp. 635–639

[Davis88]       A. Davis, A taxonomy for early stages of the software development life
                cycle, *J. Syst. Software*, 8(4):297–311, 1988

[Davis02]       W. R. Davis et al., A design environment for high-throughput low-
                power dedicated signal processing systems, *IEEE J. Solid-State
                Circuits*, 37(3):420–431, 2002

[Debaillie01a]  B. Debaillie, B. Côme, W. Eberle, S. Donnay, and M. Engels, Impact of
                front-end filters on bit error rate performances in WLAN–OFDM
                transceivers, in *Proceedings of IEEE Radio and Wireless Conference*,
                Boston, USA, August 2001, pp. 193–196

[Debaillie01b]  B. Debaillie, H. Minami, B. Côme, W. Eberle, and S. Donnay, Filter
                design methodology controlling the impact on bit error rate perfor-
                mances in WLAN–OFDM transceivers, in *Proceedings of International
                Workshop on Multi-Carrier Spread Spectrum*, Oberpfaffenhofen,
                Germany, September 2001

[Debaillie02]   B. Debaillie, H. Minami, B. Côme, W. Eberle, and S. Donnay, System-
                level filter design methodology for WLAN–OFDM transceivers,
                *Microwave J.*, 45(5):268–279, 2002

[DeLocht05]     L. De Locht et al., Identification of contributions to nonlinear circuit
                behavior caused by multitone excitation, in *Proceedings of ARFTG
                Conference*, Florida, USA, November 2005, pp. 75–84

[DeMan00]       H. De Man, System design challenges in the post-PC era, in
                *Proceedings of ACM/IEEE Design Automation Conference*, Los
                Angeles, USA, June 2000

[Deneire00a]    L. Deneire, W. Eberle, M. Engels, B. Gyselinckx, S. Thoen, P.
                Vandenameele, and L. Van der Perre, Broadband wireless OFDM
                communication beyond standards, in *Proceedings of International
                Symposium on Mobile Multimedia Systems and Applications*, Delft, The
                Netherlands, December 2000, pp. 71–78

[Deneire00b]    L. Deneire, W. Eberle, M. Engels, B. Gyselinckx, S. Thoen, P. Vandenameele, and L. Van der Perre, Broadband wireless OFDM communication, *Revue HF*, 4:30–38, 2000

[Deneire00c]    L. Deneire, B. Gyselinckx, and M. Engels, Training sequence vs. cyclic prefix – A new look on single carrier communication, in *Proceedings of IEEE Global Telecommunications Conference*, San Francisco, USA, November 2000, pp. 1056–1060

[Deneire03]     L. Deneire, P. Vandenameele, L. Van der Perre, B. Gyselinckx, and M. Engels, A low complexity ML channel estimator for OFDM, *IEEE Trans. Commun.*, 51(2):135–140, 2003

[Despain79]     A. M. Despain, Very Fast Fourier Transform algorithms hardware for implementation, *IEEE Trans. Comput.*, 28(5):333–341, 1979

[Diesing94]     N. Diesing, Why has industry been slow to embrace mixed analog–digital simulation tools? in *Proceedings of IEEE International Symposium on Circuits and Systems*, London, UK, May 1994, pp. 269–274

[Dobrovolný01]  P. Dobrovolný, P. Wambacq, G. Vandersteen, D. Hauspie, S. Donnay, M. Engels, and I. Bolsens, The effective high-level modeling of a 5-GHz RF variable gain amplifier, in *Proceedings of Workshop on Nonlinear Dynamics of Electronic Systems*, Delft, The Netherlands, June 2001

[Doelz57]       M. L. Doelz, F. T. Heald, and D. L. Martin, Binary data transmission techniques for linear systems, *Proc. IRE*, 45:656–661, 1957

[Donnay94]      S. Donnay, K. Swings, G. Gielen, W. Sansen, W. Kruiskamp, and D. Leenaerts, A methodology for analog design automation in mixed-signal ASICs, in *Proceedings of European Design and Test Conference*, February 1994, pp. 530–534

[Donnay00]      S. Donnay et al., Chip-package co-design of a low-power 5-GHz RF front-end, *Proc. IEEE*, 88(10):1583–1597, 2000

[Doufexi02]     A. Doufexi, S. Armour, M. Butler, A. Nix, D. Bull, and J. McGeehan, A comparison of the HIPERLAN/2 and IEEE 802.11a wireless LAN standards, *IEEE Commun. Mag.*, 40(5):172–180, 2002

[Duhamel90]     P. Duhamel and M. Vetterli, Fast Fourier Transforms: A tutorial review and a state of the art, *IEEE Signal Process. Mag.*, 19:259–299, 1990

[DuttaRoy99]    A. Dutta-Roy, Networks for homes, *IEEE Spectrum*, 36(12):26–33, 1999

[DVB-H]         ETSI EN 302 304, Digital Video Broadcasting (DVB); Transmission System for Handheld Terminals (DVB-H), V1.1.1, November 2004

[DVB-T]         ETSI EN 300 744, Digital Video Broadcasting (DVB); Framing Structure, Channel Coding and Modulation for Digital Terrestrial Television, V1.4.1, January 2001 (created: February 2000)

---

[Eberle97a]     W. Eberle, Wireless LAN activities including ASIC architecture and design methodology, presented at *Workshop of the Innovationskolleg Kommunikationssysteme*, Dresden, Germany, December 1997

[Eberle97b]     W. Eberle, L. Van der Perre, B. Gyselinckx, M. Engels, and I. Bolsens, Design aspects of an OFDM-based wireless LAN with regard to ASIC integration, in *Proceedings of International OFDM Workshop*, Braunschweig, Germany, September 1997

**[Eberle99a]**      W. Eberle, M. Badaroglu, V. Derudder, S. Thoen, P. Vandenameele, L. Van der Perre, M. Vergara, B. Gyselinckx, M. Engels, and I. Bolsens, Flexible OFDM transceiver for high-speed WLAN, in *Proceedings of IEEE Vehicular Technology Conference*, Amsterdam, The Netherlands, September 1999, pp. 2677–2681

**[Eberle99b]**      W. Eberle, L. Van der Perre, B. Gyselinckx, M. Engels, and S. Thoen, European Patent EP1030489 Multicarrier transceiver based on European Patent EP1083721, filed 1999 and granted on 6 July 2004

**[Eberle00a]**      W. Eberle, M. Badaroglu, V. Derudder, S. Thoen, P. Vandenameele, L. Van der Perre, M. Vergara, B. Gyselinckx, M. Engels, and I. Bolsens, A digital 80 Mb/s OFDM transceiver IC for wireless LAN in the 5-GHz band, in *IEEE International Solid-State Circuits Conference Digest*, San Francisco, USA, February 2000, pp. 74–75 and 448

**[Eberle00b]**      W. Eberle, M. Badaroglu, V. Derudder, L. Van der Perre, M. Vergara, B. Gyselinckx, M. Engels, I. Bolsens, and H. De Man, A flexible OFDM transceiver ASIC for high-speed wireless local networks, in *Proceedings of International Conference on Telecommunications*, Acapulco, Mexico, May 2000, pp. 1122–1128

**[Eberle00c]**      W. Eberle, L. Deneire, H. De Man, B. Gyselinckx, and M. Engels, Automatic gain control for OFDM-based wireless burst receivers, in *Proceedings of International OFDM Workshop*, Hamburg, Germany, September 2000

**[Eberle01a]**      W. Eberle, V. Derudder, L. Van der Perre, G. Vanwijnsberghe, M. Vergara, L. Deneire, B. Gyselinckx, M. Engels, I. Bolsens, and H. De Man, A digital 72 Mb/s 64-QAM OFDM transceiver for 5-GHz wireless LAN in 0.18 μm CMOS, in *IEEE International Solid-State Circuits Conference Digest*, San Francisco, USA, February 2001, pp. 336–337 and 462

**[Eberle01b]**      W. Eberle, V. Derudder, G. Vanwijnsberghe, M. Vergara, L. Deneire, L. Van der Perre, M. Engels, I. Bolsens, and H. De Man, 80 Mb/s QPSK and 72 Mb/s 64-QAM flexible and scalable digital OFDM transceiver ASICs for wireless local area networks in the 5-GHz band, *IEEE J. Solid-State Circuits*, 36(11):1829–1838, 2001

**[Eberle02a]**      W. Eberle, J. Tubbax, B. Côme, S. Donnay, G. Gielen, and H. De Man, OFDM–WLAN receiver performance improvement using digital compensation techniques, in *Proceedings of IEEE Radio and Wireless Conference*, Boston, USA, August 2002, pp. 111–114

**[Eberle02b]**      W. Eberle, B. Côme, S. Donnay, G. Gielen, and H. De Man, Mixed-signal compensation techniques for low-cost 802.11a receiver front-ends, in *Proceedings of Communications Design Conference*, San Jose, USA, September 2002

**[Eberle02c]**      W. Eberle, Putting it all together, in *Wireless OFDM Systems: How to Make Them Work?* M. Engels, Ed. Boston: Kluwer, 2002, pp. 151–189

**[Eberle02d]**      W. Eberle, European Patent EP03447080.7 Device with front-end reconfiguration, based on US Patent US60/370,642 A Wireless Communication Device, filed 2002

**[Eberle02e]**      W. Eberle, Flexible devices without shipping the engineer along with it: Panel on 'WLAN, WPAN and IP: The PACWOMAN approach', presented at *International Symposium on Wireless Personal Multimedia Communications*, Honolulu, USA, 27–30 October 2002

| [Eberle03] | W. Eberle, G. Vandersteen, P. Wambacq, S. Donnay, G. Gielen, and H. De Man, Behavioral modeling and simulation of a mixed analog/ digital automatic gain control loop in a 5-GHz WLAN receiver, in *Proceedings of Design, Automation and Test in Europe Conference*, München, Germany, March 2003, pp. 642–647 |
|---|---|
| [Ebert99] | J.-P. Ebert and A. Wolisz, Combined tuning of RF power and medium access control for WLANs, in *Proceedings of IEEE International Workshop on Mobile Multimedia Communications*, November 1999, pp. 74–82 |
| [Edenfeld04] | D. Edenfeld, A. B. Kahng, M. Rodgers, and Y. Zorian, 2003 technology roadmap for semiconductors, *IEEE Comput. Mag.*, 37(1):47–56, 2004 |
| [Edfors98] | O. Edfors, M. Sandell, J. J. van de Beek, S. K. Wilson, and P. O. Börjesson, OFDM channel estimation by singular value decomposition, *IEEE Trans. Commun.*, 46(7):931–939, 1998 |
| [Eklund02] | C. Eklund, R. B. Marks, K. L. Stanwood, and S. Wang, IEEE Standard 802.16: A technical overview of the WirelessMAN air interface for broadband wireless access, *IEEE Commun. Mag.*, 40(6):98–107, 2002 |
| [Elliott82] | D. F. Elliott and K. R. Rao, *Fast Transforms – Algorithms, Analyses, Applications*. New York: Academic, 1982 |
| [Elwan98] | H. O. Elwan, T. B. Tarim, and M. Ismail, A digitally controlled dB-linear CMOS AGC for low voltage mixed signal applications, in *Proceedings of Midwest Symposium on Circuits and Systems*, Notre Dame, USA, August 1998, pp. 423–425 |
| [Engels98] | M. Engels, W. Eberle, and B. Gyselinckx, Design of a 100-Mbps wireless local area network, in *Proceedings of URSI International Symposium on Signals, Systems, and Electronics*, Pisa, Italy, September/October 1998, pp. 253–256 |
| [Englund97] | C. Englund, Future directions of personal multimedia communication space: Multimedia over IP, Technical Report, MIT Internet & Telecoms Convergence Consortium, 15 August 1997 |

| [Favalli96] | M. Favalli, L. Benini, and G. De Micheli, Design for testability of gated-clock FSMs, in *Proceedings of European Design and Test Conference*, Paris, France, March 1996, pp. 589–596 |
|---|---|
| [Ferguson68] | M. J. Ferguson, Communication at low data rates – Spectral analysis receivers, *IEEE Trans. Commun. Technol.*, 16(5):657–668, 1968 |
| [Ferrari99] | A. Ferrari and A. Sangiovanni-Vincentelli, System design: Traditional concepts and new paradigms, in *International Conference on Computer Design*, Austin, USA, pp. 2–12, October 1999 |
| [Fertner97] | A. Fertner and C. Sölve, An adaptive gain control with a variable step size for use in high-speed data communication systems, *IEEE Trans. Circuits Syst. II*, 44(11):962–966, 1997 |
| [Flament02] | M. Flament, Broadband wireless OFDM systems, Ph.D. Dissertation, Chalmers University of Technology, Göteborg, Sweden, November 2002 |
| [Fluckiger95] | F. Fluckiger, *Understanding Networked Multimedia*. Englewood Cliffs, NJ: Prentice-Hall, 1995 |

| | |
|---|---|
| **[Fodor03]** | G. Fodor, A. Eriksson, and A. Tuoriniemi, Providing quality of service in always best connected networks, *IEEE Commun. Mag.*, 40(7):154–163, 2003 |
| **[Fort03a]** | A. Fort, J.-W. Weijers, V. Derudder, W. Eberle, and A. Bourdoux, A performance and complexity comparison of auto-correlation and cross-correlation for OFDM burst synchronization, in *Proceedings of IEEE International Conference on Acoustics, Speech, and Signal Processing*, Hong Kong, China, April 2003, pp. 341–344 |
| **[Fort03b]** | A. Fort and W. Eberle, Synchronization and AGC proposal for IEEE 802.11a burst OFDM systems, in *Proceedings of IEEE Global Telecommunications Conference,* San Francisco, USA, December 2003, pp. 1335–1338 |
| **[Fujisawa03]** | T. Fujisawa et al., A single-chip 802.11a MAC/PHY with a 32-b RISC processor, *IEEE J. Solid-State Circuits*, 38(11):2001–2009, 2003 |

| | |
|---|---|
| **[Gajski83]** | D. Gajski and R. H. Kuhn, New VLSI tools, *IEEE Comput.*, 16(2):11–14, 1983 |
| **[Gardner96]** | F. M. Gardner and J. D. Baker, *Simulation Techniques: Models of Communication Signals and Processes.* New York: Wiley, 1996 |
| **[Getreu90]** | I. E. Getreu, Behavioral modeling of analog blocks using the Saber simulator, in *Proceedings of Midwest Symposium on Circuits and Systems*, August 1989, pp. 977–980 |
| **[Ghosh96]** | M. Ghosh, Analysis of the effect of impulsive noise on MC and SC QAM systems, *IEEE Trans. Commun.*, 44(2):145–147, 1996 |
| **[Gielen00]** | G. Gielen and R. Rutenbar, Computer-aided design of analog and mixed-signal integrated circuits, *Proc. IEEE*, 88(12):1825–1854, 2000 |
| **[Gielen02]** | G. Gielen, Modeling and analysis techniques for system-level architectural design of telecom front-ends, *IEEE Trans. Microwave Theory Techn.*, 50(1):360–368, 2002 |
| **[Givargis02]** | T. Givargis, F. Vahid, and J. Henkel, System-level exploration for Pareto-optimal configurations in parameterized system-on-a-chip, *IEEE Trans. VLSI Syst.*, 10(4):416–422, 2002 |
| **[Goering03]** | R. Goering, Show time for ESL design? *EEdesign*, 30 June 2003 |
| **[Goffioul02]** | M. I. Goffioul, P. Wambacq, G. Vandersteen, and S. Donnay, Analysis of nonlinearities in RF front-end architectures using a modified Volterra series approach, in *Proceedings of Design, Automation and Test in Europe Conference*, Paris, France, March 2002, pp. 352–356 |
| **[Goldberg62]** | B. Goldberg, Applications of statistical communications theory, in *Proceedings of West Point Army Conference*, 1962; reprinted in *IEEE Commun. Mag.*, 26–33, 1981 |
| **[Gordon02]** | R. Gordon, A silicon virtual prototype is key in achieving design closure, EE Times, 19 August 2002 |
| **[Gozdecki03]** | J. Gozdecki, A. Jajszczyk, and R. Stankiewicz, Quality of service terminology in IP networks, *IEEE Commun. Mag.*, 41(3):153–159, 2003 |
| **[Gupta01]** | R. Gupta and S. Rawat, The next HDL: If C++ is the answer, what was the question? in *Proceedings of ACM/IEEE Design Automation Conference*, June 2001, pp. 71–72 |

[Gutierrez01]  J. A. Gutierrez, M. Naeve, E. Callaway, M. Bourgeois, V. Mitter, and B. Heile, IEEE 802.15.4: A developing standard for low-power low-cost wireless personal area networks, *IEEE Network*, 15(5):12–19, 2001

[Gyselinckx98]  B. Gyselinckx, W. Eberle, M. Engels, C. Schurgers, S. Thoen, P. Vandenameele, and L. Van der Perre, A flexible architecture for future wireless local area networks, in *Proceedings of International Conference on Telecommunications*, Chalkidiki, Greece, June 1998, pp. 115–119

[Gyselinckx99]  B. Gyselinckx, W. Eberle, M. Engels, and M. Vergara, A 256-point FFT/IFFT for a 100 Mbit/s orthogonal frequency division multiplex modem, in *Proceedings of International Conference on VLSI Design*, Goa, India, January 1999

---

[H1]  ETSI ETS 300 652, HIPERLAN Type 1; Functional Specification, October 1996 (created: November 1991)

[H2-MAC]  ETSI ETS 101 761-1, Broadband Radio Access Networks (BRAN); HIPERLAN Type 2; Data Link Control (DLC) Layer; Part 1: Basic Data Transport Functions, April 2000 (created: May 1999)

[H2-PHY]  ETSI ETS 101 475, Broadband Radio Access Networks (BRAN); HIPERLAN Type 2; Physical (PHY) Layer, April 2000 (created: December 1997)

[Hajimiri98]  A. Hajimiri and T. Lee, A general theory of phase noise in electrical oscillators, *IEEE J. Solid-State Circuits*, 33(2):179–194, 1998

[Halim94]  R. Y. Halim, J. Harris, M. Chadwick, T. Quan, N. Diesing, and E. MacRobbie, Mixed analog–digital simulation: The tools are here…is anyone really using them? in *Proceedings of IEEE International Symposium on Circuits and Systems*, May/June 1994, pp. 269–274

[Harame03]  D. L. Harame et al., Design automation methodology and rf/analog modeling for rf CMOS and SiGe BiCMOS technologies, *IBM J. Res. Dev.*, 47(23):139–175, 2003

[Haroun03]  I. Haroun and F. Gouin, WLANs meet fiber optics – Evaluating 802.11a WLANs over fiber optics links, *RF Des. Mag.*, 36–39, 2003

[Harris96]  Harris Semiconductor Corp., *1996 Wireless Communications Design Seminar Handbook*, March 1996

[Hashemi93]  H. Hashemi, The indoor radio propagation channel, *Proc. IEEE*, 81(7):943–968, 1993

[Hazy97]  L. Hazy and M. El-Tanany, Synchronization of OFDM systems over frequency selective fading channels, in *Proceedings of IEEE Vehicular Technology Conference*, 1997, pp. 2094–2098

[He96]  S. He and M. Torkelson, A new approach to pipeline FFT processor, in *Proceedings of IEEE International Parallel Processing Symposium*, Honolulu, USA, April 1996, pp. 766–770

[He98]  S. He and M. Torkelson, Design and implementation of a 1024-point pipeline FFT processor, in *Proceedings of Custom Integrated Circuits Conference*, Santa Clara, USA, May 1998, pp. 131–134

[Heideman84]  M. T. Heideman, D. H. Johnson, and C. S. Burrus, Gauss and the history of the fast Fourier transform, *IEEE Signal Proc. Mag.*, 1(4):14–21, 1984

[Henry02]  P. S. Henry and H. Luo, WiFi: What's next? *IEEE Commun. Mag.*, 40(12):66–72, 2002

[Hirosaki81]   B. Hirosaki, An orthogonally multiplexed QAM system using the Discrete Fourier Transform, *IEEE Trans. Commun.*, 29:982–989, 1981

[Hollemans94]   W. Hollemans and A. Verschoor, Performance study of WaveLAN and Altair radio-LANs, in *Proceedings of IEEE International Symposium on Personal, Indoor, and Mobile Radio Communications*, September 1994, pp. 831–837

[Honcharenko97]   W. Honcharenko, J. P. Kruys, D. Y. Lee, and N. J. Shah, Broadband wireless access, *IEEE Commun. Mag.*, 35(1):20–26, 1997

[Honkasalo02]   H. Honkasalo, K. Pehkonen, M. T. Niemi, and A. T. Leino, WCDMA and WLAN for 3G and beyond, *IEEE Wireless Commun.*, 9(2):14–18, 2002

[Huisken98]   J. Huisken, F. van de Laar, M. Bekooij, G. Gielis, P. Gruijters, and F. Welten, A power-efficient single-chip OFDM demodulator and Channel decoder for multimedia broadcasting, *IEEE J. Solid-State Circuits*, 33(11):1793–1798, 1998

[Hulbert96]   A. P. Hulbert, A general purpose digital demodulator for VSAT and mobile spread spectrum CDMA signal reception, in *Proceedings of ESA Workshop on DSP Techniques Applied to Space Communications*, Barcelona, Spain, September 1996, pp. 11.45–11.59

[Huys03]   R. Huys, Optimalisatie van het acquisiteproces in laagvermogen draadloze communicatiesystemen, M.S. Thesis, Katholieke Universiteit Leuven, Belgium, May 2003

[IEEE802.11a]   IEEE Std. 802.11a, Amendment 1 to 802.11: High-speed physical layer in the 5-GHz band, 1999

[IEEE802.11b]   IEEE Std. 802.11b, Supplement to 802.11: Higher speed physical layer (PHY) extension in the 2.4-GHz band, 1999

[IEEE802.11g]   IEEE Std. 802.11g, Supplement to 802.11: Further higher data rate extension in the 2.4-GHz band, June 2003

[IEEE802.11n]   IEEE Std. 802.11n, Supplement to 802.11: Enhancements for higher throughput, draft proposal, January 2006

[IEEE802.16]   IEEE Std. 802.16-2004, Standard for local and metropolitan area networks Part 16: Air Interface for Fixed Broadband Wireless Access Systems, 2004

[IEEE802.16e]   IEEE Std. 802.16E-2005, Standard for local and metropolitan area networks Part 16: Air Interface for Fixed and Mobile Broadband Wireless Access Systems Amendment for Physical and Medium Access Control Layers for Combined Fixed and Mobile Operation in Licensed Bands, January 2006

[IEEE830]   IEEE Std. 830, Recommended practice for software requirements specifications, June 1998

[IEEE1233]   IEEE Std. 1233, Guide for developing system requirements specifications, 1998

[Ikeda02]   N. Ikeda, The spectrum as commons, RIETI[167] Discussion Paper Series 02-E002, March 2002

---

[167] RIETI is the Research Institute of Economy, Trade, and Industry located in Tokyo, Japan.

[Jacome96]     M. F. Jacome and S. W. Director, A formal basis for design process planning and management, *IEEE Trans. CAD*, 15(10):1197–1211, 1996

[Jakes93]      W. C. Jakes, *Microwave Mobile Communications*. New York: IEEE, 1993

[Janssen96]    G. J. M. Janssen, P. A. Stigter, and R. Prasad, Wideband indoor channel measurements and BER analysis of frequency selective multipath channels at 2.4, 4.75, and 11.5 GHz, *IEEE Trans. Commun.*, 44(10):1272–1288, 1996

[Jeruchim92]   M. C. Jeruchim, P. Balaban, and K. S. Shanmugan, *Simulation of Communication Systems*. New York: Plenum, 1992

[Jia00]        Q.-W. Jia and G. Mathew, A novel AGC scheme for DFE read channels, *IEEE Trans. Magn.*, 36(5):2210–2212, 2000

[Johansson99]  S. Johansson, D. Landström, and P. Nilsson, Silicon realization of an OFDM synchronization algorithm, in *IEEE International Conference on Electronics, Circuits, and Systems*, 1999, pp. 319–322

[Jones98]      V. K. Jones and G. C. Raleigh, Channel estimation for wireless OFDM systems, in *Proceedings of IEEE Global Telecommunications Conference*, Sydney, Australia, November 1998, pp. 980–985

[Kabal86]      P. Kabal and B. Sayar, Performance of fixed-point FFT's: Rounding and scaling considerations, in *Proceedings of IEEE International Conference on Acoustics, Speech, and Signal Processing*, Tokyo, Japan, March 1986, pp. 221–224

[Kabulepa01]   L. D. Kabulepa, M. Glesner, and T. Kella, Finite-precision analysis of an OFDM burst synchronization scheme, in *Proceedings of IEEE Global Telecommunications Conference*, San Antonio, USA, November 2001, pp. 310–314

[Kaleh95]      G. K. Kaleh, Channel equalization for block transmission systems, *IEEE J. Sel. Areas Commun.*, 13(1):110–121, 1995

[Kalet89]      I. Kalet, The multitone channel, *IEEE Trans. Commun.*, 37(2):119–124, 1989

[Kalliokulju01] J. Kalliokulju, P. Meche, M. J. Rinne, J. Vallström, P. Varshney, and S.-G. Häggman, Radio access selection for multistandard terminals, *IEEE Commun. Mag.*, 39(10):116–124, 2001

[Kandukuri02]  S. Kandukuri and S. Boyd, Optimal power control in interference-limited fading wireless channels with outage-probability specifications, *IEEE Trans. Wireless Commun.*, 1(1):46–55, 2002

[Karaoguz01]   J. Karaoguz, High-rate wireless personal area networks, *IEEE Commun. Mag.*, 39(12):96–102, 2001

[Keutzer00]    K. Keutzer, S. Malik, A. R. Newton, J. M. Rabaey, and A. Sangiovanni-Vincentelli, System-level design: Orthogonalization of concerns and platform-based design, *IEEE Trans. Comput. Aid. Des.*, 19(12):1523–1543, 2000

[Khaled05]     N. Khaled, S. Thoen, and L. Deneire, Optimizing the joint transmit and receive MMSE design using mode selection, *IEEE Trans. Commun.*, 53(4):730–737, 2005

[Khoury98]     J. M. Khoury, On the design of constant settling time AGC circuits, *IEEE Trans. Circuits Syst. II*, 45(3):283–294, 1998

[Kienhuis99]     A. C. J. Kienhuis, Design space exploration of stream-based dataflow
                 architectures, Ph.D. Dissertation, Delft University of Technology, The
                 Netherlands, January 1999

[Kirsch69]       A. L. Kirsch, P. R. Gray, and D. W. Hanna, Field-test results of the
                 AN/GSC-10 (KATHRYN) digital data terminal, *IEEE Trans. Commun.
                 Technol.*, 17(2):118–128, 1969

[Kneip02]        J. Kneip et al., Single chip programmable baseband ASSP for 5 GHz
                 wireless LAN applications, *IEICE Trans. Electron.*, 85-c(2):359–367,
                 2002

[Kumar92]        K. Kumar and R. J. Welke, Methodology engineering: A proposal
                 for situation-specific methodology construction, in *Challenges and
                 Strategies for Research in Systems Development*. New York: Wiley,
                 1992, pp. 257–269

---

[Lambrette97]    U. Lambrette, M. Speth, and H. Meyr, OFDM burst frequency
                 synchronization by single carrier training data, *IEEE Commun.
                 Lett.*, 1(2):46–48, 1997

[Lampinen00]     J. Lampinen, Multiobjective nonlinear Pareto-optimization. Pre-
                 investigation Report, Lappeenranta University of Technology

Lanschützer03]   C. Lanschützer, A. Springer, L. Maurer, Z. Boos, and R. Weigel,
                 Integrated adaptive LO leakage cancellation for W-CDMA direct
                 upconversion transmitters, in *Proceedings of IEEE Radio Frequency
                 Integrated Circuits Symposium*, Philadelphia, USA, June 2003, pp. 19–
                 22

[Lansford01]     J. Lansford, A. Stephens, and R. Nevo, Wi-Fi (802.11b) and Bluetooth:
                 Enabling coexistence, *IEEE Network*, 15:20–27, 2001

[Lee96]          E. Lee and A. Sangiovanni-Vincentelli, Comparing models of
                 computation, in *Proceedings of International Conference on Computer-
                 Aided Design*, San Jose, USA, 1996, pp. 234–241

[Leenaerts01]    D. Leenaerts, G. Gielen, and R. A. Rutenbar, CAD solutions and
                 outstanding challenges for mixed-signal and RF IC design, in
                 *Proceedings of IEEE/ACM International Conference on Computer-
                 Aided Design*, San Jose, USA, 4–8 November 2001, pp. 270–277

[Leeson66]       D. B. Leeson, A simple model of feedback oscillator noise spectrum,
                 *Proc. IEEE*, 54(2):329–330, 1966

[LeFloch89]      B. Le Floch, R. Halbert-Lassalle, and D. Castelain, Digital sound
                 broadcasting to mobile receivers, *IEEE Trans. Consum. Electron.*,
                 35(3):493–503, 1989

[Lennard00]      C. K. Lennard, P. Schaumont, G. de Jong, A. Haverinen, and P. Hardee,
                 Standards for system-level design: Practical reality or solution in search
                 of a question, in *Proceedings of Design, Automation and Test in
                 Europe Conference*, Paris, France, March 2000, pp. 576–583

[Lieverse99]     P. Lieverse, P. van der Wolf, E. Deprettere, and K. Vissers, A
                 methodology for architecture exploration of heterogeneous signal
                 processing systems, in *Proceedings of IEEE Workshop on Signal
                 Processing Systems*, Taipei, Taiwan, October 1999, pp. 181–190

[Lindoff00]      B. Lindoff, Using a direct conversion receiver in EDGE terminals – A
                 new DC offset compensation algorithm, in *Proceedings of IEEE
                 International Symposium on Personal, Indoor, and Mobile Radio
                 Communications*, London, UK, September 2000, pp. 959–963

| | |
|---|---|
| [Lovrich88] | A. Lovrich, G. Troullinos, and R. Chirayil, An all digital automatic gain control, in *Proceedings of IEEE International Conference on Acoustics, Speech, and Signal Processing*, April 1988, pp. 1734–1737 |
| [Lucky03a] | R. W. Lucky, Down into darkness or up into fog, *IEEE Spectrum*, 40(3):76, 2003 |
| [Luise96] | M. Luise and R. Reggiannnini, Carrier frequency acquisition and tracking for OFDM systems, *IEEE Trans. Commun.*, 44(11):1590–1598, 1996 |
| [Luo03] | H. Luo, Z. Jiang, B.-J. Kim, N. K. Shankaranarayanan, and P. Henry, Integrating wireless LAN and cellular data for the enterprise, *IEEE Internet Comput.*, 7(2):25–33, 2003 |

---

| | |
|---|---|
| [Malmgren96] | G. Malmgren, Impact of carrier frequency offset, Doppler spread and time synchronization errors in OFDM based single frequency networks, in *Proceedings of IEEE Global Telecommunications Conference*, 1996, pp. 729–733 |
| [Mandl00] | C. Mandl, M. Bacher, G. Krampl, and F. Kuttner, 0.35 µm COFDM receiver chip for DVB-T, in *IEEE International Solid-State Circuits Conference Digest*, San Francisco, USA, February 2000, pp. 76–77 |
| [Mannion03] | P. Mannion, Smart antenna boost IQ of WLANs, startup says, in *EETimes/CommsDesign*, 18 August 2003 |
| [Martone00] | M. Martone, On the necessity of high performance RF front-ends in broadband wireless access employing multicarrier modulations (OFDM), in *Proceedings of IEEE Global Telecommunications Conference*, San Francisco, USA, November 2000, pp. 1407–1411 |
| [Mayaram00] | K. Mayaram, D. C. Lee, S. Moinian, D. A. Rich, and J. Roychowdhury, Computer-aided circuit analysis tools for RFIC simulation: Algorithms, features, and limitations, *IEEE Trans. Circuits Syst. II*, 47(4):274–286, 2000 |
| [McCain] | D. McCain and G. Xu, Rapid prototyping for a high speed wireless local area network radio (http://ww.ednc.com/products/aptix/db/Rapid_prototype.pdf) |
| [McDermott97] | T. McDermott, P. Ryan, M. Shand, D. Skellern, T. Percival, and N. Weste, A wireless LAN demodulator in a Pamette: Design and experience, in *Proceedings of IEEE Symposium on FPGA-Based Custom Computing Machines*, Napa Valley, USA, April 1997, pp. 40–45 |
| [Medbo98] | J. Medbo and P. Schramm, Channel models for HIPERLAN/2 in different indoor scenarios, ETSI EP BRAN – 3ERI085B, March 1998 |
| [Medbo99] | J. Medbo, H. Hallenberg, and J.-E. Berg, Propagation characteristics at 5 GHz in typical radio-LAN scenarios, in *Proceedings of IEEE Vehicular Technology Conference Spring*, Houston, USA, May 1999, pp. 185–189 |
| [Medbo00] | J. Medbo and J.-E. Berg, Simple and accurate path loss modeling at 5 GHz in indoor environments with corridors, *Proc. IEEE VTC*, September 2000, pp. 30–36 |
| [MEDEA02] | MEDEA+, *The MEDEA+ Design Automation Roadmap*, 3rd release, 2002 |
| [MEDEA05] | MEDEA+, *The MEDEA+ Design Automation Roadmap*, 5th release, 2005 |

[Mehrotra03]     A. Mehrotra, L. van Ginneken, and Y. Trivedi, Design flow and methodology for 50M gate ASIC, in *Proceedings of Asia and South Pacific Design Automation Conference*, January 2003, pp. 640–647

[Mehta01]        M. Mehta, N. Drew, and C. Niedermeier, Reconfigurable terminals: An overview of architectural solutions, *IEEE Commun. Mag.*, 39(8):82–89, 2001

[Melander96]     J. Melander, T. Widhe, and L. Wanhammar, Design of an 128-point FFT processor for OFDM applications, in *Proceedings of IEEE International Conference on Electronics, Circuits, and Systems*, Rhodos, Greece, October 1996, pp. 828–831

[Melgaard94]     H. Melgaard, Identification of physical models, Ph.D. Dissertation, Technical University of Denmark, Lyngby, 1994

[Melsa96]        P. J. W. Melsa, R. C. Younce, and C. E. Rohrs, Impulse response shortening for discrete multitone transceivers, *IEEE Trans. Commun.*, 44(12):1662–1672, 1996

[Merritt03]      R. Merritt, Wi-Fi prices fall, *EE Times*, 29 August, 2003

[Meyr98]         H. Meyr, M. Moeneclay, and S. A. Fechtel, *Digital Communication Receivers: Synchronization, Channel Estimation and Signal Processing*. New York: Wiley, 1998

[Meyr01]         H. Meyr, Why we need all these MIPS in future wireless communication systems, presented at *IEEE Workshop on Signal Processing Systems*, Antwerp, Belgium, September 2001

[Miliozzi00]     P. Miliozzi, K. Kundert, K. Lampaert, P. Good, and M. Chian, A design system for RFIC: Challenges and solutions, *Proc. IEEE*, 88(10):1613–1632, 2000

[Minn00a]        H. Minn, M. Zeng, and V. K. Bhargava, On timing offset estimation for OFDM systems, *IEEE Commun. Lett.*, 4:242–244, 2000

[Minn00b]        H. Minn and V. K. Bhargava, A simple and efficient timing offset estimation for OFDM, *in Proceedings of IEEE Vehicular Technology Conference*, Tokyo, Japan, 2000, pp. 51–55

[Minnis03]       B. J. Minnis and P. A. Moore, A highly digitized multimode receiver architecture for 3G mobiles, *IEEE Trans. Veh. Technol.*, 52(3):637–653, 2003

[Mitola95]       J. Mitola, The software radio architecture, *IEEE Commun. Mag.*, 33(5):26–38, 1995

[Mohr00]         W. Mohr and W. Konhauser, Access network evolution beyond third generation mobile communications, *IEEE Commun. Mag.*, 38(12):122–133, 2000; also: R. Becher, M. Dillinger, M. Haardt, and W. Mohr, Broad-band wireless access and future communication networks, *Proc. IEEE*, 89(1):58–75, 2001

[Moore65]        G. E. Moore, Cramming more components onto integrated circuits, *Electronics*, 38(8):82–85, 1965

[Moose94]        P. H. Moose, A technique for orthogonal frequency division multiplexing frequency offset correction, *IEEE Trans. Commun.*, 42(10):2908–2914, 1994

[Morelli99]      M. Morelli and U. Mengali, An improved frequency offset estimator for OFDM applications, *IEEE Commun. Lett.*, 3(3):75–77, 1999

[Moretti03b]     G. Moretti, Tight squeeze: RF design, in *EDN*, 27 November 2003

[Morgan75]       D. R. Morgan, On discrete-time AGC amplifiers, *IEEE Trans. Circuits Syst.*, 22(2):135–146, 1975

| | |
|---|---|
| **[Moult98]** | L. Moult and J. E. Chen, The K-model: RF IC modelling for communication systems simulation, in *IEE Colloquium on Analog Signal Processing*, October 1998 |
| **[Muller03]** | G. Muller, The arisal of a system architect, 25 April 2003 (www.extra.research.philips.com/natlab/sysarch) |
| **[MüllerW98]** | S. H. Müller-Weinfurtner et al., Analysis of frame and frequency synchronizer for (bursty) OFDM, in *Proceedings of IEEE Global Telecommunications Conference*, November 1998, pp. 201–206 |
| **[MüllerW01]** | S. H. Müller-Weinfurtner, Burst frame and frequency synchronization with a sandwich preamble, in *Proceedings of IEEE Global Telecommunications Conference*, San Antonio, USA, November 2001, pp. 1366–1370 |
| **[Münch00]** | M. Münch, B. Wurth, R. Mehra, J. Sproch, and N. When, Automating RT-level operand isolation to minimize power consumption in datapaths, in *Proceedings of Design, Automation and Test in Europe Conference*, March 2000, pp. 624–631 |
| **[Murthy01]** | P. K. Murthy, E. G. Cohen, and S. Rowland, System Canvas: A new design environment for embedded DSP and telecommunications systems, in *Proceedings of International Symposium on Hardware/Software Codesign*, Copenhagen, Denmark, April 2001, pp. 54–59 |
| **[Muschallik95]** | C. Muschallik, Influence of RF oscillators on an OFDM signal, *IEEE Trans. Consum. Electron.*, 41(3):592–603, 1995 |
| **[Muschallik00]** | C. Muschallik, Ein Beitrag zur Optimierung der Empfangbarkeit von Orthogonal-Frequency-Division-Multiplexing (OFDM) – Signalen, Ph.D. Dissertation, Technische Universität Braunschweig, Germany, 2000 |

---

| | |
|---|---|
| **[Nakagawa03]** | M. Nakagawa, H. Zhang, and H. Sato, Ubiquitous homelinks based on IEEE 1394 and ultra wideband solutions, *IEEE Commun. Mag.*, 41(4):74–82, 2003 |
| **[Ness99]** | R. Ness, S. Thoen, L. Van der Perre, B. Gyselinckx, and M. Engels, Interference mitigation in OFDM-based WLANs, in *Proceedings of Multi-Carrier Spread Spectrum Workshop*, Oberpfaffenhofen, Germany, September 1999 |
| **[Nicolay02]** | T. Nicolay, Theoretische und experimentelle Untersuchung eines breitbandigen, direktmischenden Empfängers unter besonderer Berücksichtigung des Einsatzes von Fuzzy Logic in der Leistungsverstärkung, Ph.D. Dissertation, Universität des Saarlandes, Germany |
| **[Niemann98]** | R. Niemann, *Hardware/Software Co-Design for Data Flow Dominated Embedded Systems*. Boston: Kluwer, 1998 |
| **[Note91]** | S. Note, W. Geurts, F. Catthoor, and H. De Man, Cathedral III: Architecture driven high-level synthesis for high throughput DSP applications, in *Proceedings of ACM/IEEE Design Automation Conference*, San Francisco, USA, June 1991, pp. 597–602 |

---

| | |
|---|---|
| **[O'Brien89]** | J. O'Brien, J. Mather, and B. Holland, A 200 MIPS single-chip 1K FFT processor, in *IEEE International Solid-State Circuits Conference Digest*, San Francisco, USA, pp. 166–167, 1989 |

[Ohlson74]      J. E. Ohlson, Exact dynamics of automatic gain control, *IEEE Trans. Commun.*, 22(1):72–75, 1974

[Ohr04]         S. Ohr, Analog age pronounced live and well, *EE Times*, 18 February 2004

[Ojanpera98]    T. Ojanpera and R. Prasad, An overview of third-generation wireless personal communications: A European perspective, *IEEE Pers. Commun.*, 5(6):59–65, 1998

[Osgood97]      K. Osgood et al., A flexible approach to 5-GHz U-NII band WLAN radio development, in *Proceedings of Workshop on Applications of Radio Science*, Australia, 21–23 September 1997, pp. 175–180

---

[PageJ88]       M. Page-Jones, *Practical Guide to Structured Systems Design*, 2nd edition. Englewood Cliffs, NJ: Prentice-Hall, 1988

[Palicot03]     J. Palicot and C. Roland, A new concept for wireless reconfigurable receivers, *IEEE Commun. Mag.*, 41(7):124–132, 2003

[Panda01]       P. R. Panda, SystemC – A modeling platform supporting multiple design abstractions, in *Proceedings of International Symposium on System Synthesis*, Montreal, Canada, September 2001, pp. 75–80

[Papalambros00] P. Y. Papalambros and D. J. Wilde, *Principles of Optimal Design – Modeling and Computation*. Cambridge: Cambridge University Press, 2000

[Parekh01]      S. N. Parekh, Evolution of wireless home networks: The role of policy-makers in a standards-based market, M.S. Thesis, Massachusetts Institute of Technology, June 2001

[Pareto06]      V. Pareto, *Manuale di economia politica con una introduzione alla scienza sociale*. Milano: Società Editrice Libraria, 1906

[Parhi92]       K. K. Parhi, Systematic synthesis of DSP data format converters using life-time analysis and forward–backward register allocation, *IEEE Trans. Circuits Syst. II*, 39(7):423–440, 1992

[Pasko00]       R. Paško, L. Rijnders, P. Schaumont, S. Vernalde, and D. Durackova, High-performance flexible all-digital quadrature up and down converter chip, in *Proceedings of Custom Integrated Circuits Conference*, Orlando, USA, May 2000, pp. 43–46

[Pasko02]       R. Paško, S. Vernalde, and P. Schaumont, Techniques to evolve a C++ based system design language, in *Proceedings of Design, Automation and Test in Europe Conference*, Paris, France, March 2002, pp. 302–309

[Peled80]       A. Peled and A. Ruiz, Frequency domain data transmission using reduced computational complexity algorithms, in *Proceedings of IEEE International Conference on Acoustics, Speech, and Signal Processing*, Denver, USA, April 1980, pp. 964–967

[Pereira01]     J. M. Pereira, Reconfigurable radio: The evolving perspectives of different players, in *Proceedings of IEEE International Symposium on Personal, Indoor, and Mobile Radio Communications*, September/ October 2001, pp. 79–84

[Perl87]        J. Perl, A. Shpigel, and A. Reichman, Adaptive receiver for digital comm-unication over HF channels, *IEEE J. Sel. Areas Commun.*, 5(2):304–308, 1987

[Peterson96]    L. Peterson and B. Davie, *Computer Networks – A System Approach*. San Francisco: Morgan Kaufman, 1996

[Piaget72]        J. Piaget, in *Proceedings of Workshop 'L'interdisciplinarité –*
                  *Problèmes d'enseignement et de recherche dans les universités'*, Nice,
                  France, September 1970 (OECD 1972), p. 144

[Pollet99]        T. Pollet and M. Peeters, Synchronization with DMT modulation, *IEEE*
                  *Commun. Mag.*, 37(4):80–86, 1999

[Pollet00]        T. Pollet, M. Peeters, M. Moonen, and L. Vandendorpe, Equalization
                  for DMT-based broadband modems, *IEEE Commun. Mag.*, 38(5):106–
                  113, 2000

[Potkonjak99]     M. Potkonjak and J. Rabaey, Algorithm selection: A quantitative
                  optimization-intensive approach, *IEEE Trans. CAD*, 18(5):524–532,
                  1999

[Prasad99]        R. Prasad, J. Schwarz DaSilva, and B. Arroyo-Fernández, Air interface
                  access schemes for wireless communications, *IEEE Commun. Mag.*,
                  37(9):104–105, 1999

[Proakis95]       J. G. Proakis, *Digital Communications*, 3rd edition. New York:
                  McGraw Hill, 1995

[Proakis96]       J. G. Proakis and D. G. Manolakis, *Digital Signal Processing*, 3rd
                  edition. Upper Saddle River, NJ: Prentice-Hall, 1996

[Prodanov01]      V. Prodanov, G. Palaskas, J. Glas, and V. Boccuzzi, A CMOS AGC-
                  less IF strip for Bluetooth, in *Proceedings of European Solid-State*
                  *Circuits Conference*, Villach, Austria, September 2001, pp. 488–491

[Prophet99]       G. Prophet, System-level design languages: To C or not to C? in *EDN*,
                  pp. 135–146, 14 October 1999

[Qu01]            G. Qu, What is the limit of energy saving by dynamic voltage scaling,
                  in *Proceedings of IEEE/ACM International Conference on Computer-*
                  *Aided Design*, San Jose, USA, November 2001, pp. 560–563

[Raab02]          F. H. Raab et al., Power amplifiers and transmitters for RF and
                  microwave, *IEEE Trans. Microwave Theory Techn.*, 50(3):814–826,
                  2002

[Rabaey96]        J. M. Rabaey, *Digital Integrated Circuits – A Design Perspective*, 1st
                  edition. Upper Saddle River, NJ: Prentice-Hall, 1996

[Raivio01]        Y. Raivio, 4G – Hype or reality, in *Proceedings of IEE 3G Mobile*
                  *Communication Technologies Conference*, March 2001, pp. 346–350

[Rappaport02]     T. S. Rappaport, A. Annamalai, R. M. Buehrer, and W. H. Tranter,
                  Wireless communications: Past events and a future perspective, *IEEE*
                  *Commun. Mag.*, 50(5):148–161, 2002

[Razavi97a]       B. Razavi, Design considerations for direct-conversion receivers, *IEEE*
                  *Trans. Circuits Syst. II*, 44(6):428–435, 1997

[Razavi97b]       B. Razavi, *RF Microelectronics*. Englewood Cliffs, NJ: Prentice-Hall,
                  1997

[Razavi99]        B. Razavi, A 2.4 GHz CMOS receiver for IEEE 802.11 wireless LANs,
                  *IEEE J. Solid-State Circuits*, 34(10):1382–1385, 1999

[Rhett02]         W. Rhett Davis et al., A design environment for high-throughput,
                  low-power dedicated signal processing systems, *IEEE J. Solid State*
                  *Circuits*, 37(3):420–431, 2002

[Riezenman01]     M. J. Riezenman, The rebirth of radio, *IEEE Spectrum*, 38(1):62–64,
                  2001

[Rijnders00]   L. Rijnders, P. Schaumont, S. Vernalde, and I. Bolsens, High-level analysis of clock regions in a C++ system description, in *IEICE Trans. Fund. Electron. Commun. Comput. Sci. – Special Section on VLSI Design and CAD Algorithms*, E83-A(12):2631–2632, 2000

[Rissone02]   P. Rissone and G. Cascini, Creativity as means for technical innovation, in *Proceedings of SEFI Annual Conference*, Firenze, Italy, September 2002

[Robles01]   T. Robles et al., QoS support for an all-IP system beyond 3G, *IEEE Commun. Mag.*, 39(8):64–72, 2001

[Rohling97]   H. Rohling, R. Grünheid, and K. Brüninghaus, Comparison of multiple access schemes for an OFDM downlink system, in *Proceedings of International Workshop on Multi-Carrier Spread Spectrum*, Oberpfaffenhofen, Germany, April 1997

[Rose01]   B. Rose, Home networks: A standards perspective, *IEEE Commun. Mag.*, 12:78–85, 2001

[Rosenfield92]   P. L. Rosenfield, The potential of transdisciplinary research for sustaining and extending linkages between the health and social sciences, *Soc. Sci. Med.*, 35(11):1343–1357

[Rowson97]   J. A. Rowson and A. Sangiovanni-Vincentelli, Interface-based design, in *Proceedings of ACM/IEEE Design Automation Conference*, Anaheim, USA, June 1997, pp. 178–183

[Rumbaugh98]   J. Rumbaugh, G. Booch, and I. Jacobson, *The Unified Modeling Language Reference Manual*. Reading, MA: Addison-Wesley, 1998

[Ryan95]   P. Ryan, T. Percival, and D. Skellern, A 16-point FFT IC for wireless communication systems, in *Workshop on Applications of Radio Science (WARS) Digest*, Canberra, Australia, June 1995

[Ryan01]   P. Ryan et al., A single chip PHY COFDM modem for IEEE 802.11a with integrated ADCs and DACs, in *IEEE International Solid-State Circuits Conference Digest*, San Francisco, USA, 2001, pp. 338–339

---

[Sakiyama03]   K. Sakiyama, P. Schaumont, and I. Verbauwhede, Finding the best system design flow for a high-speed JPEG encoder, in *Proceedings of Asia and South Pacific Design Automation Conference*, Kitakyushu, Japan, January 2003, pp. 577–578

[Sakurai03]   T. Sakurai, Perspectives on power-aware electronics, in *IEEE Solid-State Circuits Conference Digest*, San Francisco, USA, February 2003, pp. 26–29

[Saleh83]   A. A. M. Saleh and D. C. Cox, Improving the power-added efficiency of FET amplifiers operating with varying-envelope signals, *IEEE Trans. Microwave Theory Techn.*, 83(1):51–56, 1983

[Saleh87]   A. A. M. Saleh and R. A. Valenzuela, A statistical model for multipath propagation, *IEEE Trans. Sel. Areas Commun.*, 5(2):128–137, 1987

[Saltzberg67]   B. R. Salzberg, Performance of an efficient parallel data transmission system, *IEEE Trans. Commun.*, 15(6):805–813, 1967

[Sampath02]   H. Sampath, S. Talwar, J. Tellado, V. Erceg, and A. Paulraj, A fourth-generation MIMO–OFDM broadband wireless system: Design, performance, and field trial results, *IEEE Commun. Mag.*, 40(9):143–149, 2002

[Sampei92]      S. Sampei and K. Feher, Adaptive DC-offset compensation algorithm for burst mode operated direct conversion receivers, in *Proceedings of IEEE Vehicular Technology Conference*, May 1992, pp. 93–96

[Saracco03]     R. Saracco, Forecasting the future of information technology: How to make research investment more cost-effective? *IEEE Commun. Mag.*, 41(12):38–45, 2003

[Sari94]        H. Sari, G. Karam, and I. Jeanclaude, Frequency-domain equalization of mobile radio and terrestrial broadcast channels, in *Proceedings of IEEE Global Telecommunications Conference*, San Francisco, USA, November 1994, pp. 1–5

[Sari95]        H. Sari, G. Karam, and I. Jeanclaude, Transmission techniques for digital terrestrial TV broadcasting, *IEEE Commun. Mag.*, 33(2):100–109, 1995. See also comments in *IEEE Commun. Mag.*, 33(11):22–26, 1995

[Savage03]      P. Savage, The perfect handheld: Dream on, *IEEE Spectrum*, 40(1):44–46, 2003

[Scaglione99]   A. Scaglione, S. Barbarossa, and G. B. Giannakis, Filterbank transceivers optimizing information rate in block transmissions over dispersive channels, *IEEE Trans. Inform. Theory*, 45(3):1019–1032, 1999

[Schaumont98]   P. Schaumont, S. Vernalde, L. Rijnders, M. Engels, and I. Bolsens, A design environment for the design of complex high-speed ASICs, in *Proceedings of ACM/IEEE Design Automation Conference*, San Francisco, USA, June 1998, pp. 315–320

[Schaumont99a]  P. Schaumont, R. Cmar, S. Vernalde, M. Engels, and I. Bolsens, Hardware reuse at the behavioral level, in *Proceedings of ACM/IEEE Design Automation Conference*, New Orleans, USA, June 1999, pp. 784–789

[Schaumont99b]  P. Schaumont, R. Cmar, S. Vernalde, and M. Engels, A 10-Mb/s upstream cable modem with automatic equalization, in *Proceedings of ACM/IEEE Design Automation Conference*, New Orleans, USA, June 1999, pp. 337–340

[Schaumont01a]  P. Schaumont, I. Verbauwhede, and H. De Man, Post-PC systems, architectures and design challenges, presented at *CANDE 2001 Workshop*, Jackson Hole, USA, September 2001

[Schaumont01b]  P. Schaumont, I. Verbauwhede, M. Sarrafzeadeh, and K. Keutzer, A quick safari through the reconfiguration jungle, in *Proceedings of ACM/IEEE Design Automation Conference*, June 2001, pp. 18–22

[Schlebusch03]  H.-J. Schlebusch et al., Transaction based design: Another buzzword or the solution to a design problem? in *Proceedings of Design, Automation and Test in Europe Conference*, München, Germany, March 2003

[Schmidl97]     T. Schmidl and D. C. Cox, Robust frequency and timing synchronization for OFDM, *IEEE Trans. Commun.*, 45(12):1613–1621, 1997

[Schön83]       D. Schön, *The Reflective Practitioner*. New York: Basic Books, 1983

[Schurgers02a]  C. Schurgers, Energy-aware communication systems, Ph.D. Dissertation, University of California, Los Angeles, USA, November 2002

**[Schurgers02b]**     C. Schurgers, V. Tsiatsis, S. Ganeriwal, and M. Srivastava, Optimizing
                       sensor networks in the energy–latency–density design space, *IEEE
                       Trans. Mobile Comput.*, 1(1):70–80, 2002

**[Schwanenberger03]** T. Schwanenberger, M. Ipek, S. Roth, and H. Schemmann, A multi
                       standard single-chip transceiver covering 5.15 to 5.85 GHz, in *IEEE
                       International Solid-State Circuits Conference Digest*, San Francisco,
                       USA, February 2003, pp. 350–351

**[Shakkottai03]**     S. Shakkottai, T. S. Rappaport, and P. C. Karlsson, Cross-layer design
                       for wireless networks, *IEEE Commun. Mag.*, 41(10):74–80, 2003

**[Shan88]**           T. J. Shan and T. Kailath, Adaptive algorithms with an automatic gain
                       control feature, *IEEE Trans. Circuits Syst. II*, 35(1):122–127, 1988. See
                       also S. Karni and G. Zeng, Comments, with Reply, on 'Adaptive
                       algorithms with an automatic gain control feature' by T. J. Shan and
                       T. Kailath, *IEEE Trans. Circuits Syst. II*, 37(7):974–975, 1990

**[Shen02]**           M. Shen, L.-R. Zheng, and H. Tenhunen, Cost and performance
                       analysis for mixed-signal system implementation: System-on-chip or
                       system-on-package? *IEEE Trans. Electron. Packag. Manufact.*,
                       25(4):262–272, 2002

**[Sheng92]**          S. Sheng, A. Chandrakasan, and R. W. Brodersen, A portable
                       multimedia terminal, *IEEE Commun. Mag.*, 30(12):64–75, 1992

**[Sherif02]**         M. H. Sherif, Intelligent homes: A new challenge in telecommu-
                       nications standardization, *IEEE Commun. Mag.*, 40(1):8, 2002

**[Shibutani91]**      M. Shibutani, T. Kanai, K. Emura, and J. Namiki, Feasibility studies on
                       an optical fiber feeder system for microcellular mobile communication
                       systems, in *Proceedings of IEEE International Conference on
                       Communications*, June 1991, pp. 1176–1181

**[Shimozawa96]**      M. Shimozawa, K. Kawakami, K. Itoh, A. Iida, and O. Ishida, A novel
                       sub-harmonic pumping direct conversion receiver with high
                       instantaneous dynamic range, in *IEEE MTT-S International Microwave
                       Symposium Digest*, San Francisco, USA, June 1996, pp. 819–822

**[Shiue98]**          M.-T. Shiue, K.-H. Huang, C.-C. Lu, C.-K. Wang, and W. I. Way, A
                       VLSI design of dual-loop automatic gain control for dual-mode
                       QAM/VSB CATV modem, in *Proceedings of IEEE International
                       Symposium on Circuits and Systems*, Monterey, USA, May/June 1998,
                       pp. 490–493

**[Siegmund01]**       R. Siegmund and D. Müller, SystemC$^{SV}$: An extension of SystemC for
                       mixed multi-level communication modeling and interface-based system
                       design, in *Proceedings of Design, Automation and Test in Europe
                       Conference*, München, Germany, March 2001, pp. 26–32

**[Siegmund02]**       R. Siegmund and D. Müller, Automatic synthesis of communication
                       controller hardware from protocol specifications, *IEEE Des. Test
                       Comput.*, 19(4):84–95, 2002

**[Simoens02]**        S. Simoens, M. de Courville, F. Bourzeix, and P. de Champs, New I/Q
                       imbalance modeling and compensation in OFDM systems with
                       frequency offset, in *Proceedings of International Symposium on
                       Personal, Indoor, and Mobile Radio Communications*, Lisboa,
                       Portugal, September 2002, pp. 561–566

**[Singh95]**          D. Singh et al., Power conscious CAD tools and methodologies: A
                       perspective, *Proc. IEEE*, 83(4):570–594, 1995

[Sinyanskiy98]   V. Sinyanskiy, J. Cukier, A. Davidson, and T. Poon, Front-end of a digital ATV receiver, *IEEE Trans. Consum. Electron.*, 44(3):817–822, 1998

[Skellern97]   D. J. Skellern et al., A high-speed wireless LAN, *IEEE Micro*, 17(1):40–47, 1997

[Smith97]   P. J. Smith, M. Shafi, and H. Gao, Quick simulation: A review of importance sampling techniques in communications systems, *IEEE J. Sel. Areas Commun.*, 15(4):597–613, 1997

[Smulders02]   P. Smulders, Exploiting the 60-GHz band for local wireless multimedia access: Prospects and future directions, *IEEE Commun. Mag.*, 40(1):140–147, 2002

[Sony99]   Sony International Europe, Preamble structures for Hiperlan type 2 systems, ETSI BRAN, Technical Report, HL13SON1a, 1999

[Souza03]   C. Souza, Intersil, a leading WLAN chip player, exits a market ripe for consolidation, in *EE Times*, 22 July 2003

[Sperling03]   E. Sperling, Is Moore's law irrelevant? in *Electronic News*, 14 August 2003

[Speth97]   M. Speth, F. Classen, and H. Meyr, Frame synchronization in OFDM systems in frequency selective fading channels, in *Proceedings of IEEE Vehicular Technology Conference*, Phoenix, USA, May 1997, pp. 1807–1811

[Speth98]   M. Speth, D. Daecke, and H. Meyr, Minimum overhead burst synchronization for OFDM-based broadband transmission, in *Proceedings of IEEE Global Telecommunications Conference*, Sydney, Australia, November 1998, pp. 2777–2782

[Speth01]   M. Speth, S. A. Fechtel, G. Fock, and H. Meyr, Optimum receiver design for OFDM-based broadband transmission: Part II, *IEEE Trans. Commun.*, 49(4):571–578, 2001

[SRC00]   Semiconductor Research Corporation (SRC), Research needs for mixed-signal technologies. Report of the 2000 Mixed-Signal Task Force, 25 October 2000

[Srikanteswara03]   S. Srikanteswara, R. C. Palat, J. H. Reed, and P. Athanas, An overview of configurable computing machines for software radio handsets, *IEEE Commun. Mag.*, 41(7):134–141, 2003

[Stantchev99]   B. Stantchev and G. Fettweis, Optimum frame synchronization for orthogonal FSK in flat fading channels and one burst application, *Proc. IEEE WCNC*, September 1999, pp. 1070–1074

[Steele00]   R. Steele, Beyond 3G, in *Proceedings of IEEE International Zurich Seminar on Broadband Communications*, 2000, pp. 1–7

[Stoltermann92]   E. Stolterman, How system designers think – About design and methods, *Scand. J. Inform. Syst.*, 4:137–150, 1992

[Struhsaker03]   T. Struhsaker, Trends in WLAN silicon integration, panel on Key technologies enabling success of Wi-Fi public access networks, presented at *IEEE Wireless Communications and Networking Conference*, New Orleans, USA, March 2003

[Swartzlander84]   E. Swartzlander and G. Hallnor, High speed FFT processor implementation, in *Proceedings of IEEE Workshop on VLSI Signal Processing*, November 1984, pp. 27–34

[SystemC]   The Open SystemC Initiative (OSCI) (http://www.systemc.org)

[Taura96]        K. Taura, M. Tsujishita, M. Takeda, H. Kato, M. Ishida, and Y. Ishida,
                 A digital audio broadcasting (DAB) receiver, *IEEE Trans. Consum.
                 Electron.*, 42(3):322–327, 1996

[Teger02]        S. Teger and D. J. Waks, End-user perspectives on home networking,
                 *IEEE Commun. Mag.*, 40(4):114–119, 2002

[Tellado99]      J. Tellado-Mourelo, Peak to average power reduction for multicarrier
                 modulation, Ph.D. Dissertation, Stanford University, USA, September
                 1999

[Thoen02a]       S. Thoen, Transmit optimization for OFDM/SDMA-based wireless
                 local area networks, Ph.D. Dissertation, Katholieke Universiteit
                 Leuven, Leuven, Belgium, May 2002

[Thoen02b]       S. Thoen, L. Van der Perre, and M. Engels, Modeling the channel time-
                 variance for fixed wireless communications, *IEEE Commun. Lett.*,
                 6(8):331–333, 2002

[Thompson83]     C. D. Thompson, Fourier transforms in VLSI, *IEEE Trans. Comput.*,
                 32(11):1047–1057, 1983

[Thomson02]      J. Thomson et al., An integrated 802.11a baseband and MAC processor,
                 in *IEEE International Solid-State Circuits Conference Digest*, San
                 Francisco, USA, 2002, pp. 126–127

[Tolvanen98]     J.-P. Tolvanen, Incremental method engineering with modeling tools,
                 Ph.D. Dissertation, University of Jyväskylä, 1998

[Trân-Thông76]   Trân-Thông and B. Liu, Fixed-point fast Fourier transform error
                 analysis, *IEEE Trans. Acoust. Speech Signal Process.*, 24(6):563–573,
                 1976

[Truman98]       T. E. Truman, T. Pering, R. Doering, and R. W. Brodersen, The
                 InfoPad multimedia terminal: A portable device for wireless
                 information access, *IEEE Trans. Comput.*, 47(10):1073–1087, 1998

[Tubbax04]       J. Tubbax, A digital approach to low-cost low-power broadband radios,
                 Ph.D. Dissertation, Katholieke Universiteit Leuven, Leuven, Belgium,
                 April 2004

[Tufvesson99]    F. Tufvesson, O. Edfors, and M. Faulkner, Time and frequency
                 synchronization for OFDM using PN-sequence preambles, in
                 *Proceedings of IEEE Vehicular Technology Conference*, 1999, pp.
                 2203–2207


[UML]            OMG, Unified Modeling Language (http://www.omg.org/uml)


[Vanassche01]    P. Vanassche, G. Gielen, and W. Sansen, Efficient time-domain
                 simulation of telecom front-ends using a complex damped exponential
                 signal model, in *Proceedings of Design, Automation and Test in Europe
                 (DATE) Conference*, March 2001, pp. 169–175

[vandeBeek95]    J. J. van de Beek, O. Edfors, M. Sandell, S. K. Wilson, and P. O.
                 Börjesson, On channel estimation in OFDM systems, in *Proceedings of
                 IEEE Vehicular Technology Conference*, Chicago, USA, 1995, pp.
                 815–819

[vandeBeek97]    J. J. van de Beek, M. Sandell, and P. O. Börjesson, ML estimation of
                 time and frequency offset in OFDM systems, *IEEE Trans. Signal
                 Process.*, 45:1800–1805, 1997

[Vandenameele00] P. Vandenameele, Space division multiple access for wireless local area networks, Ph.D. Dissertation, Katholieke Universiteit Leuven, Leuven, Belgium, October 2000

[VanderPerre98] L. Van der Perre, S. Thoen, P. Vandenameele, B. Gyselinckx, and M. Engels, Adaptive loading strategy for a high speed OFDM-based WLAN, in *Proceedings of IEEE Global Communications Conference*, Sydney, Australia, November 1998, pp. 1936–1940

[Vandersteen00] G. Vandersteen et al., A methodology for efficient high-level dataflow simulation of mixed-signal front-ends of digital telecom transceivers, in *Proceedings of ACM/IEEE Design Automation Conference*, Los Angeles, USA, June 2000, pp. 440–445

[Vandersteen01] G. Vandersteen, P. Wambacq, S. Donnay, W. Eberle, and Y. Rolain, FAST – An efficient high-level dataflow simulator of mixed-signal front-ends of digital telecom transceivers, in *Low-Power Design Techniques and CAD Tools for Analog and RF Integrated Circuits*, P. Wambacq, G. Gielen, and J. Gerrits, Eds. Boston: Kluwer, 2001, pp. 43–59

[VanDriessche03] J. Van Driessche, G. Cantone, W. Eberle, B. Côme, and S. Donnay, Transmitter cost/efficiency exploration for 5-GHz WLAN, in *Proceedings of IEEE Radio and Wireless Conference*, Boston, USA, August 2003, pp. 35–38

[vanLamsweerde00] A. van Lamsweerde, Requirements engineering in the year 00: A research perspective, in *Proceedings of International Conference on Software Engineering*, Limerick, Ireland, June 2000

[vanLamsweerde03] A. van Lamsweerde and E. Letier, From object orientation to goal orientation: A paradigm shift for requirements engineering, in *Proceedings of Workshop on Radical Innovations of Software and Systems Engineering*, Venice, Italy, 2003

[Vanmeerbeeck01] G. Vanmeerbeeck, P. Schaumont, S. Vernalde, M. Engels, and I. Bolsens, Hardware/software partitioning of embedded system in OCAPI-xl, in *Proceedings of International Symposium on Hardware/Software Codesign*, April 2001, pp. 30–35

[vanNee99] R. van Nee, G. Awater, M. Morikura, H. Takanashi, M. Webster, and K. W. Halford, New high-rate wireless LAN standards, *IEEE Commun. Mag.*, 37(12):82–88, 1999

[VanRompaey96] K. Van Rompaey, D. Verkest, I. Bolsens, and H. De Man, CoWare – A design environment for heterogeneous hardware/software systems, in *Proceedings of European Design Automation Conference*, September 1996, pp. 252–257

[Vassiliou99] I. Vassiliou, Design methodologies for RF and mixed-signal systems, Ph.D. Dissertation, University of California at Berkeley, USA, 1999

[Veithen99] D. Veithen et al., A 70 Mb/s variable-rate DMT-based modem for VSDL, in *IEEE International Solid-State Circuits Conference Digest*, San Francisco, USA, February 1999, pp. 248–249

[Velez02] F. J. Velez and L. M. Correia, Mobile broadband services: Classification, characterization, and deployment scenarios, *IEEE Commun. Mag.*, 40(4):142–150, 2002

[Vergara98a] M. Vergara, M. Strum, W. Eberle, and B. Gyselinckx, A 195 kFFT/s 256-point high performance FFT/IFFT processor for OFDM applications, in *Proceedings of SBT/IEEE International Telecommunications Symposium*, São Paulo, Brazil, August 1998, pp. 273–278

[Vergara98b]    M. Vergara, Projeto de uma macro-célula FFT/IFFT para aplicações sem fio, M.S. Thesis, Universidade de São Paulo, Brazil, 1998

[Verilog]    IEEE Std. 1364–1995, Verilog Language, 1995[168]

[Verkest00]    D. Verkest, J. Kunkel, and F. Schirrmeister, System level design using C++, in *Proceedings of Design, Automation and Test in Europe Conference*, Paris, France, March 2000, pp. 74–81

[Verkest01a]    D. Verkest, W. Eberle, P. Schaumont, B. Gyselinckx, and S. Vernalde, C++ based system design of a 72 Mb/s OFDM transceiver for wireless LAN, in *Proceedings of Custom Integrated Circuits Conference*, San Diego, USA, May 2001, pp. 433–439

[Vermeulen00]    F. Vermeulen, F. Catthoor, D. Verkest, and H. De Man, Formalized three-layer system-level model and reuse methodology for embedded data-dominated applications, *IEEE Trans. VLSI Syst.*, 8(2):207–216, 2000

[Vermeulen02]    F. Vermeulen, Reuse of system-level design components in data-dominated digital systems, Ph.D. Dissertation, Katholieke Universiteit Leuven, December 2002

[Vernalde99]    S. Vernalde, P. Schaumont, and I. Bolsens, An object-oriented programming approach for hardware design, in *Proceedings of IEEE Computer Society Workshop on VLSI*, Orlando, USA, April 1999, pp. 68–73

[VHDL]    IEEE Std. 1076-1993, VHDL Language, 1993[168]

[Victor60]    W. K. Victor and M. H. Brockman, The application of linear servo theory to the design of AGC loops, *Proc. IRE*, 48:234–238, 1960

[Walker94]    S. Walker, A high speed feed forward pseudo automatic gain control circuit for an amplifier cascade, in *IEEE MTT-S International Microwave Symposium Digest*, San Diego, USA, May 1994, pp. 941–944

[Walzman73]    T. Walzman and M. Schwartz, Automatic equalization using the discrete frequency domain, *IEEE Trans. Inform. Theory*, 19(1):59–68, 1973

[Wambacq98]    P. Wambacq and W. M. Sansen, *Distortion Analysis of Analog Integrated Circuits*. Boston: Kluwer, 1998

[Wambacq00]    P. Wambacq, P. Dobrovolný, S. Donnay, M. Engels, and I. Bolsens, Compact modeling of nonlinear distortion in analog communication circuits, in *Proceedings of Design, Automation and Test in Europe Conference*, Paris, France, March 2000, pp. 350–354

[Wambacq01]    P. Wambacq, G. Vandersteen, J. Phillips, J. Roychowdhury, W. Eberle, B. Yang, D. Long, and A. Demir, CAD for RF circuits, in *Proceedings of Design, Automation and Test in Europe Conference*, München, Germany, March 2001, pp. 520–526

[Wambacq02a]    P. Wambacq, G. Vandersteen, P. Dobrovolny, M. Goffioul, W. Eberle, M. Badaroglu, and S. Donnay, High-level simulation and modeling tools for mixed-signal front-ends of wireless systems, in *Proceedings of Workshop on Advances in Analog Circuit Design (AACD)*, Spa, Belgium, March 2002

---

[168] Years given are indicative only; we do not refer to a particular feature here.

| | |
|---|---|
| [Wambacq02b] | P. Wambacq, G. Vandersteen. P. Dobrovolny, M. Goffioul, W. Eberle, M. Badaroglu, and S. Donnay, High-level simulation and modeling tools for mixed-signal front-ends of wireless systems, in *Analog Circuit Design: Structured Mixed-Mode Design, Multi-Bit Sigma Delta Converters, Short Range RF Circuits*, M. Steyaert, A. H. M. van Roermund, J. H. Huijsing, Eds. Boston: Kluwer, 2002, pp. 77–94 |
| [Wang89] | P. C. Wang and C. R. Ward, A software AGC scheme for integrated communication receivers, in *Proceedings of IEEE National Aerospace and Electronics Conference*, 1989, pp. 2085–2091 |
| [Weber75] | W. J. Weber, Decision-directed automatic gain control for MAPSK systems, *IEEE Trans. Commun.*, 23(5):510–517, 1975 |
| [Weinstein71] | S. B. Weinstein and P. M. Ebert, Data transmission by frequency division multiplexing using the Discrete Fourier Transform, *IEEE Trans. Commun.*, 19(5):628–634, 1971 |
| [Weiser91] | M. Weiser, The computer for the 21st century, *Sci. Am.*, 265(3):94–104, 1991 |
| [Weste97] | N. Weste et al., A 50-MHz 16-point FFT processor for WLAN applications, in *Proceedings of IEEE Custom Integrated Circuits Conference*, Santa Clara, USA, May 1997, pp. 457–460 |
| [Wheeler03] | B. Wheeler and L. Gwennap, *A Guide to Wireless LAN Chip Sets*, 1st edition. Sunnyvale, CA: The Linley Group, 2003 |
| [Wickelgren96] | I. J. Wickelgren, Local area networks go wireless, *IEEE Spectrum*, 33(9):34–40, 1996 |
| [Wiesler01] | A. Wiesler, Parametergesteuertes Software Radio für Mobilfunksysteme, Ph.D. Dissertation, Universität Karlsruhe (TH), Germany, May 2001 |
| [Williams03] | R. Williams, Improving efficiency when detecting WLAN preambles, in *CommsDesign*, 18 November 2003 (http://www.commsdesign.com/story/OEG20031118S0024) |
| [Wittmann03] | R. Wittmann, J. Hartung, H.-J. Wassener, G. Tränkle, and M. Schröter, Hot topic session: RF design technology for highly integrated communication systems, in *Proceedings of Design, Automation and Test in Europe Conference*, Munich, Germany, 2003, pp. 842–847 |
| [Wojituk05] | J. Wojituk, Analysis of frequency conversion for M-QAM and M-PSK modems, *Dissertation, Chalmers Univ. of Tech.*, June 2005 |
| [Wouters02b] | M. Wouters, G. Vanwijnsberghe, P. Van Wesemael, T. Huybrechts, and S. Thoen, Real-time implementation on FPGA of an OFDM based wireless LAN modem extended with adaptive loading, in *Proceedings of European Solid-State Circuits Conference*, Firenze, Italy, September 2002, pp. 531–534 |

| | |
|---|---|
| [Yang00] | B. Yang, K. Letaief, R. Cheng, and C. Zhigang, Timing recovery for OFDM transmission, *IEEE J. Sel. Areas in Commun.*, 18(11):2278–2291, 2000 |
| [Yee01] | D. G. W. Yee, A design methodology for highly-integrated low-power receivers for wireless communications, Ph.D. Dissertation, University of California at Berkeley, USA, Spring 2001 |
| [Yeh01] | W. C. Yeh, Arithmetic module design and its application to FFT, Ph.D. Dissertation, National Chiao-Tung University, China, October 2001 |

[Yun99]            K. Y. Yun and A. E. Dooply, Pausible clocking-based heterogeneous
                   systems, *IEEE Trans. VLSI Syst.*, 7(4):482–488, 1999

[Zander00]         J. Zander, Trends in resource management future wireless networks, in
                   *Proceedings of IEEE Wireless Communications and Networking
                   Conference,* Chicago, USA, September 2000, pp. 159–163
[Zargari02]        M. Zargari et al., A 5-GHz CMOS transceiver for IEEE 802.11a
                   wireless LAN systems, *IEEE J. Solid-State Circuits*, 37(12):1688–
                   1694, 2002
[Zhang01]          N. Zhang, Algorithm/architecture co-design for wireless communi-
                   cations systems, Ph.D. Dissertation, University of California at
                   Berkeley, USA, Fall 2001
[Zhao03]           J. Zhao, Z. Guo, and W. Zhu, Power efficiency in IEEE 802.11a
                   WLAN with cross-layer adaptation, in *Proceedings of IEEE
                   International Conference on Communications*, May 2003, pp. 2030–
                   2034
[Zhou02]           S. Zhou, G. B. Giannakis, and C. Le Martret, Chip-interleaved block-
                   spread code division multiple access, *IEEE Trans. Commun.*,
                   50(2):235–248, 2002
[Zimmerman67]      M. S. Zimmerman and A. L. Kirsch, The AN/GSC-10 (KATHRYN)
                   variable rate data modem for HF radio, *IEEE Trans. Commun.
                   Technol.*, 15(2):197–204, 1967
[Zimmermann99]     T. G. Zimmermann, Wireless networked digital devices: A new
                   paradigm for computing and communication, *IBM Syst. J.*, 38:566–574,
                   1999
[Zuberek92]        W. M. Zuberek, Flexible circuit simulation with mixed-domain and
                   mixed-mode applications, in *Proceedings of IEEE International
                   Symposium on Circuits and Systems*, San Diego, USA, May 1992, pp.
                   81–84

# Index

Printed in the United States of America